Debating Design

From Darwin to DNA

William Dembski, Michael Ruse, and other prominent philosophers provide here a comprehensive, balanced overview of the debate concerning biological origins – a controversial dialectic since Darwin published *The Origin of Species* in 1859. Invariably, the source of controversy has been "design."

Is the appearance of design in organisms (as exhibited in their functional complexity) the result of purely natural forces acting without prevision or teleology? Or, does the appearance of design signify genuine prevision and teleology, and, if so, is that design empirically detectable and thus open to scientific inquiry? Four main positions have emerged in response to these questions: Darwinism, self-organization, theistic evolution, and intelligent design. The contributors to this volume define their respective positions in an accessible style, inviting readers to draw their own conclusions. Two introductory essays furnish a historical overview of the debate.

William A. Dembski is an associate research professor in the conceptual foundations of science at Baylor University as well as a Senior Fellow with Seattle's Discovery Institute. His most important books are *The Design Inference* (Cambridge University Press, 1998) and *No Free Lunch* (2002).

Michael Ruse is Lucyle T. Wekmeister Professor of Philosophy at Florida State University. He is the author of many books, including *Darwinism and Its Discontents* (Cambridge University Press, 2006).

Debating Design

From Darwin to DNA

Edited by

WILLIAM A. DEMBSKI

Baylor University

MICHAEL RUSE

Florida State University

CAMBRIDGE
UNIVERSITY PRESS

CAMBRIDGE UNIVERSITY PRESS
Cambridge, New York, Melbourne, Madrid, Cape Town,
Singapore, São Paulo, Delhi, Mexico City

Cambridge University Press
32 Avenue of the Americas, New York, NY 10013-2473, USA

www.cambridge.org
Information on this title: www.cambridge.org/9780521709903

First published 2004
Reprinted 2006 (twice)
First paperback edition 2007
Reprinted 2009 (twice), 2011, 2013

A catalog record for this publication is available from the British Library.

Library of Congress Cataloging in Publication Data

Debating design : from Darwin to DNA / edited by William A. Dembski, Michael Ruse.
 p. cm.
Papers from a conference, entitled Design and its Critics, held at
Concordia University, Mequon, Wis., June 2–24, 2000.
Includes bibliographical references and index.
ISBN 0-521-82949-6 (hb)
1. Evolution (Biology) – Philosophy – Congresses. 2. Intelligent design
(Teleology) – Congresses. 3. Evolution (Biology) – Religious aspects –
Christianity – Congresses. 4. Self-organizing systems – Congresses.
I. Dembski, William A., 1960– II. Ruse, Michael.
QH360.5.D42   2004
576.8–dc22   2004047363

ISBN 978-0-521-82949-6 Hardback
ISBN 978-0-521-70990-3 Paperback

Contents

Notes on Contributors

Francisco J. Ayala was born in Madrid, Spain, and has been a U.S. citizen since 1971. Ayala has been president and chairman of the board of the American Association for the Advancement of Science (1993–96) and was a member of the President's Committee of Advisors on Science and Technology (1994–2001). Ayala is currently Donald Bren Professor of Biological Sciences and of Philosophy at the University of California at Irvine. He is a recipient of the National Medal of Science for 2001. Other honors include election to the National Academy of Sciences, the American Academy of Arts and Sciences, the American Philosophical Society, and numerous foreign academies, including the Russian Academy of Sciences and the Accademia Nazionale dei Lincei (Rome). He has received numerous prizes and honorary degrees. His scientific research focuses on population and evolutionary genetics, including the origin of species, genetic diversity of populations, the origin of malaria, the population structure of parasitic protozoa, and the molecular clock of evolution. He also writes about the interface between religion and science and on philosophical issues concerning epistemology, ethics, and the philosophy of biology. He is author of more than 750 articles and of 18 books.

James Barham was trained in classics at the University of Texas at Austin and in the history of science at Harvard University. He is an independent scholar who has published some dozen articles on evolutionary epistemology, the philosophy of mind, and the philosophy of biology in both print and electronic journals, including *BioSystems, Evolution and Cognition, Rivista di Biologia,* and Metanexus.net. His work consists of a critique of the mechanistic and Darwinian images of life and mind, as well as an exploration of alternative means of understanding value, purpose, and meaning as objectively real, natural phenomena, in both their human and their universal biological manifestations. He is working on a book to be called *Neither Ghost nor Machine.*

Michael J. Behe graduated from Drexel University in Philadelphia in 1974, with a B.S. degree in chemistry. He did graduate studies in biochemistry at the University of Pennsylvania and was awarded a Ph.D. in 1978 for his dissertation research on sickle-cell disease. From 1978 to 1982, he did post-doctoral work on DNA structure at the National Institutes of Health. From 1982 to 1985, he was an assistant professor of chemistry at Queens College in New York City. In 1985 he moved to Lehigh University, where he is currently a professor of biochemistry. In his career he has authored more than forty technical papers and one book, *Darwin's Black Box: The Biochemical Challenge to Evolution*, which argues that living systems at the molecular level are best explained as being the result of deliberate intelligent design. *Darwin's Black Box* has been reviewed by the *New York Times, Nature, Philosophy of Science, Christianity Today*, and more than eighty other publications and has been translated into eight languages. He and his wife reside near Bethlehem, Pennsylvania, with their eight children.

Walter L. Bradley, Ph.D., P.E., received his B.S. in engineering science and his Ph.D. in materials science, both from the University of Texas at Austin. He taught for eight years as an assistant and associate professor at the Colorado School of Mines in its Metallurgical Engineering Department before assuming a position as professor of mechanical engineering at Texas A&M University in 1976. He served as head of his department of 67 professors and 1,500 students from 1989 to 1993. He also served as the director of the Texas A&M University Polymer Technology Center from 1986 to 1990 and from 1994 to 2000. He has received more than $5 million in research contracts from government agencies such as NSF, NASA, DOE, and AFOSR and from major corporations such as Dupont, Exxon, Shell, Phillips, Equistar, Texas Eastman, Union Carbide, and 3M. He has published more than 125 technical articles in archival journals, conference proceedings, and as book chapters. He was honored by being elected a Fellow of the American Society for Materials in 1992. He has received one national and five local research awards and two local teaching awards. He coauthored a seminal work on the origin of life entitled *The Mystery of Life's Origin: Reassessing Current Theories* in 1984, has published several book chapters and journal articles related to the origin of life, and has spoken on more than sixty university campuses on this topic over the past ten years. He took early retirement from Texas A&M University in 2000 and now holds the title of Professor Emeritus of Mechanical Engineering.

Paul Davies was born in London in 1946 and obtained a doctorate from University College, London, in 1970. He held academic appointments at Cambridge and London Universities until, at the age of thirty-four, he was appointed professor of theoretical physics at the University of Newcastle upon Tyne. From 1990 until 1996 he was professor of mathematical physics, and later of natural philosophy, at the University of Adelaide. He currently holds the positions of visiting professor at Imperial College, London, and honorary professor at the University of Queensland, although he remains based in south Australia, where he runs a science, media, and publishing consultancy called Orion Productions. Professor Davies has published more than 100 research papers in specialist journals in the areas of cosmology, gravitation, and quantum field theory, with particular emphasis on black holes and the origin of the universe. In addition to his research, Professor Davies is well known as an author, broadcaster, and public lecturer. He has written more than twenty-five books, including *God and the New Physics, The Cosmic Blueprint, The Mind of God, The Last Three Minutes, About Time, Are We Alone?* and *The Fifth Miracle.* Davies's commitment to bringing science to the wider public includes a heavy program of public lecturing in Australia, Europe, and the United States. In addition to addressing scientific topics, Davies lectures to religious organizations around the world and has had meetings with the Pope and the Dalai Lama. He frequently debates science and religion with theologians. Paul Davies is married and has four children.

William A. Dembski is an associate research professor in the conceptual foundations of science at Baylor University and a senior Fellow with Discovery Institute's Center for Science and Culture in Seattle. He is also the executive director of the International Society for Complexity, Information, and Design <www.iscid.org>, a professional society that explores complex systems apart from programmatic constraints such as naturalism. Dr. Dembski previously taught at Northwestern University, the University of Notre Dame, and the University of Dallas. He has done postdoctoral work in mathematics at MIT, in physics at the University of Chicago, and in computer science at Princeton University. A graduate of the University of Illinois at Chicago, where he earned a B.A. in psychology, an M.S. in statistics, and a Ph.D. in philosophy, he also received a doctorate in mathematics from the University of Chicago in 1988 and a master of divinity degree from Princeton Theological Seminary in 1996. He has held National Science Foundation graduate and postdoctoral fellowships. Dr. Dembski has published articles in mathematics, philosophy, and theology journals and is the author of several books. In *The Design Inference: Eliminating Chance through Small Probabilities* (Cambridge University Press, 1998), he examines the design argument in a post-Darwinian context and analyzes the connections linking chance, probability, and intelligent causation.

David J. Depew is professor of communication studies and rhetoric of inquiry at the University of Iowa. He is the coauthor, with Bruce H. Weber, of *Darwinism Evolving: Systems Dynamics and the Genealogy of Natural Selection* (1994). He is currently at work, with Marjorie Grene, on a history of the philosophy of biology to be published by Cambridge University Press.

John F. Haught is the Landegger Distinguished Professor of Theology at Georgetown University. Dr. Haught received his Ph.D. from Catholic University of America. He served as chair of the Georgetown Department of Theology from 1990 to 1995. He is now also director of the Georgetown Center for the Study of Science and Religion. Dr. Haught has published many articles and lectured widely, especially on topics related to religion and science, cosmology and theology, and ecology and theology. He is the author of many books, including *Responses to 101 Questions on God and Evolution* (2001), *God After Darwin* (2000), *Science and Religion: From Conflict to Conversation* (1995), *The Promise of Nature: Ecology and Cosmic Purpose* (1993), *Mystery and Promise: A Theology of Revelation* (1993), *What Is Religion?* (1990), *The Revelation of God in History* (1988), *What Is God?* (1986), *The Cosmic Adventure* (1984), *Nature and Purpose* (1980), and *Religion and Self-Acceptance* (1976), and he is the editor of *Science and Religion in Search of Cosmic Purpose* (2000).

Stuart Kauffman is an external professor for the Santa Fe Institute in New Mexico. He received his M.D. degree from the University of California at San Francisco in 1968 and was a professor in biochemistry and biophysics at the University of Pennsylvania until 1995. Since 1985, he has been a consultant for Los Alamos National Laboratory, and from 1986 to 1997 he was a professor at the Santa Fe Institute. Dr. Kauffman is also a founding general partner of the Bios Group in Santa Fe. He has served on the editorial boards of numerous scientific journals, including the *Journal of Theoretical Biology*. He is the author or coauthor of more than 100 scientific articles and the author of three books: *Origins of Order: Self-Organization and Selection in Evolution* (1993), *At Home in the Universe* (1995), and *Investigations* (2000).

Angus Menuge is associate professor of philosophy and program associate of the Cranach Institute at Concordia University, Wisconsin <www.cuw.edu/institutes/Cranach/>. He received his B.A. in philosophy from the University of Warwick and his Ph.D., on action explanation, from the University of Wisconsin–Madison. Dr. Menuge is editor of three books – *C. S. Lewis: Lightbearer in the Shadowlands* (1997), *Christ and Culture in Dialogue* (1999), and *Reading God's World: The Vocation of Scientist* (forthcoming). With the help of William Dembski, Menuge hosted the Design and Its Critics conference in June 2000, which inspired the present volume. Dr. Menuge has written a number of recent articles on Intelligent Design and is currently writing a book defending a robust notion of agency against reductionist theories, entitled *Agents Under Fire: Materialism and the Rationality of Science*.

Stephen C. Meyer is director of the Discovery Institute's Center for Science and Culture in Seattle, Washington, and serves as University Professor, Conceptual Foundations of Science, at Palm Beach Atlantic University in West Palm Beach, Florida. He received his Ph.D. in the history and

philosophy of science from Cambridge University, where he did a disser-
tation on the history of origin-of-life biology and the methodology of the
historical sciences. Meyer worked previously as a geophysicist for the Atlantic
Richfield Company and as a professor of philosophy at Whitworth College.
He is coauthor of the book *Science and Evidence of Design in the Universe*
(Ignatius 2002) and coeditor of the book *Darwinism, Design and Public
Education* (Michigan State University Press 2003). Meyer has contributed
scientific and philosophical articles to numerous scholarly books and journals and has published opinion-editorial columns for major newspapers and
magazines such as *The Wall Street Journal, The Los Angeles Times, The Chicago
Tribune, National Review,* and *First Things.* He has appeared on national television and radio programs such as Fox News, PBS's *TechnoPolitics* and *Freedom
Speaks,* MSNBC's *Hardball with Chris Matthews,* and NPR's *Talk of the Nation*
and *Science Friday.* He coauthored the film *Unlocking the Mystery of Life,* which
recently aired on PBS stations around the country.

Kenneth R. Miller is professor of biology at Brown University. Dr. Miller has
a Sc.B. in biology from Brown University (1970) and a Ph.D. in biology
from the University of Colorado (1974). He has taught at the University
of Colorado, Harvard University, and Brown University, where he has been
full professor since 1986. He is the recipient of numerous honors for teaching excellence. Dr. Miller is a member of the American Association for the
Advancement of Science and the American Society for Cell Biology, and he
has been chairman and council member of the ASCB, editor of *The Journal of Cell Science,* and general editor of *Advances in Cell Biology.* Dr. Miller's
scientific interests include the structure, composition, and function of biological membranes, electron microscopy and associated techniques, and
photosynthetic energy conversion. He has published a large number of technical scientific papers and essays, edited three volumes of *Advances in Cell
Biology,* and is author or coauthor of several high school and college biology
textbooks, including *Biology: The Living Science* and *Biology: Discovering Life.*
Recently, Dr. Miller has produced a general-audience work defending evolution and its compatibility with Christian faith and critiquing Intelligent
Design: *Finding Darwin's God: A Scientist's Search for Common Ground between
God and Evolution* (1999).

Robert T. Pennock is associate professor of science and technology studies
and philosophy at Michigan State University's Lyman Briggs School and
in the Philosophy Department. He is also on the faculty of MSU's Ecology
and Evolutionary Biology and Behavior program. He has published numerous articles that critique Intelligent Design creationism, including one that
won a Templeton Prize for Exemplary Paper in Theology and the Natural Sciences. He is the author of *Tower of Babel: The Evidence against the New
Creationism* (1999).

John Polkinghorne was professor of mathematical physics at Cambridge University, working in theoretical elementary particle physics. He was elected a Fellow of the Royal Society in 1974. In 1982 he was ordained to the priesthood in the Church of England. He is the author of a number of books about science and theology. In 1996 he retired as president of Queens' College, Cambridge, and in 1997 he was made a Knight of the British Empire.

Michael Roberts studied geology at Oxford and spent three years in Africa as an exploration geologist. He studied theology at Durham and was ordained into the Anglican Church in 1974 (along with Peter Toon). He is now vicar of Chirk, near Llangollen in North Wales. He is a keen mountain walker and has written articles on science and religion (one, on Darwin and design, received a Templeton Award in 1997) and on Darwin's British geology. In June 2000 he was a plenary speaker at the conference on Intelligent Design at Concordia University Wisconsin. He is married to Andrea, and they have two almost-grown-up children.

Michael Ruse is Lucyle T. Werkmeister Professor of Philosophy at Florida State University. He received his B.A. in philosophy and mathematics from Bristol University, an M.A. in philosophy from McMaster University, and his Ph.D. from Bristol University. He was full professor of philosophy at Guelph from 1974 to 2000. Dr. Ruse is a Fellow of the Royal Society of Canada and of the American Association for the Advancement of Science. He has received numerous visiting professorships, fellowships, and grants. Michael Ruse's many publications include *The Philosophy of Biology; Sociobiology: Sense or Nonsense?; The Darwinian Revolution: Science Red in Tooth and Claw; Darwinism Defended: A Guide to the Evolution Controversies; Taking Darwin Seriously: A Naturalistic Approach to Philosophy; But Is It Science? The Philosophical Question in the Evolution/Creation Controversy;* and *Monad to Man: The Concept of Progress in Evolutionary Biology.* His most recent works include *Mystery of Mysteries: Is Evolution a Social Construction?* and *Can a Darwinian Be a Christian? The Relationship between Science and Religion* (Cambridge University Press, 2000). Michael Ruse was the founding editor of the journal *Biology and Philosophy* and is now on the editorial board of several major journals, including *Zygon, Philosophy of Science,* and the *Quarterly Review of Biology.* On a more public level, Ruse has appeared on many television programs, including *Firing Line,* and was a witness for the ACLU in the 1981 Arkansas hearings that overturned a creation science law. His latest book is *Darwin and Design: Does Evolution have a Purpose?*

Elliott Sober is Hans Reichenbach Professor of Philosophy and Henry Vilas Research Professor at the University of Wisconsin–Madison, where he has taught since 1974. His research is in the philosophy of science, especially in the philosophy of evolutionary biology. Sober's books include *The Nature of Selection: Evolutionary Theory in Philosophical Focus* (1984, 2nd ed. 1993);

Reconstructing the Past: Parsimony, Evolution, and Inference (1988); *Philosophy of Biology* (1993); *From a Biological Point of View: Essays in Evolutionary Philosophy* (Cambridge University Press, 1994); and, most recently, *Unto Others: The Evolution and Psychology of Unselfish Behavior* (with David Sloan Wilson) (1998). Sober is a past president of the American Philosophical Association Central Division and a Fellow of the American Academy of Arts and Sciences.

Richard Swinburne has been Nolloth Professor of the Philosophy of the Christian Religion at the University of Oxford since 1985. For the twelve years before that he was a professor of philosophy at the University of Keele. Since 1992, Dr. Swinburne has been a Fellow of the British Academy. His books include *Space and Time* (1968, 2nd ed. 1981), *The Concept of Miracle* (1971), *An Introduction to Confirmation Theory* (1973), *The Coherence of Theism* (1977, 2nd ed. 1993), *The Existence of God* (1979, 2nd ed. 1991), *Faith and Reason* (1981), *Personal Identity* (with Sidney Shoemaker) (1984), *The Evolution of the Soul* (1986, 2nd ed. 1997), *Responsibility and Atonement* (1989), *Revelation* (1991), *The Christian God* (1994), *Providence and the Problem of Evil* (1998), *Is There a God?*(1996), and *Epistemic Justification* (forthcoming).

Keith Ward is a philosopher and theologian. He has taught philosophy at Glasgow, St. Andrews, London, and Cambridge Universities. He was ordained in the Church of England in 1972. Dr. Ward has been dean of Trinity Hall, Cambridge; professor of moral theology, London; professor of the history and philosophy of religion, London; and is presently Regius Professor of Divinity, Oxford. His books include *God, Chance and Necessity*; *God, Faith and the New Millennium*; and *Divine Action*.

Bruce H. Weber is professor of biochemistry at California State University at Fullerton and Robert H. Woodworth Professor of Science and Natural Philosophy at Bennington College. His is coauthor (with David Depew) of *Darwinism Evolving: Systems Dynamics and the Genealogy of Natural Selection* (1995), coauthor (with John Prebble) of *Wandering in the Gardens of the Mind: Peter Mitchell and the Making of Glynn* (2003), and coeditor (with David Depew) of *Evolution and Learning: The Baldwin Effect Reconsidered* (2003). He is also director of the Los Angeles Basin California State University Minority International Research Training Program.

Debating Design

From Darwin to DNA

INTRODUCTION

1

General Introduction

William A. Dembski and Michael Ruse

Intelligent Design is the hypothesis that in order to explain life it is necessary to suppose the action of an unevolved intelligence. One simply cannot explain organisms, those living and those long gone, by reference to normal natural causes or material mechanisms, be these straightforwardly evolutionary or a consequence of evolution, such as an evolved extraterrestrial intelligence. Although most supporters of Intelligent Design are theists of some sort (many of them Christian), it is not necessarily the case that a commitment to Intelligent Design implies a commitment to a personal God or indeed to any God that would be acceptable to the world's major religions. The claim is simply that there must be something more than ordinary natural causes or material mechanisms, and moreover, that something must be intelligent and capable of bringing about organisms.

Intelligent Design does not speculate about the nature of such a designing intelligence. Some supporters of Intelligent Design think that this intelligence works in tandem with a limited form of evolution, perhaps even Darwinian evolution (for instance, natural selection might work on variations that are not truly random). Other supporters deny evolution any role except perhaps a limited amount of success at lower taxonomic levels – new species of birds on the Galapagos, for instance. But these disagreements are minor compared to the shared belief that we must accept that nature, operating by material mechanisms and governed by unbroken natural laws, is not enough.

To say that Intelligent Design is controversial is to offer a truism. It is opposed, often bitterly, by the scientific establishment. Journals such as *Science* and *Nature* would as soon publish an article using or favourable to Intelligent Design as they would an article favourable to phrenology or mesmerism – or, to use an analogy that would be comfortable to the editors of those journals, an article favourable to the claims of the Mormons about Joseph Smith and the tablets of gold, or favourable to the scientific creationists' claims about the coexistence of humans and dinosaurs. Recently, indeed, the American Association for the Advancement of Science (the organization that publishes *Science*) has declared officially that in its opinion Intelligent Design is not so much bad science as no science at all and accordingly has no legitimate place in the science classrooms of the United States.

Once one leaves the establishment and moves into the more popular domain, however, one finds that the level of interest in and sympathy for Intelligent Design rises rapidly. Many people think that there may well be something to it, and even those who are not entirely sure about its merits think that possibly (or probably) it is something that should be taught in schools, alongside more conventional, purely naturalistic accounts of origins. Students should be exposed to all sides of the debate and given a choice. That, after all, is the American Way – open debate and personal decision.

The editors of this volume, *Debating Design: Darwin to DNA*, fall at opposite ends of the spectrum on the Intelligent Design debate. William Dembski, a philosopher and a mathematician, has been one of the major contributors to the articulation and theory of Intelligent Design. He has offered analyses of design itself and has argued that no undirected natural process can account for the information-rich structures exhibited by living matter. Moreover, he has argued that the very features of living matter that place it beyond the remit of undirected natural causes also provide a reliable signature of design. Michael Ruse, a philosopher and historian of science, has long been an advocate of Darwinian evolution, and has devoted many years to fighting against those who argue that one must appeal to non-natural origins for plants and animals. He has appeared in court as an expert witness on behalf of Darwinism and has written many books on the subject.

For all their differences, the editors share the belief that – if only culturally – Intelligent Design is a significant factor on the contemporary landscape and should not be ignored. For the Intelligent Design proponents, it is a major breakthrough in our understanding about the world. For the Intelligent Design opponents, it is at the least a major threat to the status quo and something with a real chance of finding its way into classrooms. The editors also share the belief that, in a dispute such as this, it is important that the two sides have a real grasp of the opinions of those that they oppose. Ignorance is never the way to fight error.

There are of course already books that deal with Intelligent Design and with the arguments of the critics. The editors have themselves contributed to this literature. We believe, however, that there is virtue in producing one volume, containing arguments from both sides, in which each side puts forward its strongest case (previous volumes have tended to bias discussion toward one side over the other). The reader then can quickly and readily start to grasp the fundamental claims and counterclaims being made. Of course – and this is obviously an argument that comes more from the establishment – even doing something like this can be seen as giving one's opponents some kind of status and legitimacy. And there is probably truth in this. But we do live in a democracy, and we are committed to working things out without resort to violence or to underhanded strategies, and so, despite the worries and fears, we have come together hoping that the merits of such an enterprise will outweigh the negative factors. Those who know how to do things better will of course follow their own principles.

The collection is divided into four main sections, with a shorter introductory section. The aim of the introductory section is simply to give the reader some background, and hence that section contains an overall historical essay by one of the editors, Michael Ruse, on the general history of design arguments – "The Argument from Design: A Brief History," and then a second essay by Angus Menuge on the specific history of the Intelligent Design movement – "Who's Afraid of ID? A Survey of the Intelligent Design Movement." Although the first author has very strongly negative views on Intelligent Design and, as it happens, the second author has views no less strongly favourable, the intent in this introductory section is to present a background of information without intruding value commentary. The essays are written, deliberately, in a nonpartisan fashion; they are intended to set the scene and to help the reader in evaluating the discussions of the rest of the volume.

Michael Ruse traces design arguments back to the Greeks and shows that they flourished in biology down to the eighteenth century, despite the rethinking of issues in the physical sciences. Then David Hume made his devastating attack, but still it was not until Charles Darwin in his *Origin of Species* (1859) offered a naturalistic explanation of organisms that the design argument was truly rejected by many. The essay concludes with a discussion of the post-Darwinian period, showing that many religious people today endorse a "theology of nature" over natural theology. Most important in Ruse's discussion is the distinction he draws between the argument to complexity – the argument that there is something distinctive about the organic world – and the argument to design – the argument that this complexity demands reference to a (conscious) designer to provide a full explanation. These are the issues that define the concerns of this collection.

Next, Angus Menuge provides a short history of the contemporary Intelligent Design movement and considers its future prospects. He notes that some, such as Barbara Forrest, dismiss the movement as stealth creationism. Menuge, however, finds this designation to be misleading. He argues that Intelligent Design is significantly different from typical creationist approaches in its aims, methods, and scope, and that scientists became interested in design apart from political or religious motivations. Thus he traces the roots of the Intelligent Design movement not to the political and religious zeal of anti-evolutionists but to the legitimate scientific critiques of evolution and origin-of-life studies in the mid-eighties by scientists such as Michael Denton and Walter Bradley. Yet because criticism by itself rarely threatens a dominant paradigm, the Intelligent Design movement did not gain prominence until the work of Michael Behe (*Darwin's Black Box*, The Free Press, 1996) and William Dembski (*The Design Inference*, Cambridge University Press, 1998). These works outlined a positive program for understanding design in the sciences. Mengue concludes his essay by noting that regardless of whether Intelligent Design succeeds in becoming mainstream science, it is helping scientists to think more clearly about the causal pathways that account for the emergence of biological complexity.

We move now to the main sections, each of which has four or five contributions. We go from discussions favourable to evolution and critical of Intelligent Design, to discussions favourable to Intelligent Design and critical at least of unbroken evolution. The first such section, *Darwinism*, starts with a piece by the leading evolutionary biologist Francisco J. Ayala, a former Catholic priest and a person with great sensitivity to and sympathy for the religious attitude. In "Design without Designer: Darwin's Greatest Discovery," Ayala makes three claims. First, he claims that Darwin successfully brought the question of organic origins into the realm of science; second, that Darwin spoke to and solved successfully the question of complexity or adaptation; and third, that nevertheless there is something distinctive (something "teleological") about biological understanding even in the post-Darwinian world. The reader should refer back to the introductory essay of Michael Ruse to fit what Ayala is claiming into the division drawn between the argument to complexity (that Ayala thinks Darwin addresses and solves scientifically) and the argument to design (that Ayala thinks is now out of science but still carrying a form of argumentation that transfers over to modern science). Ayala concludes that science is not the only way of knowing.

Kenneth R. Miller, a scientist and a practicing Roman Catholic, is one of the strongest critics of Intelligent Design. In his contribution, "The Flagellum Unspun: The Collapse of 'Irreducible Complexity,' " Miller takes aim at one of the most important concepts promoted by Intelligent Design supporters, namely that of *irreducible complexity*. Introduced by Michael Behe in his *Darwin's Black Box*, this is a property possessed by certain aspects of organisms that supposedly could not be produced by unguided natural causes. It denotes something so overwhelmingly intricate and complex that it defies normal natural understanding and demands an explanation in terms of intelligence. Behe's prime biological example is of certain motorlike processes in microorganisms, and Miller's intent is to show that Behe is mistaken in his claims (as is Dembski in his support). Note that Miller explicitly asserts that his naturalistic position is more theologically satisfactory than that of his opponents.

Elliott Sober is a well-known philosopher whose piece – "The Design Argument" – is of a general nature. He is concerned to give a theoretical analysis of design arguments and particularly of arguments of the kind offered by Archdeacon William Paley (see Ruse's introductory chapter). He analyses matters in terms of likelihood, that is, the idea of which of two hypotheses is more likely given a particular observation – in Paley's case, an intelligence or blind chance given the discovery of a watch. Although Sober does not want to go all the way with Paley to the inference of a God (certainly not the Christian God), given his analysis he is more critical than most philosophers are of Hume's arguments (especially inasmuch as they are analogical), but he is also not convinced that one can simply dismiss design arguments once Darwin appears on the scene. Having said this, however, Sober has little time for Intelligent Design, which he thinks fails as genuine science with respect to important properties such as prediction.

Finally in this section we have Robert Pennock, a well-known philosopher and critic of creationism (the author of *The Tower of Babel*) who argues that the Intelligent Design movement is built upon problematic religious assumptions. Considering the writings of Stephen Meyer (one of the contributors to this collection), Pennock takes up the claim that human dignity (and morality generally) can be justified only if the assumption that man is created in the image of God is factual. Pennock's aim is to criticize not the belief in "the God hypothesis," but rather the claim to have established it scientifically as an alternative to evolution. His essay critiques the theological presuppositions that he finds hidden in Intelligent Design, as well as the proposition that the design inference, interpreted as a scientific inference to the best explanation, confirms not just theism, but specifically the Judeo-Christian God. Along the way, Pennock points out problems with the recurring arguments that supporters of Intelligent Design use in their lobbying to get their view taught in the public schools.

The second section, *Complex Self-Organization*, contains pieces by those who believe that nature itself, simply obeying the laws of physics and chemistry without the aid of selection (or with, at best, a very limited contribution by selection), can produce entities showing the kind of complexity that Darwinians think can be produced only by their mechanism. This idea of "order for free" (as it has been termed by Stuart Kauffman) has a long history; its most notable exponent was the early twentieth-century Scottish morphologist D'Arcy Wentworth Thompson in his *On Growth and Form*.

The first piece in this section is by Stuart Kauffman himself. Here Kauffman tries to imagine what it would be like for biologists to develop what he calls a "general biology." By a general biology Kauffman means a general theory of what it means to be alive and of how things that are alive originated. Kauffman concedes that we don't at this time possess a general biology. According to Kauffman, a general biology would consist in principles that are applicable to all possible forms of life and that uncover their deep structure. The problem with natural selection, for Kauffman, is not that it is false or even that it is less than universally applicable. The problem is that natural selection cannot account for its own success (or, as he states it more precisely, cannot account for the "smooth fitness landscapes" that enable it to be a "successful search strategy"). Kauffman's essay attempts not to provide solutions but to ask the right questions. Implicit throughout the essay, therefore, is the admission that biology's key conceptual problems remain to be solved. Kauffman thus differs from Darwinists who think that Darwin is the "Newton of the blade of grass." At the same time, Kauffman does not think that Intelligent Design holds the solution to a general biology.

Next comes a chapter written jointly by the biologist Bruce H. Weber and the philosopher David J. Depew. In "Darwinism, Design, and Complex Systems Dynamics," they argue that both strict Darwinians and Intelligent Design theorists are at fault for putting too heavy an emphasis on the design-like nature of the organic world (the argument to complexity, in the sense given earlier). They stress that it is possible to have functioning systems with many components that are far from perfect. The aim in nature is not to achieve some ideal standard, but simply to get things working at all. In this light, they feel that the natural processes of physics and chemistry can do far more than is often realized, and the authors make their case through a detailed discussion of the origin of life, something often downplayed in scientific discussions (especially those of Darwinians). As practicing Christians, Weber and Depew have a more-than-casual interest in the Intelligent Design debate, and their important concluding discussion points to the lack of a uniform Christian tradition giving unambiguous support for natural theology – that part of theology that stresses reason over faith and that focuses on arguments for the existence of God, such as the design argument (the second part of the distinction drawn earlier).

Paul Davies is one of the best known of all writers on the science–religion interface. His *God and the New Physics* is rightfully considered a classic. He is ever keen to show that the world works according to law, and yet for some time now he has been a critic of strict Darwinism, thinking that more is needed to explain life and its complexity. Mere selection will not do. (Unlike Weber and Depew, Davies has no trouble with the argument to complexity as such.) In "Emergent Complexity, Teleology and the Arrow of Time," Davies explores the question of whether, balancing the negative downgrading effects of the Second Law of Thermodynamics, there is some kind of cosmic law of increasing complexity. He raises the contentious question of progress, something that has been much debated by evolutionists. Although it is not directly related to the question of possible design in the universe (in his *The Blind Watchmaker*, the arch-atheist Richard Dawkins argues for biological progress), for many thinkers (as Ruse notes in his introductory chapter) progress provides a new argument for God's existence to replace the one (they believe to have been) destroyed by Darwin.

Finally in this section we have James Barham's piece on the emergence of biological value. In it, he critiques what he calls the Mechanistic Consensus in contemporary scientific and philosophical thought. According to the Mechanistic Consensus, the theory of natural selection and molecular biology suffice to explain the appearance of design in living things. This, he argues, is a mistake, because these disciplines make use of primitive concepts that are themselves normative and teleological in character. Furthermore, he argues, the widespread belief that the teleological language of biology is only "as if" and can be "cashed out" through reduction to lower-level physical theories is a mistake based on an outdated conception of physics itself. According to Barham, the best way to make scientific sense of bio-logical design is not by looking to natural selection, and not by looking to an intelligent designer, but by looking to an emergent, purposive dynamics within living matter that allows organisms globally to coordinate their own activity so as to maintain themselves in existence as organized wholes. In this way, the living state of matter may be viewed as having intrinsic value. Some recent developments in nonlinear dynamics and condensed matter physics tending to support this view are briefly surveyed.

We come next to the section on *Theistic Evolution.* Here we find committed Christian believers who nevertheless want to find some place for evolution, although perhaps boosted by some kind of divine forethought or ongoing concern. John F. Haught is a distinguished Catholic writer on the science–religion relationship. His thinking is marked (he would say informed) by a sensitivity to the ongoing, unfurling nature of the world, something he finds explicable thanks not only to Christian theology but also to the philosophical thinking of Alfred North Whitehead, where the creation is not a once-and-for-all event, but rather something that is continuous and that God can try to influence and direct but cannot command. His chapter, "Evolution, Design and the Idea of Providence," finds fault with both Darwinians and Intelligent Design supporters, feeling that both overstress the significance of design for an understanding of the Christian God (that is, overstress the significance of the second part of the twofold argument, the argument to design). Haught suggests that a God who is working in an ongoing fashion in the creation is truer to the Christian message than one conceived solely in the terms of traditional natural theology.

John Polkinghorne is both a distinguished physical scientist and an Anglican priest, which dual roles and interests have led him to be one of the most prolific writers in recent years on the science–religion relationship. He has long been an enthusiast for the "Anthropic Principle," where it is the constants of the universe coming together in such a remarkable way to produce intelligent life that is the true mark of design. In the present essay, "The Inbuilt Potentiality of Creation," Polkinghorne explores these ideas. His intent is positive rather than negative, but in a way his approach could be taken as implying that neither Darwinians nor Intelligent Design enthusiasts are focusing on the most important issues for understanding the Creator. In terms of the division of the argument for design into an argu-ment for complexity and then an argument to design, Polkinghorne seems to accept the latter move but to feel that the real issue of complexity is not biological adaptation but rather the specific physical phenomena that allow life – especially intelligent life – to exist at all.

Keith Ward is Regius Professor of Religion at the University of Oxford and another who has written extensively on theological issues in the light of modern science. In "Theistic Evolution," he faces the issue that the world – the world of evolutionary life – cannot be something that simply occurred by chance. At least, for the Christian it cannot be something that simply occurred by chance. In some way, we must find space for purpose, for God's intentions. Ward explores various ways in which this might be done. It is clear that (by implication) he would not look favourably on an Intelligent Design approach, for this would put God too directly into His creation. Ward wants God creating through the natural processes of law, and to this end he invites us to look sympathetically to the progressivist thinking of the French Jesuit paleontologist Pierre Teilhard de Chardin. But rather than Teilhard's kind of vitalistic thinking, Ward inclines toward the idea that somehow God's influence on nature stands in the same relationship as does the mind to the body. The two are intertwined, but in some way separate.

The Anglican priest and geologist Michael Roberts is interested in historical issues. Although he denies that he is committed to any kind of theistic evolution, he has as little sympathy for the hard-line materialistic Darwinian as he has for the Intelligent Design theorist. Through a study of the thinking of earlier scientists, particularly those interested in geology, he concludes that neither side has the true picture and that both are seduced by the rhetoric of their language and thinking. In some way, Roberts wants to break down the distinction between things working according to blind law (the stance of many Darwinians) and things working through miraculous intervention (the stance he attributes to Intelligent Design supporters). God is at work all of the time, through His laws. This means that He is never absent from the world – something that Darwinians are free to suppose is always true and that Intelligent Design proponents suppose, by default, is generally true.

Finally in this section we have Richard Swinburne of Oxford University. He, like Polkinghorne, sides with a version of the Anthropic Principle, although he approaches the issue from a more philosophical basis than does Polkinghorne, a physicist. In Swinburne's thinking, it is all a matter of probabilities. Which is the more likely? That everything was set up to work by design, or that everything simply came together by chance? In Swinburne's opinion, there is no doubt but that the intention-based explanation is better, from a simplicity perspective. In other words, like Polkinghorne, for Swinburne the Darwinian–Intelligent Design debate takes second place to an argument from design that begins with an argument to complexity that is not biologically based.

The final section turns to the proponents of *Intelligent Design*. The first piece is by William A. Dembski, in which he outlines his method of design detection. If Intelligent Design is going to stand a chance of entering mainstream science, it must provide scientists with some rigorous way of identifying the effects of intelligent causation and of distinguishing them from the effects of undirected natural causes. Dembski claims to have provided design theorists with such a method in his criterion of specified complexity. The most common criticism made against using specified complexity to detect design is that it commits an argument from ignorance. In his chapter, Dembski answers this criticism by analyzing the logic by which specified complexity detects design. But he goes futher, arguing that to reject specified complexity as a reliable empirical marker of actual design renders naturalistic theories of life's origin and history invulnerable to refutation, even in principle. His essay therefore attempts to level the playing field on which theories of biological origins are decided.

Walter L. Bradley is an engineer who specializes in polymers. In the mid-1980s, he coauthored what supporters consider a seminal critique of origin-of-life studies in *The Mystery of Life's Origin*. In that book, he distinguished between thermal and configurational entropy. That distinction has proven to be essential in relating the Second Law of Thermodynamics to the origin of life. The Second Law of Thermodynamics is widely abused. Some creationists have used it to provide a one-line refutation of any naturalistic attempt to account for life's origin. Alternatively, some evolutionists have treated the Second Law as a creative principle that provides a one-line solution to life's origin. Bradley is more careful and in his chapter delineates exactly how the Second Law applies to the origin of life. In particular, he updates his work on configurational entropy and clarifies how the crucial form of entropy that life must overcome is not thermal but configurational entropy.

The biochemist Michael J. Behe is the best-known scientific proponent of Intelligent Design. His chief claim to fame in his widely cited book published in 1996, *Darwin's Black Box: The Biochemical Challenge to Evolution*. In that book, he argued that the irreducible complexity of certain biochemical systems convincingly confirms their actual design. In this essay, Behe briefly explains the concept of irreducible complexity and reviews why he thinks that it poses such a severe problem for Darwinian gradualism. In addition, he addresses several misconceptions about how the theory of Intelligent Design applies to biochemistry. In particular, he discusses several putative counterexamples that some scientists have advanced against his claim that irreducibly complex biochemical systems demonstrate design. Behe turns the tables on these counterexamples, arguing that these examples in fact underscore the barrier that irreducible complexity poses to Darwinian explanations and, if anything, show the need for design explanations.

Finally, the philosopher of biology Stephen C. Meyer argues for design on the basis of the Cambrian explosion – the geologically sudden appearance of new animal body plans during the Cambrian period. Meyer notes that this episode in the history of life represents a dramatic and discontinuous increase in the complex specified information (or specified complexity) of the biological world. He argues that neither the neo-Darwinian mechanism of natural selection acting on random mutations nor alternative self-organizational mechanisms are sufficient to produce such an increase in information in the time allowed by the fossil evidence. Instead, he suggests that such discrete increases in complex specified information are invariably associated with conscious and rational activity – that is, with design. Thus, he argues that design can be inferred as the best, most causally adequate explanation of what he calls the Cambrian Information Explosion.

It remains only for the editors to express their thanks to those who helped to make possible the creation and publication of this collection. Above all, we are indebted to Angus Menuge, who organized a conference bringing together the different sides in the Intelligent Design debate. This conference, entitled Design and Its Critics, was held at Concordia University in Mequon, Wisconsin, June 22–24, 2000, and was sponsored by the Cranach Institute, *Touchstone* magazine, and the Sir John Templeton Foundation. The conference presented a scholarly debate on the merits of Intelligent Design and included many of the strongest proponents and critics of the theory. Many of the sessions had a debate format, and there were presentations and exchanges among twenty-six plenary speakers from around the world: Larry Arnhart, Michael Behe, Robin Collins, Ted Davis, William Dembski, David DeWolf, Brian Josephson, John Leslie, Stephen Meyer, Kenneth Miller, Scott Minnich, Lenny Moss, Paul Nelson, Warren Nord, Ronald Numbers, Robert O' Connor, Del Ratzsch, Patrick Henry Reardon, Michael Roberts, Michael Ruse, Michael Shermer, Kelly Smith, Walter and Lawrence Starkey, Jean Staune, and Mike Thrush. In addition, thirty-five papers were presented in the concurrent sessions.

It was Angus Menuge who suggested to the editors that there might be a potential volume worth pursuing, and since then he has acted as our general assistant and organizer, far beyond the call of duty. We are grateful also to Terence Moore and Stephanie Achard of Cambridge University Press, who had to get reports and to steer the project through the Press (and we are grateful also to the eight anonymous referees who gave us much good advice and seven and a half positive recommendations). Finally, we are truly grateful to our contributors, who, often despite serious doubts, allowed us to talk them into contributing to a project such as this, agreeing with us that the way to solve contentious issues is by open and frank discussion among those at different ends of the spectrum.

2

The Argument from Design

A Brief History

Michael Ruse

The argument from design for the existence of God – sometimes known as the teleological argument – claims that there are aspects of the world that cannot be explained except by reference to a Creator. It is not a Christian argument as such, but it has been appropriated by Christians. Indeed, it forms one of the major pillars of the natural-theological approach to belief – that is, the approach that stresses reason, as opposed to the revealed-theological approach that stresses faith and (in the case of Catholics) authority. This chapter is a very brief history of the argument from design, paying particular attention to the impact of Charles Darwin's theory of evolution through natural selection, as presented in his *Origin of Species*, published in 1859.[1]

FROM THE GREEKS TO CHRISTIANITY

According to Xenophon (*Memorabilia*, I, 4.2–18), it was Socrates who first introduced the argument to Western thought, but it is Plato who gives the earliest full discussion, in his great dialogue about the death of Socrates (the *Phaedo*) and then in later dialogues (the *Timaeus*, especially). Drawing a distinction between causes that simply function and those that seem to reveal some sort of plan, Plato wrote about the growth of a human being:

I had formerly thought that it was clear to everyone that he grew through eating and drinking; that when, through food, new flesh and bones came into being to supplement the old, and thus in the same way each kind of thing was supplemented by new substances proper to it, only then did the mass which was small become large, and in the same way the small man big. (*Phaedo*, 96 d, quoted in Cooper 1997, 83–4)

But then, Plato argued that this kind of explanation will not do. It is not wrong, but it is incomplete. One must address the question of why someone would grow. Here one must (said Plato) bring in a thinking mind, for without this, one has no way of relating the growth to the end result, the reason for the growth:

The ordering Mind ordered everything and place each thing severally as it was best that it should be; so that if anyone wanted to discover the cause of anything, how it came into being or perished or existed, he simply needed to discover what kind of existence was *best* for it, or what it was best that it should do or have done to it. (97 b–c)

Note that here there is a two-stage argument. First, there is the claim that there is something special about the world that needs explaining – the fact of growth and development, in Plato's example. To use modern language, there is the claim that things exist for certain desired ends, that there is something "teleological" about the world. Then, second, there is the claim that this special nature of the world needs a special kind of cause, namely, one dependent on intelligence or thinking. Sometimes the first stage of the argument is known as the argument *to* design, and the second stage as the argument *from* design, but this seems to me to suppose what is to be proven, namely, that the world demands a designer. Although not unaware of the anthropomorphic undertones, I shall refer to the first stage of the argument as the "argument to (seemingly) organized complexity." Here I am using the language of a notorious atheist, the English biologist Richard Dawkins, who speaks in terms of "organized complexity" or "adaptive complexity," following his fellow English evolutionist John Maynard Smith (1969) in thinking that this is "the same set of facts" that the religious "have used as evidence of a Creator" (Dawkins 1983, 404). Then for the second stage of the argument, I shall speak of the "argument to design." Obviously, for Socrates and Plato this did not prove the Christian God, but it did prove a being whose magnificence is reflected in the results – namely, the wonderful world about us. For the two stages taken together, I shall continue to speak of the argument from design.

For Plato, it was the second stage of the argument – the argument to design – that really mattered. He was not that interested in the world as such, and clearly thought that design could be inferred from the inorganic and the organic indifferently. His student Aristotle, who for part of his life was a working biologist, emphasized things rather differently. Although, in a classic discussion of causation, he argued that all things require understanding in terms of ends or plans – in terms of "final causes," to use his language – in fact it was in the organic world exclusively that he found what I am calling organic complexity. Aristotle asked: "What are the forces by which the hand or the body was fashioned into its shape?" A woodcarver (speaking of a model) might say that it was made as it is by tools such as an axe or an auger. But note that simply referring to the tools and their effects is not enough. One must bring in desired ends. The woodcarver "must state the reasons why he struck his blow in such a way as to effect this, and for the sake of what he did so; namely, that the piece of wood should develop eventually into this or that shape." Likewise, against the physiologists he argued that "the true method is to state what the characters are that distinguish the animal – to explain what it is and what are its qualities – and to deal after the same fashion with its several parts; in fact, to proceed in exactly the same way as we should do, were we dealing with the form of a couch" (*Parts of Animals*, 641a 7–17, quoted in Barnes 1984, 997).

Aristotle certainly believed in a god or gods, but these "unmoved movers" spend their time contemplating their own perfection, indifferent to human fate. For this reason, whereas Plato's teleology is sometimes spoken of as "external," meaning that the emphasis is on the designer, Aristotle's teleology is sometimes spoken of as "internal," meaning that the emphasis is on the way that the world – the organic world, particularly – seems to have an end-directed nature. Stones fall. Rivers run. Volcanoes erupt. But hands are for grasping. Eyes are for seeing. Teeth are for biting and chewing. Aristotle emphasizes the first part of the argument from design. Plato emphasizes the second part. And these different emphases show in the uses made of the argument from design in the two millennia following the great Greek philosophers. Someone like the physician Galen was interested in the argument to organized complexity. The hand, for instance, has fingers because "if the hand remained undivided, it would lay hold only on the things in contact with it that were of the same size that it happened to be itself, whereas, being subdivided into many members, it could easily grasp masses much larger than itself, and fasten accurately upon the smallest objects" (Galen 1968, 1, 72). Someone like the great Christian thinker Augustine was interested in the argument to design.

The world itself, by the perfect order of its changes and motions, and by the great beauty of all things visible, proclaims by a kind of silent testimony of its own both that it has been created, and also that it could not have been made other than by a God ineffable and invisible in greatness, and ineffable and invisible in beauty. (Augustine 1998, 452–3)

As every student of philosophy and religion knows well, it was Saint Thomas Aquinas who put the official seal of approval on the argument from design, integrating it firmly within the Christian *Weltanschauung*, highlighting it as one of the five valid proofs for the existence of God.

The fifth way is taken from the governance of the world. We see that things that lack intelligence, such as natural bodies, act for an end, and this is evident from their acting always, or nearly always, in the same way, so as to obtain the best result. Hence it is plain that not fortuitously, but designedly do they [things of this world] achieve their end.

Then from this premise (equivalent of the argument to organization) – more claimed than defended – we move to the Creator behind things (argument to design). "Now whatever lacks knowledge cannot move towards an end, unless it be directed by some being endowed with knowledge and intelligence; as the arrow is shot to its mark by the archer. Therefore some intelligent being exists by which all natural things are directed to their end; and this being we call God" (Aquinas 1952, 26–7).

AFTER THE REFORMATION

Famous though this "Thomistic" argument has become, one should never-theless note that for Aquinas (as for Augustine before him) natural theology could never take the primary place of revealed theology. Faith first, and then reason. It is not until the Reformation that one starts to see natural theology being promoted to the status of revealed theology. In a way, this is some-what paradoxical. The great reformers – Luther and Calvin, particularly – had in some respects less time for natural theology than the Catholics from which they were breaking. One finds God by faith alone (*sola fide*), and one is guided to Him by scripture alone (*sola scriptura*). They were putting pressure on the second part of the argument from design. At the same time, scien-tists were putting pressure on the first part of the argument. Francis Bacon (1561–1626), the English philosopher of scientific theory and methodol-ogy, led the attack on Greek thinking, wittily likening final causes to vestal virgins: dedicated to God but barren! He did not want to deny that God stands behind His design, but Bacon did want to keep this kind of thinking out of his science. The argument to complexity is not very useful in science; certainly the argument to complexity in the nonliving context is not useful in science. And whatever one might want to say about the argument to com-plexity for the living world, inferences from this to or for design (a Mind, that is) have no place in science. Harshly, Bacon judged: "For the handling of final causes mixed with the rest in physical inquiries, hath intercepted the severe and diligent inquiry of all real and physical causes, and given men the occasion to stay upon these satisfactory and specious causes, to the great arrest and prejudice of further discovery" (Bacon 1605, 119).

But there was another side, in England particularly. Caught in the six-teenth century between the Scylla of Catholicism on the continent and the Charybdis of Calvinism at home, the central Protestants – the members of the Church of England, or the Anglicans – turned with some relief to nat-ural theology as a middle way between the authority of the Pope and the Catholic tradition and the authority of the Bible read in a Puritan fashion. This was especially the strategy of the Oxford-trained cleric Richard Hooker, in his *The Laws of Ecclesiastical Polity*. If one turned to reason and evidence, one did not need to rely on Catholic authority and tradition. The truth was there for all to see, given good will and reason and observation. Nor, against the other extreme, did one need to rely on the unaided word of scripture. Indeed, it is an error to think that "the only law which God has appointed unto men" is the word of the Bible (Hooker, *Works*, I, 224, quoted in Olson 1987, 8). In fact, natural theology is not just a prop but an essential part of the Christian's argument. "Nature and Scripture do serve in such full sort that they both jointly and not severally either of them be so complete that unto everlasting felicity we need not the knowledge of anything more than these two may easily furnish.... "(Hooker, *Works*, I, 216, quoted in Olson 1987, 8)

With the argument to design given firm backing by the authorities, the way was now open to exploit the argument to organized complexity – if not in the inorganic world, then in the world of plants and animals. William Harvey's whole approach to the problem of circulation – valves in the veins, the functions of the parts of the heart, and so forth – was teleological through and through, with a total stress on what was best for, or of most value for, an organism and its parts. And then, at the end of the seventeenth century, there was the clergyman-naturalist John Ray (1628–1705) and his *Wisdom of God, Manifested in the Words of Creation* (1691, 5th ed. 1709). First, the argument to adaptive complexity:

Whatever is natural, beheld through [the microscope] appears exquisitely formed, and adorned with all imaginable Elegancy and Beauty. There are such inimitable gildings in the smallest Seeds of Plants, but especially in the parts of Animals, in the Lead or Eye of a small Fry; Such accuracy, Order and Symmetry in the frame of the most minute Creatures, a Louse, for example, or a Mite, as no man were able to conceive without seeming of them.

Everything that we humans do and produce is just crude and amateurish compared to what we find in nature. Then, the argument to design: "There is no greater, at least no more palpable and convincing argument of the Existence of a Deity, than the admirable Art and Wisdom that discovers itself in the Make and Constitution, the Order and Disposition, the Ends and uses of all the parts and members of this stately fabric of Heaven and Earth" (Ray 1709, 32–3).

At the end of the eighteenth century, this happy harmony between science and religion was drowned out by the cymbals clashed together by he who has been described wittily as "God's greatest gift to the infidel." In his *Dialogues Concerning Natural Religion*, David Hume tore into the argument from design.

If we survey a ship, what an exalted idea must we form of the ingenuity of the carpenter, who framed so complicated, useful, and beautiful a machine? And what surprise must we feel, when we find him a stupid mechanic, who imitated others, and copied an art, which, through a long succession of ages, after multiplied trials, mistakes, corrections, deliberations, and controversies, had been gradually improving?

More generally:

Many worlds might have been botched and bungled, throughout an eternity, ere this system was struck out: much labour lost: many fruitless trials made: and a slow, but continued improvement carried on during infinite ages in the art of world-making.

In such subjects, who can determine, where the truth; nay, who can conjecture where the probability, lies; amidst a great number of hypotheses which may be proposed, and a still greater number which may be imagined? (Hume 1779, 140)

This is a counter to the second phase of the argument – against the argument from complexity to a Creator that we might want to take seriously. Hume also went after the argument to complexity itself, that which suggests that there is something special or in need of explanation. In Hume's opinion, we should be careful about making any such inference. We might question whether the world really does have marks of organized, adaptive complexity. For instance, is it like a machine, or is it more like an animal or a vegetable, in which case the whole argument collapses into some kind of circularity or regression? It is certainly true that we seem to have a balance of nature, with change in one part affecting and being compensated by change in another part, just as we have in organisms. But this seems to imply a kind of non-Christian pantheism. "The world, therefore, I infer, is an animal, and the Deity is the SOUL of the world, actuating it, and actuated by it" (pp. 143–4). And if this is not enough – going back again to the argument for a Designer – there is the problem of evil. This is something that apparently belies the optimistic conclusions – drawn by enthusiasts for the design argument from Socrates on – about the Designer. As Hume asked, if God did design and create the world, how do you account for all that is wrong within it? If God is all-powerful, He could prevent evil. If God is all-loving, He would prevent evil. Why then does it exist? Speaking with some feeling of life in the eighteenth century, Hume asked meaningfully, "what racking pains, on the other hand, arise from gouts, gravels, megrims, tooth-aches, rheumatisms; where the injury to the animal-machinery is either small or incurable?" (p. 172) Not much "divine benevolence" displaying itself here, I am afraid.

For students in philosophy classes today, this tends to be the end of matters. The argument from design is finished, and it is time to move on. For people at the end of the eighteenth century, this was anything but the end of matters. Indeed, even Hume himself, at the end of his *Dialogues*, rather admitted that he had proven too much. The argument still has some force. If the proposition before us is that "*the cause or causes of order in the universe probably bear some remote analogy to human intelligence,*" then "what can the most inquisitive, contemplative, and religious man do more than give a plain, philosophical assent to the proposition, as often as it occurs; and believe that the arguments, on which it is established, exceed the objections, which lie against it?" (Hume 1779, 203–4, his italics) The official counterblast came from a Christian apologist, Archdeacon William Paley of Carlisle. Warming up for the argument to complexity:

In crossing a heath suppose I pitched my foot against a stone, and were asked how the stone came to be there, I might possibly answer, that for any thing I knew to the contrary it had lain there for ever; nor would it, perhaps, be very easy to show the absurdity of this answer. But supposing I had found a watch upon the ground, and it should be inquired how the watch happened to be in that place, I should hardly think of the answer which I had before given, that for any thing I knew the watch might have always been there. Yet why should not this answer serve for the watch as well as for the stone; why is it not as admissible in the second case as in the first? For this reason, and for no other, namely, that when we come to inspect the watch, we perceive – what we could not discover in the stone – that its several parts are framed and put together for a purpose, e.g. that they are so formed and adjusted as to produce motion, and that motion so regulated as to point out the hour of the day; that if the different parts had been shaped different from what they are, or placed after any other manner or in any other order than that in which they are placed, either no motion at all would have been carried on in the machine, or none which would have answered the use that is now served by it. (Paley 1819, 1)

A watch implies a watchmaker. Likewise, the adaptations of the living world imply an adaptation maker, a Deity. The argument to design. You cannot argue otherwise without falling into absurdity. "This is atheism; for every indication of contrivance, every manifestation of design which existed in the watch, exists in the works of nature, with the difference on the side of nature of being greater and more, and that in a degree which exceeds all computation" (p. 14).

After Hume, how was Paley able to get away with it? More pertinently, after Hume, how did Paley manage to influence so many of his readers? Do logic and philosophy have so little effect? The philosopher Elliott Sober (2000) points to the answer. Prima facie, Paley is offering an analogical argument. The world is like a machine. Machines have designers/makers. Hence, the world has a designer/maker. Hume had roughed this up by suggesting that the world is not much like a machine, and that even if it is, one cannot then argue to the kind of machine-maker/designer usually identified with the Christian God. But this is not really Paley's argument. He is offering what is known as an "inference to the best explanation." There has to be some causal explanation of the world. All explanations other than one supposing a designing mind – or rather a Designing Mind – are clearly inadequate. Hence, whatever the problems, the causal explanation of the world has to be a Designing Mind. If design remains the only explanation that can do the job, then at one level all of the counterarguments put forth by Hume fall away. As Sherlock Holmes, speaking to his friend Dr. Watson, put it so well: "How often have I told you that when you have eliminated the impossible, whatever remains, *however improbable*, must be the truth." This is not to say that Hume's critical work was wasted. The believer who was prepared to face up to what Hume had argued would now know (or should now know) that the Designer is a lot less humanlike than most confidently suppose. But for the critical work to be fatal to the existence of the Designer, it would be necessary to wait until another viable hypothesis presented itself. Then, inasmuch as it rendered the design hypothesis improbable, it could come into play.

"Another viable hypothesis." There's the rub! For the first half of the nineteenth century, no one had such a hypothesis, and so the argument from design flourished as never before. The eight Bridgewater Treatises – "On the power, wisdom and goodness of God as manifested in the creation" – were taken as definitive. The Reverend William Buckland, professor of geology at the University of Oxford, drew the reader's attention to the world's distribution of coal, ores, and other minerals. He concluded that this showed not only the designing nature of wise Providence, but also the especially favoured status of a small island off the coast of mainland Europe. "We need no further evidence to shew that the presence of coal is, in an especial degree, the foundation of increasing population, riches, and power, and of improvement in almost every Art which administers to the necessities and comforts of Mankind." It took much time and forethought to lay down those strata, but their very existence, lying there for "the future uses of Man, formed part of the design, with which they were, ages ago, disposed in a manner so admirably adapted to the benefit of the Human Race" (Buckland 1836, I, 535–8). It helps, of course, that God is an Englishman. The location of vital minerals "expresses the most clear design of Providence to make the inhabitants of the British Isles, by means of this gift, the most powerful and the richest nation on earth" (Gordon 1894, 82).

CHARLES DARWIN

Darwin was not the first evolutionist. His grandfather Erasmus Darwin put forward ideas sympathetic to the transmutation of species at the end of the eighteenth century, and the Frenchman Jean Baptiste de Lamarck did the same at the beginning of the nineteenth. But it was Charles Darwin who made the fact of evolution secure and who proposed the mechanism – natural selection – that is today generally considered by scientists to be the key factor behind the development of organisms (Ruse 1979): a development by a slow natural process from a few simple forms, and perhaps indeed ultimately from inorganic substances. In the *Origin*, after first stressing the analogy between the world of the breeder and the world of nature, and after showing how much variation exists between organisms in the wild, Darwin was ready for the key inferences. First, an argument to the struggle for existence and, following on this, an argument to the mechanism of natural selection.

A struggle for existence inevitably follows from the high rate at which all organic beings tend to increase. Every being, which during its natural lifetime produces several eggs or seeds, must suffer destruction during some period of its life, and during some season or occasional year, otherwise, on the principle of geometrical increase, its numbers would quickly become so inordinately great that no country could support the product. Hence, as more individuals are produced than can possibly survive, there must in every case be a struggle for existence, either one individual with another of the same species, or with the individuals of distinct species, or with the physical conditions of life. It is the doctrine of Malthus applied with manifold force to the whole animal and vegetable kingdoms; for in this case there can be no artificial increase of food, and no prudential restraint from marriage. (Darwin 1859, 63)

Now, natural selection follows at once.

Let it be borne in mind in what an endless number of strange peculiarities our domestic productions, and, in a lesser degree, those under nature, vary; and how strong the hereditary tendency is. Under domestication, it may be truly said that the whole organization becomes in some degree plastic. Let it be borne in mind how infinitely complex and close-fitting are the mutual relations of all organic beings to each other and to their physical conditions of life. Can it, then, be thought improbable, seeing that variations useful to man have undoubtedly occurred, that other variations useful in some way to each being in the great and complex battle of life, should sometimes occur in the course of thousands of generations? If such do occur, can we doubt (remembering that many more individuals are born than can possibly survive) that individuals having any advantage, however slight, over others, would have the best chance of surviving and of procreating their kind? On the other hand we may feel sure that any variation in the least degree injurious would be rigidly destroyed. This preservation of favourable variations and the rejection of injurious variations, I call Natural Selection. (80–1)

With the mechanism in place, Darwin now turned to a general survey of the biological world, offering what the philosopher William Whewell (1840) had dubbed a "consilience of inductions." Each area was explained by evolution through natural selection, and in turn each area contributed to the support of the mechanism of evolution through natural selection. Geographical distribution (biogeography) was a triumph, as Darwin explained just why it is that one finds the various patterns of animal and plant life around the globe. Why, for instance, does one have the strange sorts of distributions and patterns that are exhibited by the Galapagos Archipelago and other island groups? It is simply that the founders of these isolated island denizens came by chance from the mainlands and, once established, started to evolve and diversify under the new selective pressures to which they were now subject. Embryology, likewise, was a particular point of pride for Darwin. Why is it that the embryos of some different species are very similar – man and the dog, for instance – whereas the adults are very different? Darwin argued that this follows from the fact that in the womb the selective forces on the two embryos would be very similar – they would not therefore be torn apart – whereas the selective forces on the two adults would be very different – they would be torn apart. Here, as always in his discussions of evolution, Darwin turned to the analogy with the world of the breeders in order to clarify and support the point at hand. "Fanciers select their horses, dogs, and pigeons, for breeding, when they are nearly grown up: they are indifferent whether the desired qualities and structures have been acquired earlier or later in life, if the full-grown animal possesses them" (Darwin 1859, 446).

All of this led to that famous passage at the end of the *Origin*: "There is a grandeur in this view of life, with its several powers, having been orig-inally breathed into a few forms or into one; and that, whilst this planet has gone cycling on according to the fixed law of gravity, from so simple a beginning endless forms, most beautiful and most wonderful have been, and are being, evolved" (Darwin 1859, 490). But what of the argument from design? What of organized complexity? What of the inference to design? Darwin's evolutionism impinged significantly on both of these stages of the main argument. With respect to organic complexity, at one level no one could have accepted it or have regarded it as a significant aspect of living nature more fully than Darwin. To use Aristotle's language, no one could have bought into the idea of final cause more than the author of the *Ori-gin of Species*. This was Darwin's starting point. He accepted completely that the eye is for seeing and the hand is for grasping. These are the adapta-tions that make life possible. And more than this, it is these adaptations that natural selection is supplied to explain. Organisms with good adaptations survive and reproduce. Organisms without such adaptations wither and die without issue. Darwin had read Paley and agreed completely about the dis-tinctive nature of plants and animals. At another level, Darwin obviously pushed adaptive complexity sideways somewhat. It was very much part of his evolutionism that not everything works perfectly all of the time. And some features of the living world have little or no direct adaptive value. Homol-ogy, for instance – the isomorphisms between organisms of very different natures and lifestyles – is clearly a mark of common descent, but it has no direct utilitarian value. What end does it serve that there are similarities between the arm of humans, the forelimb of horses, the paw of moles, the flipper of seals, the wings of birds and bats? There is adaptive complexity, and it is very important. It is not universal.

What about the argument to design? Darwin was never an atheist, and although he died an agnostic, at the time of the writing of the *Origin* he was a believer of some kind – a deist, probably, believing in a God as unmoved mover, who had set the world in motion and then stood back from the creation as all unfurled through unbroken law. So Darwin certainly did not see his theory as proving there is no God. But he certainly saw his theory as taking God out of science and as making nonbelief a possibility – as supplying that missing hypothesis on the absence of which Paley had relied. And more than that. Darwin saw the presence of natural evil brought on by natural selection as threatening to the Christian conception of God. To the

American botanist Asa Gray (a firm believer), Darwin wrote:

With respect to the theological view of the question; this is always painful to me. – I am bewildered. – I had no intention to write atheistically. But I own that I cannot see, as plainly as others do, & as I sh^d. wish to do, evidence of design & beneficence on all sides of us. There seems to me too much misery in the world. I cannot persuade myself that a beneficent & omnipotent God would have designedly created the Ichneumonidae with the express intention of their feeding within the living bodies of caterpillars, or that a cat should play with mice. Not believing this, I see no necessity in the belief that the eye was expressly designed. On the other hand I cannot anyhow be contented to view this wonderful universe & especially the nature of man, & to conclude that everything is the result of brute force. I am inclined to look at everything as resulting from designed laws, with the details, whether good or bad, left to the working out of what we may call chance. Not that this notion *at all* satisfies me. I feel most deeply that the whole subject is too profound for the human intellect. A dog might as well speculate on the mind of Newton. – Let each man hope & believe what he can. (Letter to Asa Gray, May 22, 1860)

Then later:

One more word on "designed laws" & "undesigned results." I see a bird which I want for food, take my gun & kill it, I do this *designedly*. – An innocent & good man stands under tree and is killed by flash of lightning. Do you believe (& I really shd. like to hear) that God *designedly* killed this man? Many or most persons do believe this; I can't and don't. – If you believe so, do you believe that when a swallow snaps up a gnat that God designed that that particular sparrow shd. snap up that particular gnat at that particular instant? I believe that the man & the gnat are in the same predicament. – If the death of neither man or gnat are designed, I see no good reason to believe that their *first* birth or production shd. be necessarily designed. Yet, as I said before, I cannot persuade myself that electricity acts, that the tree grows, that man aspires to loftiest conceptions all from blind, brute force. (Darwin 1985–, VIII, 275)

To sum up: Darwin stressed the argument to adaptive complexity, even as he turned it to his own evolutionary ends. He made – or, if you prefer, claimed to make – the argument to design redundant by offering his own naturalistic solution, evolution through natural selection. In other words, Darwin endorsed internal teleology to the full. He pushed external teleology out of science. He did not, and did not claim to, destroy all religious belief, but he did think that his theory exacerbated the traditional problem (for the Christian) of design. How could a good God allow such a painful process of development?

FROM DARWIN TO DAWKINS

Scientifically speaking, after the *Origin*, evolution was a great success. Natural selection was not. In certain respects, it was Herbert Spencer rather than Charles Darwin who came to epitomize the evolutionary perspective. Adaptation was downplayed, and selection was sidelined. In its stead, evolution was linked to the popular ideology of the day – progress, from blob to vertebrate, from ape to human, from primitive to civilized, from savage to Englishman. From what Spencer termed the uniform "homogenous" to what he termed the differentiated "heterogenous."

Now we propose in the first place to show, that this law of organic progress is the law of all progress. Whether it be in the development of the Earth, in the development of Life upon its surface, in the development of Society, of Government, of Manufactures, of Commerce, of Language, Literature, Science, Art, this same evolution of the simple into the complex, through successive differentiations, hold throughout. From the earliest traceable cosmical changes down to the latest results of civilization, we shall find that the transformation of the homogeneous into the heterogeneous, is that in which Progress essentially consists. (Spencer 1857, 244)

It was not until the 1930s and the coming of Mendelian genetics (as generalized across populations) that Darwinism in the strict sense really took off as a functioning professional science (Ruse 1996). People such as E. B. Ford in England and his school of ecological genetics, and Theodosius Dobzhansky in America and his fellow supporters of the synthetic theory (a synthesis of Darwin and Mendel) used natural selection as a tool of inquiry. They built up a paradigm (to use a term) that had adaptation as the central focus – as the problem to be solved. Classic work was done in Britain on snail shell colours (showing the camouflage value of different patterns against different backgrounds) by such people as Philip Sheppard and Arthur Cain. Equally classic work was done in America on issues to do with variation in natural and artificial populations, especially as found in fruitflies (*Drosophila*), by Dobzhansky and his students, such as Bruce Wallace and Richard Lewontin. In the 1960s and 1970s, the Darwinian programme became, if anything, even more intense and successful as researchers turned increasingly to problems to do with behaviour, especially social behaviour as exhibited by such organisms as the *Hymenoptera* (the ants, the bees, and the wasps). New ways of using selection were developed – for instance, through models of "kin selection" as devised by the late William Hamilton, where selection is seen to be something that can promote adaptations that aid the reproductive efforts of close relatives. At the same time, the molecular revolution came to evolutionary biology, and, through such techniques as gel electrophoresis, it became possible to study the effects of natural selection as they are recorded in heredity, especially as they show themselves in the molecular gene, the DNA molecule.

The argument to adaptive complexity has thriven, although one should not think that ultra-Darwinians have always had it their own way. A sizeable minority of evolutionary biologists – the most notable being the late Stephen Jay Gould – has always warned against seeing function and purpose in every last feature of organic nature. Gould (2002), particularly, argued that if one looks at the long-term results of evolution as shown in the fossil record, one finds good reason to think that selection is but one player in the causal band. He is well known for having formulated, with fellow paleontologist Niles Eldredge, the theory of punctuated equilibria, where change goes in jumps and where selection and adaptation have but lesser roles (Eldredge and Gould 1972; Ruse 1999). In recent years, those who warn against seeing too much (natural) design in nature have been joined by people working on development, trying to integrate this area of biological inquiry fully into the evolutionary picture (so-called "evo-devo"). The mantra of these enthusiasts is that of constraint. It is argued that nature is constrained in various ways – physical and biological – and that, hence, full and efficient adaptation is rarely, if ever, possible. Some of the most remarkable of recent findings are of homologies that exist at the micro (even the genetic) level between organisms of very different kinds (humans and fruitflies, for instance). This leads to such conclusions as:

The homologies of process within morphogenetic fields provide some of the best evidence for evolution – just as skeletal and organ homologies did earlier. Thus, the evidence for evolution is better than ever. The role of natural selection in evolution, however, is seen to play less an important role. It is merely a filter for unsuccessful morphologies generated by development. Population genetics is destined to change if it is not to become as irrelevant to evolution as Newtonian mechanics is to contemporary physics. (Gilbert, Opitz, and Raff 1996, 368)

Now is not the time or place to decide such issues, some of which are perhaps as much verbal as substantive. No Darwinian denies the nonadaptive. No critic (certainly not Gould) denies the adaptive. It is more a question of where one draws lines and what one thinks these lines signify. What cannot be denied – and would not be denied by any evolutionist today – is that these are exciting times and that natural selection and adaption have some role – an important role – to play in the whole story. The question is: how big is important? And the question for us here is: how relevant is any of this to traditional theological concerns, particularly to the second-stage argument *to* design? Did Darwin end, once and for all, that line of thought, or is there hope after the *Origin*? And is such hope predicated on (or even further compromised by) the success of the scientific critics of Darwinism, those who deny that selection and adaptation are all-important (or greatly important)? Finishing the historical story, and going back to the history of natural theology after the *Origin*, what we find is that no one was untouched by Darwin's writings. Some put up the barriers. They agreed with Darwin that organized complexity is *the* significant feature of the organic world. Yet they could not see how a process of blind law – and (for all that Darwin argued otherwise) this included Darwinism – could lead to adaptation or contrivance. They took organized complexity as basic and argued that its denial – a denial that followed necessarily from an evolutionary approach – destroyed the best of all arguments to God's existence. So that was that. To the day of his death, this was the position of Charles Hodge, a professor at Princeton Seminary and a leading systematic theologian of his day. Final causes, evidences of design, exist, and they are definitive.

The doctrine of final causes in nature must stand or fall with the doctrine of a personal God. The one cannot be denied without denying the other. And the admission of the one, involves the admission of the other. By final cause is not meant a mere tendency, or the end to which events either actually or apparently tend; but the end contemplated in the use of means adapted to attain it. (Hodge 1872, 227)

Rhetorically, the title of one of Hodge's books asked: "What is Darwinism?" Came the stern reply: "It is atheism."

It is . . . neither evolution nor natural selection, which give Darwinism its peculiar character and importance. It is that Darwin rejects all teleology, or the doctrine of final causes. He denies design in any of the organisms in the vegetable or animal world. . . . As it is this feature of his system which brings it into conflict not only with Christianity, but with the fundamental principles of natural religion, it should be clearly established. (Hodge 1874, 48–52)

More liberal theologians, however, took a somewhat different tack. Remember, even though people were moving toward evolution, this was a time when natural selection was not generally accepted. This was a time when progress took off as the doctrine of the day – a doctrine of the day supposedly firmly backed by evolutionary theorizing. And so the argument from design was modified. The argument to complexity was ignored, or rather transformed into an (evolution-backed) argument to progress. Then this was taken as the premise for a revitalized argument to design. In England, Frederick Temple, a future archbishop of Canterbury, writing in the 1880s, was quite explicit on the need to make the shift from an old-style, natural theology to a new-style, evolution-informed substitute. He stressed that, after Darwin, if one were rewriting the *Natural Theology* of Paley, the emphasis would be different. "Instead of insisting wholly or mainly on the wonderful adaptation of means to ends in the structure of living animals and plants, we should look rather to the original properties impressed on matter from the beginning, and on the beneficent consequences that have flowed from those properties" (Temple 1884, 118–19). And a similar philosophy was pushed at the more popular level. Consider the Reverend Henry Ward Beecher, brother of the novelist, charismatic preacher, and successful adulterer. Progress is his theme, if not his obsession:

If single acts would evince design, how much more a vast universe, that by inherent laws gradually builded itself, and then created its own plants and animals, a universe so adjusted that it left by the way the poorest things, and steadily wrought toward more complex, ingenious, and beautiful results! Who designed this mighty machine, created matter, gave to it its laws, and impressed upon it that tendency which has brought forth the almost infinite results on the globe, and wrought them into a perfect system? Design by wholesale is grander than design by retail. (Beecher 1885, 113)

There is much more to the story. Any full history, from the years after Darwin down to the present, would need to extend the story from Protestants to Catholics, although in fact in this case probably (for all that the Catholic Church gives a special place to the philosophy of Saint Thomas) the tale is not so very different. Certainly, the most important reconciler of science and religion in the twentieth century – the French Jesuit paleontologist Pierre Teilhard de Chardin (1955) – made progress the backbone of his evolution-based natural theology. Again, any full history would need to acknowledge the fact that in the early twentieth century – thanks to Karl Barth (1933), the greatest theologian of the century – natural theology as a very enterprise in itself took a severe beating. Barth complained, with some reason, that the God of the philosophers bears but scant resemblance to the God of the gospels. And although natural theology has recovered somewhat, there seems to be general recognition among theologians that old-fashioned approaches – supposedly proving God's existence beyond doubt – are no longer viable enterprises. What one must now aim for is a theology of nature, where the world informs and enriches belief, even though it cannot substitute for faith.

If the god of the Bible is the creator of the universe, then it is not possible to understand fully or even appropriately the processes of nature without any reference to that God. If, on the contrary, nature can be appropriately understood without reference to the God of the Bible, then that God cannot be the creator of the universe, and consequently he cannot be truly God and be trusted as a source of moral teaching either. (Pannenberg 1993, 16)

Yet for all the reservations and qualifications – perhaps because of the reservations and qualifications – in major respects natural theology (or whatever you call it now), as represented by the argument from design, has not moved that significantly from where it was in the years after Darwin. On the one hand, we have those (in the tradition of Hodge) who want to maintain the good old-fashioned emphasis on organized complexity, on adaptation, and who think that this leads to a definitive proof of a Creator – potentially, at least, the Christian Creator – and who think that evolution by blind law (especially Darwinian evolution by natural selection) denies all of this. These are the people who have embraced, with some enthusiasm, the position known today as Intelligent Design (Behe 1996; Dembski 1998). The flip side to these believers are those (in the tradition of Darwin himself) like Richard Dawkins (1995) and the philosopher Dan Dennett (1995), who are no less ardent about the argument to organized complexity, yet who feel that Darwin made any further inference unnecessary and who feel that the problem of pain makes any further inference of a theological variety impossible.

In a universe of blind physical forces and genetic replication, some people are going to get hurt, other people are going to get lucky, and you won't find any rhyme or reason in it, nor any justice. The universe we observe has precisely the properties we should expect if there is, at bottom, no design, no purpose, no evil and no good, nothing but blind, pitiless indifference. (Dawkins 1995, 133)

Then, on the other hand, we also have today the more mainstream Christians (in the tradition of Temple and Beecher), who accept evolution, who continue to be wary of natural selection, who downplay adaptation and play up progress, and who argue that this mix gives at least an adequate theology of nature, even if a traditional natural theology is no longer possible. Entirely typical is the Colorado State University professor Holmes Rolston III, a philosopher of the environment, an ordained Presbyterian, a trained physicist, and a recent Gifford Lecturer (Rolston 1999). He takes as his text a line from one of my earlier books: "The secret of the organic world is evolution caused by natural selection working on small, undirected variation." And he makes clear that (whatever today's scientists might say), although one must opt for science rather than miracles, a Darwinian approach is not adequate. The direct problem is that of design; but in a more important sense, the issue is one of direction.

The troublesome words are these: "chance," "accident," "blind," "struggle," "violent," "ruthless." Darwin exclaimed that the process was "clumsy, wasteful, blundering, low and horribly cruel." None of these words has any intelligibility in it. They leave the world, and all the life rising out of it, a surd, absurd. The process is ungodly; it only simulates design. Aristotle found a balance of material, efficient, formal, and final causes; and this was congenial to monotheism. Newton's mechanistic nature pushed that theism toward deism. But now, after Darwin, nature is more of a jungle than a paradise, and this forbids any theism at all. True, evolutionary theory, being a science, only explains *how* things happened. But the character of this *how* seems to imply that there is no *why*. Darwin seems antitheological, not merely nontheological. (Rolston 1987, 91)

Continuing, Rolston focusses on the need for a mechanism of progress. We have to have some way in which nature guarantees order and upward direction. Apparently we need (and have) a kind of (highly non-Darwinian) rachet effect, whereby molecules spontaneously organize themselves into more complex configurations, that then get incorporated into the living being and thus move the living being up a notch, from which it seems there is no return. Rolston (1987) writes:

But to have life assemble this way, there must be a sort of push-up, lock-up effect by which inorganic energy input, radiated over matter, can spontaneously happen to synthesize negentropic amino acid subunits, complex but partial protoprotein sequences, which would be degraded by entropy, except that by spiraling and folding they make themselves relatively resistant to degradation. They are metastable, locked uphill by a ratchet effect, so to speak, with such folded chains much upgraded over surrounding environmental entropic levels. Once elevated there, they enjoy a thermodynamic niche that conserves them, at least inside a felicitous microspherical environment. (111–12).

This yields a kind of upward progress to life. For all that every "form of life does not trend upslope," finally, as expected, we move up to humankind, God's special creation (116–17). At this point, Rolston (in fact echoing many earlier Christian progressionists) ties in the upward struggle of evolution with the upward struggle of humankind.

What theologians once termed an established order of creation is rather a natural order that dynamically creates, an order for creating. The order and newer accounts both concur that living creatures now exist where once they did not. But the manner of their coming into being has to be reassessed. The notion of a Newtonian Architect who from the outside designs his machines, borrowed by Paley for his Watchmaker God, has to be replaced (at least in biology, if not also in physics) by a continuous creation, a developmental struggle in self-education, where the creatures through "experience" becomes increasingly "expert" at life.

Do not be perturbed by this.

This increased autonomy, though it might first be thought uncaring, is not wholly unlike that self-finding that parents allow their children. It is a richer organic model of creation just because it is not architectural-mechanical. It accounts for the "hit and miss" aspects of evolution. Like a psychotherapist, God sets the context for self-actualizing. God allows persons to be imperfect in their struggle toward fuller lives . . . , and there seems to be a biological analogue of this. It is a part of, not a flaw in, the creative process. (131)

CONCLUSION

I draw to the end of my survey of twenty-five hundred years of the argument from design. Deliberately, I have tried to be nonjudgmental, merely telling the story of the ideas as they appeared in history. But, as I conclude, I cannot resist drawing an obvious inference from my history. Intelligent Design theorists and atheistical Darwinians cannot both be right, but they are both surely right in thinking that they are more in tune with modern evolutionary biology than are the mainstream reconcilers of science and religion. For all the qualifications, adaptation – organized complexity – is a central aspect of the living world, a central aspect of the work and attentions of the professional evolutionist today. Any adequate natural theology – or theology of nature – must start with this fact. One must indeed cherish this fact and make it a strength of one's position rather than something to be acknowledged quickly and rather guiltily, and then ignored. I say this irrespective of related questions about whether or not progress is truly the theme of life's history, although I should note that its existence is a highly contentious question in evolutionary circles. Highly contentious in theological circles also, for many believers deny the idea that unaided humans can in some sense improve their lot. Progress is seen to be in conflict with the Christian notion of Providence, where God's unmerited Grace alone gives salvation. But these are topics for another discussion. Here, I simply leave you with the reflection that the argument from design has had a long and (I would say) honorable history, and that this history seems still to be unfinished.

Note

1. This essay draws heavily on my full-length discussion of the topic in *Darwin and Design: Does Evolution Have a Purpose?*

References

Aquinas, T. 1952. *Summa Theologica, I.* London: Burns, Oates and Washbourne.

Augustine. 1998. *The City of God against the Pagans*, ed. and trans. R. W. Dyson. Cambridge: Cambridge University Press.

Bacon, F. [1605] 1868. *The Advancement of Learning.* Oxford: Clarendon Press.

Barnes, J. (ed.) 1984. *The Complete Works of Aristotle, Volume I.* Princeton, NJ: Princeton University Press.

Barth, K. 1933. *The Epistle to the Romans.* Oxford: Oxford University Press.

Beecher, H. W. 1885. *Evolution and Religion.* New York: Fords, Howard, and Hulbert.

Behe, M. 1996. *Darwin's Black Box: The Biochemical Challenge to Evolution.* New York: The Free Press.

Buckland, W. 1836. *Geology and Mineralogy* (Bridgewater Treatise, 6). London: William Pickering.

Cooper, J. M. (ed.) 1997. *Plato: Complete Works.* Indianapolis: Hackett.

Darwin, C. 1859. *On the Origin of Species.* London: John Murray.

      1985-. *The Correspondence of Charles Darwin.* Cambridge: Cambridge University Press.

Dawkins, R. 1983. Universal Darwinism. In *Molecules to Men*, ed. D. S. Bendall. Cambridge: Cambridge University Press.

      A River Out of Eden. New York: Basic Books.

Dembski, W. 1998. *The Design Inference: Eliminating Chance through Small Probabilities.* Cambridge: Cambridge University Press.

Dennett, D. C. 1995. *Darwin's Dangerous Idea.* New York: Simon and Schuster.

Eldredge, N., and S. J. Gould. 1972. Punctuated equilibria: An alternative to phyletic gradualism. In *Models in Paleobiology*, ed. T. J. M. Schopf. San Francisco: Freeman, Cooper, pp. 82–115.

Galen. 1968. On the usefulness of the parts of the body, trans. M. T. May. Ithaca, NY: Cornell University Press.

Gilbert, S. F., J. M. Opitz, and R. A. Raff. 1996. Resynthesizing evolutionary and developmental biology. *Developmental Biology* 173: 357–72.

Gordon, E. O. 1894. *The Life and Correspondence of William Buckland.* London: John Murray.

Gould, S. J. 2002. *The Structure of Evolutionary Theory*. Cambridge, MA.: Harvard University Press.

Hodge, C. 1872. *Systematic Theology*. London and Edinburgh: Nelson.

1874. *What Is Darwinism?* New York: Scribner's.

Hooker, R. 1845. *The Collected Works*. Oxford: Oxford University Press.

Hume, D. [1779] 1963. Dialogues concerning natural religion. In *Hume on Religion*, ed. R. Wollheim. London: Fontana, pp. 99–204.

Maynard Smith, J. 1969. The status of neo-Darwinism. In *Towards a Theoretical Biology*, ed. C. H. Waddington. Edinburgh: Edinburgh University Press.

Olson, R. 1987. On the nature of God's existence, wisdom, and power: The interplay between organic and mechanistic imagery in Anglican natural theology – 1640–1740. In *Approaches to Organic Form: Permutations in Science and Culture*, ed. F. Burwick. Dordrecht: Reidel, pp. 1–48.

Paley, W. [1802]1819. *Natural Theology (Collected Works*, IV). London: Rivington.

Pannenberg, W. 1993. *Towards a Theology of Nature*. Louisville: Westminster/John Knox Press.

Ray, J. 1709. *The Wisdom of God, Manifested in the Works of Creation*, 5th ed. London: Samuel Smith.

Rolston, III. H. 1987. *Science and Religion*. New York: Random House.

1999. *Genes, Genesis and God: Values and Their Origins in Natural and Human History*. Cambridge: Cambridge University Press.

Ruse, M. 1979. *The Darwinian Revolution: Science Red in Tooth and Claw*. Chicago: University of Chicago Press.

1996. *Monad to Man: The Concept of Progress in Evolutionary Biology*. Cambridge, MA.: Harvard University Press.

1999. *Mystery of Mysteries: Is Evolution a Social Construction?* Cambridge, MA.: Harvard University Press.

2003. *Darwin and Design: Does Evolution Have a Purpose?* Cambridge, MA.: Harvard University Press.

Sober, E. 2000. *Philosophy of Biology*, 2nd ed. Boulder, CO.: Westview.

Spencer, H. 1857. Progress: Its law and cause. *Westminster Review* 67: 244–67.

Teilhard de Chardin, P. 1955. *Le Phénomème Humaine*. Paris: Editions de Seuil.

Temple, F. 1884. *The Relations between Religion and Science*. London: Macmillan.

Whewell, W. 1840. *The Philosophy of the Inductive Sciences*. 2 vols. London: Parker.

Xenophon. 1994. *Memorabilia*, trans. A. L. Bonnette. Ithaca, NY: Cornell University Press.

3

Who's Afraid of ID?

Angus Menuge

Intelligent Design (ID) argues that intelligent causes are capable of leaving empirically detectable marks in the natural world. Aspiring to be a scientific research program, ID purports to study the effects of intelligent causes in biology and cosmology. It claims that the best explanation for at least some of the appearance of design in nature is that this design is actual. Specifically, certain kinds of complex information found in the natural world are said to point convincingly to the work of an intelligent agency. Yet for many scientists, any appearance of design in nature ultimately derives from the interplay of undirected natural forces. What's more, ID flies in the face of the methodological naturalism (MN) that prevails throughout so much of science. According to MN, although scientists are entitled to religious beliefs and can entertain supernatural entities in their off time, within science proper they need to proceed as if only natural causes are operative.

As compared to its distinguished colleagues – Darwinism, self-organization, and theistic evolution – Intelligent Design is the new kid on the block. Bursting onto the public scene in the 1990s, ID was greeted with both enthusiastic acceptance and strong opposition. For some, ID provides a more inclusive and open framework for knowledge that reconnects science with questions of value and purpose. For others, ID represents the latest incarnation of creationism, a confusion of religion and science that falsifies both. Between these polar reactions lie more cautious approaches, concerned that ID has not produced much scientific fruit but open to the idea that it may have something valuable to contribute. Many people, however, are just confused. They do not know what to make of ID, because they do not know where it came from nor where it is ultimately headed. The purpose of this rather modest introductory essay is to clarify the origins and goals of the ID movement.

1. THE ORIGINS OF ID

Barbara Forrest has recently published a history of ID that is well documented and informative and yet, in several important respects, inaccurate and misleading (Forrest 2001). So it is necessary to set the factual record straight, and to explain why it is that highly qualified critics of ID, such as many of the contributors to this volume, are willing to engage it in serious debate.

According to Forrest, the ID movement is "the most recent – and most dangerous – manifestation of creationism" (Forrest 2001, 5). What ID really reflects is a "wedge strategy" – a proposal due to Phillip Johnson – which aims to drive a wedge between empirical scientific practice and methodological naturalism, allowing scientists to pursue the former without commitment to the latter. Although ID claims to offer scientific proposals, Forrest argues that its origin is entirely religious.

The "Wedge," a movement – aimed at the court of public opinion – which seeks to undermine public support for teaching evolution while cultivating support for intelligent design, was not born in the mind of a scientist . . . or from any kind of scientific research, but out of personal difficulties . . . which led to Phillip Johnson's conversion to born-again Christianity (6).

Furthermore, Forrest contends, ID "really has nothing to do with science" (30). The real goal, apparently, is to make scientists think of the religious implications of their work. However, "[n]ot a single area of science has been affected in any way by intelligent design theory" (30).

In fact, Forrest thinks that the idea that ID has something to contribute to science is a deliberately cultivated deception. The real strategy, she claims, is revealed in the so-called Wedge Document. This document outlines a five-year plan for implementing the wedge strategy under the auspices of the Center for the Renewal of Science and Culture[2] (13–14). Forrest thinks that the Wedge Document reveals the hidden agenda of the Intelligent Design movement, namely "the overthrow of materialism" and the promotion of "a broadly theistic understanding of nature" (from the Introduction of the Wedge Document, quoted in Forrest 2001, 14). It is apparently this view that leads Forrest and Paul Gross to suggest that ID is a Trojan horse, with religious warriors hidden by the trappings of science (Forrest and Gross 2003).

I see no good reason to deny the existence of the Wedge Document or of Phillip Johnson's wedge strategy. Nonetheless, Forrest's account is wrong on several matters of fact, and her interpretation of those facts trades on a number of fallacious inferences.

1.1. Stealth Creationism?

According to Forrest and other critics (Coyne 2001; Pennock 1999, 2001; Ussery 2001), ID is stealth creationism. However, it can be argued that ID is significantly different from traditional varieties of creationism and that it has been quite public about its goals.

Of course, critics are entitled to argue that the creationist shoe ultimately fits, but for those who know little about ID, it is misleading to claim that ID is a creationist movement. It is not merely that proponents of ID do not refer to their movement as "Intelligent Design Creationism." There are also substantial differences between the philosophy of ID and the view historically espoused by Young Earth and Old Earth creationists. As a scientific proposal, ID does not start from the idea of an inerrant biblical text, and it does not try to find evidence that backs up specific historical claims derived from a literal (or even poetic) reading of Genesis. Further, it is false as a matter of fact that all of the current members of ID derive, by descent with modification, from earlier forms of creationism. For example, before migrating to ID, Dembski was a theistic evolutionist, and Dean Kenyon thought that he had provided a thoroughly naturalistic account of the origin of life (Kenyon and Steinman 1969). Most importantly, ID never claims that an empirically based design inference by itself establishes the identity, character, or motives of the designer. This is because the design inference as developed by Behe and Dembski depends entirely on the empirical character of the effect – its irreducible or specified complexity – and not on the presumed character of the agent that caused it.

That we can make such a distinction is shown by our experience of making design inferences in the human case. Suppose that Colonel Mustard has died in mysterious circumstances at his country home. We are confident that his death was not necessary, a consequence of his worsening gout or some other ailment. The facts make it highly unlikely that Colonel Mustard's death was the result of a chance event, such as a tragic accident while cleaning his military antiques. No, the evidence is that Mustard died because a crossbow bolt fired from twenty paces impaled him, and this has all the marks of design. Yet we do not know if the agent was Ms. Scarlet or Professor Plum, or if the motive was avarice or class warfare. We may, of course, find independent evidence that narrows down the list of suspects and homes in on the most likely motive. But none of that is necessary in order to infer design.

Now if we know how to detect design and are confident that no human could reasonably be responsible for it, there seems no reason in principle why we might not detect the marks of nonhuman (alien, artificially intelligent, or supernatural) design (Ratzsch 2001, 118–20). This is the premise of the Search for Extra Terrestrial Intelligence (SETI), of research in Strong Artificial Intelligence (which aims to make genuinely intelligent automata), and of those who think that only a being rather like God could explain the exquisite balance of the fine structure constants and the apparent fine-tuning of the cosmos for life. Nonetheless, according to ID, these suggestions about the likely identity of the designer are not necessary in order to detect design in the first place.

It is frequently replied, with a knowing nod and a wink, that proponents of ID still really think the designer is You Know Who. The suggestion is that the anonymous designer is a politically convenient fiction, a sugarcoating to make the underlying pill of creationism more palatable to those who would otherwise contest the relevance of religion to scientific practice. However, this response makes a number of doubtful assumptions. First, it assumes that all proponents of ID are religious believers, and this is false: some, such as Michael Denton, are agnostics. Besides this, as Ruse's introductory chapter points out, Aristotle accepted the design inference without the motivation of revealed religion. And we might add that Einstein thought that the success of mathematical physics depended on some ordering logos in the cosmos, even though he was far from being an orthodox Jew or Christian. But in any case, it is simply a fallacy to argue that since those proponents of ID who are believers identify the designer with God, this is what they are claiming can be inferred from the scientific evidence. Rather, this conclusion is drawn from a combination of the scientific facts and a theological and metaphysical interpretation. Theistic evolutionists and Darwinian Christians can see the fallacy in reverse when Richard Dawkins and William Provine claim to infer atheism from evolutionary theory, as if the unvarnished scientific evidence had established that atheological conclusion.

Given the clear contrasts between ID and traditional creationism, it seems plausible that the pejorative "creationist" label is used chiefly to encourage an attitude of dismissive rejection, which avoids engagement with ID's proposals.

The other main problem with Forrest's characterization is its suggestion that ID is really a conspiracy, that there is (or was, until her sleuthing uncovered it) a hidden agenda to undermine scientific materialism. This idea does not hold water, because there is nothing in the Wedge Document that has not been publicized elsewhere for quite some time. In fact, although Johnson did not use the term "wedge" in his first main book on evolution, *Darwin on Trial,* the idea of distinguishing the empirical methodology of science from a commitment to naturalism is already present in that book.

Naturalism and empiricism are often erroneously assumed to be very nearly the same thing, but they are not. In the case of Darwinism, these two foundational principles of science are in conflict. (Johnson 1993, 117)

And one cannot know Phillip Johnson and suppose that he is the sort of person who minces his words or keeps things under wraps. This is how he ends the same book.

Darwinian evolution ... makes me think of a great battleship. ... Its sides are heavily armored with philosophical barriers to criticism, and its decks are stacked with big rhetorical guns ready to intimidate would-be attackers. ... But the ship has sprung a metaphysical leak. ... (Johnson 1993, 169)

More generally, it is hard to reconcile the picture of a secret society with the fact that proponents of ID have participated in so many public conferences, presentations, and radio shows, making their opposition to scientific materialism perfectly clear. Is it really a Trojan horse if all the soldiers are on the outside waving their spears? And how secret can the wedge strategy have been after Johnson published *The Wedge of Truth: Splitting the Foundations of Naturalism* (2000)?

1.2. Life before Johnson

Forrest contends that ID is fundamentally a religious movement, not a scientific one. Part of her reason for saying this is the clear Christian orientation and motivation of Phillip Johnson after his conversion. This makes Johnson seem like a man with a religious mission to attack evolution, and lends credence to the idea that the scientists he recruited were fundamentally of the same mind. But first, it is worth pointing out that this commits the genetic fallacy, since it is erroneously claimed that ideas cannot have scientific merit if they have a religious motivation. No one thinks it is a serious argument against the scientific discoveries of Boyle, Kepler, and Newton that they all believed in divine Providence. (Newton, in fact, thought the *primary* importance of his natural philosophy was apologetics for a Creator.) And likewise, one cannot show that ID is false or fruitless by pointing to the religious (or political) beliefs of its proponents. Contemporary history of science is actually almost univocal in maintaining that religious motivations have made extremely important contributions to science (Brooke and Osler 2001; Harrison 1998; Jaki 2000; Osler 2000; Pearcey and Thaxton 1994).

Furthermore, if religious commitments did detract from the legitimacy of ideas, one could easily point out that secular humanism is also a kind of religion and that Barbara Forrest is a member of the board of directors of the New Orleans Secular Humanist Association.[3] One could then note that Forrest nowhere discloses this fact, either in her essay or in her biography (Pennock 2001, xviii). Is Forrest merely posing as a neutral investigator with the real aim of establishing secular humanism by stealth? Were one prone to conspiracy theories, one could waste quite a lot of time pursuing this line of thought. But it would be a pointless distraction from the real issue – whether or not people's proposals have any merit.

More significantly, Forrest is led to the view that ID is fundamentally a religious movement by an erroneous prior assumption, namely, that the ID movement began with the wedge strategy. This reflects the perception that Johnson, who undoubtedly helped to organize the fledgling design movement, is the intellectual father of ID. But this is simply not the case.

In fact, the contemporary conception of ID received its earliest sharp statement in a book entitled *The Mystery of Life's Origin* (Bradley, Olsen, and Thaxton 1984). This book surveys the various attempts to explain the appearance of life via naturalistic chemical evolution and finds all of them wanting. The idea that life resulted from random reactions in a primeval prebiotic soup is rejected because there is strong evidence of a reducing atmosphere, ultraviolet radiation, and a plethora of chemical cross-reactions, all of which would prevent the formation or stability of important organic molecules.

[B]oth in the atmosphere and in the various water basins of the primitive earth, many destructive interactions would have so vastly diminished, if not altogether consumed, essential precursor chemicals, that chemical evolution rates would have been negligible. (Bradley, Olsen, and Thaxton 1984, 66)

Thus it appears that simple chance is insufficient to explain the first appearance of life. This conclusion is only strengthened by an analysis of the complexity of the simplest self-replicating molecules, as many scientists who are not proponents of ID acknowledge. For example, Cairns-Smith had already noted that

Low levels of cooperation [blind chance] can produce exceedingly easily (the equivalent of small letters and small words), but [blind chance] becomes very quickly incompetent as the amount of organization increases. Very soon indeed long waiting periods and massive material resources become irrelevant. (Cairns-Smith 1971, 95)

At the same time, it is difficult to see how chemical laws could explain the complex aperiodic information found in biological molecules. Bradley, Olsen, and Thaxton argue that even in an open system, thermodynamic principles are incapable of supporting the configurational entropy work needed to account for the coding found in complex proteins and DNA molecules (Bradley, Olsen, and Thaxton 1984, Chapters 7–9). As Stephen Meyer later argued, appeal to a natural chemical affinity does not seem to help either.

[J]ust as magnetic letters can be combined and recombined in any way to form various sequences on a metal surface, so too can each of the four bases A, T, G and C attach to any site on the DNA backbone with equal facility, making all sequences equally probable (or improbable). Indeed, there are no significant affinities between any of the four bases and the binding sites on the sugar-phosphate backbone. (Meyer 2000, 86)

Considerations such as these made it seem that necessity or self-organization could not account for the origin of life either. Nor did it seem to help matters to extend Darwinism to prebiotic conditions and claim that life arose via the interaction of chance and necessity. For, as Theodosius Dobzhansky, one of the great architects of the neo-Darwinist synthesis, had long since pointed out, "prebiological natural selection is a contradiction in terms" (Dobzhansky 1965, 310), since natural selection presupposes the very kind of replicators whose emergence has to be explained.

Now suppose one thinks that there are exactly four possible explanations of the origin of life: chance, necessity, a combination of chance and necessity, and design. And suppose also that one believes one has reason to eliminate the first three candidates. However surprising or bizarre, design is then the rational inference. Along with this purely negative case for design, there is the positive observation that in our experience, intelligent agency is the only known cause of complex specified information.[4] On uniformitarian grounds, therefore, it is plausible to infer that such agency accounts for the biological complexity that appeared in the remote past. Thus according to proponents of ID, it is not some desire to rejuvenate creationism but an emerging crisis in normal, naturalistic science that points to design. It is the discovery that pursuing naturalistic science leads to an unexpected breakdown and our increasing insights into the nature and source of information that put design back on the table for discussion.

A couple of years after Bradley, Olsen, and Thaxton's seminal work, the molecular biologist Michael Denton published a sustained critique of Darwinism, *Evolution: A Theory in Crisis* (1986). Denton pointed out that many of the great biologists who aided in developing systems of morphological classification (for example, Carl Linnaeus, Georges Cuvier, Louis Aggasiz, and Richard Owen) held views antithetical to Darwin's.

The fact that so many of the founders of modern biology, those who discovered all the basic facts of comparative morphology upon which modern evolutionary biology is based, held nature to be fundamentally a discontinuum of isolated and unique types unbridged by transitional varieties . . . is obviously very difficult to reconcile with the popular notion that all the facts of biology irrefutably support an evolutionary interpretation. (Denton 1986, 100).

Denton's own position is close to Cuvier's typological view, according to which

each class of organism . . . possesses a number of unique defining characteristics which occur in fundamentally invariant form in all the species of that class but which are not found even in rudimentary form in any species outside that class. (Denton 1986, 105)

The invariance of these constraints on the biological classes argues that they did not gradually evolve by natural selection, but rather were somehow built in from the beginning. While one way of interpreting this idea is self-organization (Denton's own current position; see Denton 1998), it could also point to some form of design. At any rate, Denton defends his thesis with a number of considerations that proponents of ID have used in their critique of Darwinism. First, Denton notes that there are limits on the kinds of transformations allowed by a gradual series of small changes. Anticipating the work of Behe (1996), Denton notes that complex systems do not remain functional when subjected to local changes, because of the need for compensatory changes in the other, coadapted parts of the system. Thus in a watch,

[a]ny major functional innovation, such as the addition of a new cogwheel or an increase in the diameter of an existing cogwheel, necessarily involves simultaneous highly specific correlated changes throughout the entire cogwheel system. (Denton 1986, 90)

How are such changes to be synchronized and coordinated, if not by design?

Such theoretical considerations are buttressed by a number of empirical arguments against Darwinism. The jewel in the Darwinian crown is the argument from homology, according to which the similarity in certain structures (such as the forelimbs of mammals), despite their varied uses and adaptations, points to a common ancestor and hence to the mechanism of descent with modification. However, Denton argues that the "organs and structures considered homologous in adult vertebrates cannot be traced back to cells or regions in the earliest stages of embryogenesis" (Denton 1986, 146), a point more recently defended by Jonathan Wells (2000). Indeed, "apparently homologous structures are specified by quite different genes in different species," and "non-homologous genes are involved to some extent in the specification of homologous structures" (Denton 1986, 149). It has happened rather often that apparent cases of homology were really cases only of analogy or convergence, which cannot support common descent with modification.

Further, Darwin's theory predicts the existence of numerous transitional forms, but the evidence of their existence seems to be poorly documented by the fossil record. Denton agrees with Stanley, who writes that

[t]he known fossil record fails to document a single example of phyletic (gradual) evolution accomplishing a major morphological transition and hence offers no evidence that the gradualistic model can be valid. (Stanley 1979, 39, quoted in Denton 1986, 182)

Most telling of all, Denton thinks, is the way the coordinated complexity of biological structures makes gradualistic narratives highly implausible. For example, Denton argues against both the "from the tree down" and "from the ground up" theories of the evolution of avian flight.

The stiff impervious property of the feather which makes it so beautiful an adaptation for flight, depends basically on such a highly involved and unique system of coadapted components that it seems impossible that any transitional feather-like structure could possess even to a slight degree the crucial properties. (Denton 1986, 209)

While defenders of Darwinism complain that this is no more than an "Argument From Personal Incredulity" (Dawkins 1996, 38), proponents of ID reply that they are actually giving an argument from probability grounded in the known resources and the creative potential of gradualistic processes, and that it is the Darwinists who are guilty of an "Argument From Personal Credulity" – their belief in some poorly specified causal pathway (see, for example, Dembski 2002, 239–46).

1.3. Johnson and After

Firing a few scientific salvos at Darwinism was an important first step, but it did not by itself cause a discernible alternative movement to coalesce. Unquestionably – and here is the grain of truth in Forrest's history – the ID movement started to take shape as the result of the leadership of Phillip Johnson, a professor of law at Berkeley and an expert on legal reasoning. Forrest's mistake is analogous to supposing that the complete history of a football team starts with the moment that the coach gathers together the players, thereby ignoring the important work the players had already done. Before Johnson ever contacted them, many of the players selected for Johnson's team had independently arrived at conclusions that pointed to Intelligent Design. Michael Behe, Michael Denton, Dean Kenyon, and Henry Schaefer had established scientific careers and were already sympathetic to the idea that design lay behind the universe. Indeed, that was precisely Johnson's reason for recruiting them. It is therefore inappropriate for Forrest to insinuate that proponents of ID obtained their qualifications in order to infiltrate the academy. According to Forrest,

The CRSC creationists [sic] have taken the time and trouble to acquire legitimate degrees, providing them a degree of cover both while they are students and after they join university faculties. (Forrest 2001, 38)

Forrest gives no evidence to back this conspiratorial suggestion, and it surely constitutes an unseemly attack on the academic reputations of some senior scholars. For example, the quantum chemist Henry Schaefer is a CRSC fellow, yet he has been doing scientific research since 1969 (long before Johnson became interested in design), has over 900 science journal publications to his credit, and has been nominated five times for the Nobel Prize.[5]

Forrest is correct that Johnson's involvement with design began shortly after his conversion to Christianity at the age of thirty-eight. In 1987, Johnson was on sabbatical in England.

[H]is doubts about Darwinism had started with a visit to the British Natural History Museum, where he learned about the controversy that had raged there earlier in the 1980s. At that time, the museum paleontologist presented a display describing Darwin's theory as "one possible explanation" of origins. A furor ensued, resulting in the removal of the display, when the editors of the prestigious *Nature* magazine and others in the scientific establishment denounced the museum for its ambivalence about "established fact." (Meyer 2001, 57–8)

Johnson then read two pivotal books: the first edition of Richard Dawkins's *The Blind Watchmaker* and Denton's *Evolution: A Theory in Crisis*. (Notice that both the contents and publication date of the latter book [1986] ought to have told Forrest that design did not begin with Johnson.) After reading these books, Johnson became fascinated with evolution and devoted himself to studying evolutionary theory and using his legal skills to analyze its arguments. He also benefited from conversations with Stephen Meyer, "whose own skepticism about Darwinism had been well cemented by this time" (Meyer 2001, 57) and who happened to be in Cambridge working on a doctorate in the history and philosophy of science.

Johnson's work produced two fruits. First, there was the publication of *Darwin on Trial* in 1991 (revised edition 1993), in which Johnson argued that the scientific establishment had appropriated the word "science" in order to protect their favored naturalistic philosophy. If science is about following the evidence wherever it leads, then why should scientists rule out a priori the possibility of discovering evidence for supernatural design? As we have seen, implicit in this book's thesis was the idea of a wedge that could be driven between the empirical methods of science and the commitment of most scientists to naturalism. This idea led to a movement, whose first major event was a conference held at Southern Methodist University in 1992, featuring Phillip Johnson together with Michael Behe, Stephen Meyer, and William Dembski. Johnson was responsible for making a number of important early contacts, but the movement very soon took on a life of its own and attracted a significant cadre of scientists and philosophers. In 1996, an official organization appeared, the Center for the Renewal of Science and Culture (CRSC), operating under the umbrella of a Seattle-based think tank, the Discovery Institute, which provided fellowship support for scientists critical of Darwinism and supportive of ID. From 1996 to the present, Discovery fellows have appeared at no less than six major conferences, at Biola (1996), the University of Texas at Austin (1997), Baylor University (2000), Concordia University Wisconsin (2000), Yale University (2000), and Calvin College (2001), in addition to many other smaller presentations and symposia. The Baylor and Concordia conferences were particularly significant in that proponents of design faced their best critics in debate.

Along with these conferences have come a number of significant books. Johnson himself has continued his polemical work, with such influential books as *Reason in the Balance* (1995) and *The Wedge of Truth* (2000). There is a substantial collection of philosophical, scientific, and cultural essays drawn from the landmark Biola conference of 1996, entitled *Mere Creation: Science, Faith and Intelligent Design* (Dembski 1998b). From the field of biochemistry, Michael Behe wrote *Darwin's Black Box: The Biochemical Challenge to Evolution* (1996). In this book, Behe argued that modern biochemistry was revealing a world of irreducibly complex molecular machines, inaccessible to gradualistic pathways. Dembski followed this with a rigorous formulation of the conditions under which chance and necessity are insufficient to account for a phenomenon, *The Design Inference: Eliminating Chance through Small Probabilities* (1998a). This work was followed by Dembski's more popular exposition linking faith and science, *Intelligent Design: The Bridge between Science and Theology* (1999) and, more recently, by his rigorous attempt to show that Darwinian mechanisms are incapable of generating complex specified information, *No Free Lunch: Why Specified Complexity Cannot Be Purchased without Intelligence* (2002). Another collection, focused exclusively on the scientific case for design (in cosmology, origin-of-life studies, and biological complexity) is *Science and Evidence for Design in the Universe* (Behe, Dembski, and Meyer 2000). The textbook evidence for Darwinism is critiqued by Jonathan Wells in his *Icons of Evolution* (2000). Eugenie Scott, a well-known advocate of excluding design from biology curricula, admitted that this book would cause a lot of trouble, while strongly criticizing it (Scott 2001).

Alongside the scientific works, a number of philosophers have pressed the case that naturalism, and particularly the Darwinian variety, threatens human rationality and the very enterprise of science. Alvin Plantinga (1993, Chapter 12; 2000, Chapter 7) has suggested that evolutionary naturalism is epistemically self-defeating, because, if it were true, we could never have sufficient warrant to believe it. Michael Rea and Robert Koons (in Craig and Moreland 2000) both argue that naturalism cannot justify some assumptions required by scientific practice. For Rea, the problem is that naturalism cannot explain the modal qualities of particular physical objects. For Koons, the problem is that naturalism cannot account for the reliability of scientists' appeal to aesthetic criteria of theory choice (such as symmetry, coherence, and simplicity), and so cannot hope to resolve Nelson Goodman's famous riddle about the proper way to project observed features into the future. Here the target is not Darwinism but the assumption that naturalism is integral to scientific rationality. If, as Koons argues, theistic assumptions are necessary in order to ground the rationality of science, then, it may be argued, the possibility of empirically detectable, supernatural design can no longer be excluded in principle.

Aside from their purely academic work, proponents of ID have been quite busy with other activities. They have been instrumental in arguing for a broadened discussion of origins in the biology classroom, giving expert testimony, and developing legal briefs. Popularized versions of their work have appeared in magazines, mass circulation books, and on video, aimed at getting the basic ideas out to a wider audience and at influencing upcoming generations. Darwinists have responded at the same popular level, most notably in the recent seven-part PBS series *Evolution* (2001). There is no question that issues of cultural authority and power are important in motivating the current controversy. Some see naturalistic evolution as the very icon of progressive thinking, while others see it as a universal acid[6] that eventually eats its way through every valuable cultural institution.

2. THE FUTURE OF ID

Forrest is quite correct to suggest that the ID movement has by no means fulfilled all of its goals. In the limited space available, I will outline some of the main areas in which ID still has much work to do.

2.1. Acceptance by the Educational Mainstream

Unlike some of the more extreme creationists, who have wanted to ban the teaching of evolution, proponents of ID advance the more modest goal of having their ideas included for discussion in high school and college science classes. They argue that current legislation gives Darwinism a virtual monopoly, ruling substantial criticisms and alternative proposals out of court. According to the ID movement, this is bad for education, because teaching the controversy about a theory helps students to gain an understanding of the theory's strengths and weaknesses. Such open dialogue would also prevent the dogmatic retention of Darwinist theory in the face of strong counterarguments. If science is a critical enterprise analogous to a series of legal trials, then all relevant evidence must be allowed its day in court.

From an ID perspective, it does not help that some members of the scientific establishment appear to intimidate and censor highly qualified critics of Darwinism. In 1990, when the accomplished science writer and inventor Forrest Mims admitted that he questioned evolutionary theory, he was not hired to write the "Amateur Science" column for *Scientific American*. In the ensuing protest, which included many voices opposed to design but even more opposed to viewpoint discrimination, the journal *Science* printed the following:

Even today, some members of the scientific establishment have seemed nearly as illiberal toward religion as the church once was to science. In 1990, for instance, *Scientific American* declined to hire a columnist, Forrest Mims, after learning that he had religious doubts about evolution. (Easterbrook 1997, 891)

Similarly, in 1992, Dean Kenyon, a biology professor at San Francisco State University, was barred from teaching introductory biology classes after he shared his misgivings about evolutionary theory (including his own theory of chemical evolution) with his students.[7]

Mr. Kenyon ... had for many years made a practice of exposing students to both evolutionary theory and evidence uncongenial to it. He also discussed the philosophical controversies raised by the issue and his own view that living systems display evidence of intelligent design. . . . [H]e was yanked from teaching introductory biology and reassigned to labs. . . . Fortunately, San Francisco State University's Academic Freedom Committee ... determined that ... a clear breach of academic freedom had occurred. (Meyer 1993, A14)

The same pattern has been repeated in several other cases, including the well-known removal of William Dembski from his position as director of the Michael Polanyi Center at Baylor University in 2000 (Menuge 2001).

One of ID's long-term goals is to place at major universities more scientists whose work is explicitly shaped by an ID research program. For that reason, it is essential for the ID movement to build bridges with its opponents and to find sympathetic ears in the academy. If sound scholarship produces results that attract the interest of already-established scholars, this will start to happen. There are signs that the younger generation of scientists is more open to pursuing ID than previous generations. These scientists are to be found at ID conferences and in on-line discussion groups.

Nonetheless, there remains a great deal of hostility. Indeed, some scientists are quite willing to defend the way critics of Darwinism have been treated. For example, Arthur Caplan gave the following reasons for siding with *Scientific American* against Forrest Mims:

Forrest Mims is a competent writer and amateur scientist. But his personal beliefs about creation limit what he can and cannot tell his readers about all the nooks and crannies of science. They also distort the picture he conveys regarding what scientific methodology is all about. (Caplan 1991)

And this takes us to the heart of the matter. Most Darwinists see science as inherently committed to methodological naturalism; they argue that this approach is therefore not up for democratic debate. One does not have to accept methodological naturalism, but if one rejects it, then one is no longer viewing the world as a scientist.

In their response to this, proponents of ID argue that a residual positivism makes Darwinists identify the scientific method with endorsement of a particular epistemology and metaphysics, and note that the scientific revolution succeeded with no such commitment. If Boyle, Kepler, and Newton did superb science while believing that the success of the scientific enterprise depended on God's Providence, it does not seem absurd to suggest that science again might flourish in a non-naturalistic framework. But it will be replied that the kind of teleology that once seemed indispensable was shown to be redundant by Darwin (1859). So ultimately everything depends on whether design can be shown to do any work that cannot be reduced to undirected causes.

It is here that critics press the case that ID has not generated significant scientific journal articles or data (Forrest 2001, 23–4). If what counts as science depends on the verdict of peer review, then, it is claimed, ID has yet to establish a track record. In response, proponents of ID have made a number of points. First, they argue that it is not so much new data as the interpretation of existing data that matters. The scientists within the ID movement do perform new experiments; they have published articles in scientific journals (which do not mention ID); and they have also published peer-reviewed work (which does mention ID) outside of scientific journals. However, their main case rests on a reassessment of existing research, much of it performed by Darwinists. After all, it is fallacious to argue that scientific experiments motivated by Darwinism must always support Darwinian theory. If work that is guided by naturalistic assumptions meets with repeated failure, and if one is convinced that there is some principle to this failure, one that excludes all undirected causes, then this work may be used to support ID conclusions. As we have seen, this is precisely the reasoning used by Bradley, Olsen, and Thaxton, and by Denton and Kenyon. Similarly, Behe's claims about irreducible complexity are based in part on recently published work that has unlocked the mechanical structure of the bacterial flagellum. That is one reason that ID scientists are not impressed with the objection that the term "intelligent design" is rarely mentioned in scientific journals. Another reason is, they claim, that many scientific journal editors refuse to publish articles and even letters that explicitly defend ID (see web postings in Behe 2000a and Wells 2002). From the perspective of ID, claiming that no journal articles explicitly support ID is like pointing out that published Chinese government statistics do not support allegations of human rights abuses. Besides that, before the advent of peer review important but highly unpopular scientific work was done outside of journals. (Copernicus' *De revolutionibus*, Newton's *Principia*, and Darwin's *Origin of Species* are examples.) And finally, defenders of ID claim that Darwinists have also failed to publish in important areas; in particular, they have provided few if any causally specific reconstructions of the pathways that lead to the formation of irreducibly complex structures.

Having said all this, proponents of ID keenly feel the sting of the charge that they need more scientific publications. There are results in the pipeline. For example, there is currently research by ID scientists affiliated with the International Society for Complexity, Information and Design (ISCID). According to its web site,[8]

ISCID is a cross-disciplinary professional society that investigates complex systems apart from external programmatic constraints like materialism, naturalism, or reductionism. The society provides a forum for formulating, testing, and disseminating research on complex systems through critique, peer review, and publication. Its aim is to pursue the theoretical development, empirical application, and philosophical implications of information- and design-theoretic concepts for complex systems.

For example, ISCID scientists are studying evolutionary algorithms, aiming to show that Darwinian mechanisms are unable to generate certain kinds of information. One such project is the Monotonic Evolutionary Simulation Algorithm (MESA).[9]

Of course, critics may claim that the real reason that proponents of ID find it difficult to publish is that they are mixing science and religion. This was commonplace in the writings of Newton, but modern science believes that the objectivity of its results depends on excluding religious interpretations. Scientists can of course be religious, but their religious perspectives have no objective validity as science, since scientific statements must be amenable to public verification by everyone, regardless of religious persuasion. To this, proponents of ID reply that their claims do meet standards of public verification, because the criteria for detecting design are empirical and do not depend on a specific metaphysical interpretation. They also point out that naturalism is not a religiously neutral position, and that by excluding non-naturalistic insights, Darwinists are open to the charge of establishing their own naturalistic religion, at least for purposes of intellectual inquiry.

2.2. Theoretical Refinements

At the scientific level, proponents of ID have argued that Darwinian processes are insufficient to account for certain kinds of complexity manifested by biological systems. Behe (1996) has famously argued that some biological structures are irreducibly complex (IC), having a number of well-matched, interacting components, the removal of any one of which disrupts the structure's function. Candidates for IC systems include the cilium, the bacterial flagellum, and the blood-clotting cascade. Behe's main point is that Darwinism requires gradual increments of complexity, each one of which is sufficiently functional to be selected. Yet any supposed precursor p of an IC system s would lack one of s's components, making p nonfunctional and therefore unavailable for selection. So it would seem that irreducibly complex systems would have to be developed all at once, which is beyond the resources of the undirected bottom-up mechanism of Darwinism, but not beyond goal-driven, top-down design.

Critics have responded in a number of ways. They have pointed out that the fact that a precursor system p lacks the function of an IC system s does not show that p has no function. Perhaps p had some *other* function and was simply co-opted. After all, natural selection is a satisficer and works with the materials actually available, not ones it hopes to find later (Miller 1999, 152–8). But others, such as Allen Orr, argue that co-optation is too unlikely to account for highly complex systems with parts delicately adapted to one another: "You may as well hope that half your car's transmission will suddenly help out in the airbag department" (Orr 1996/97, 29). Orr instead prefers a solution that Dembski (2002, 256–61) has dubbed "incremental indispensability":

Some part (A) initially does some job (and not very well, perhaps). Another part (B) later gets added because it helps A. This new part isn't essential, it merely improves things. But later on A (or something else) may change in such a way that B now becomes indispensable. . . . [A]t the end of the day, many parts may all be required. (Orr 1996/97, 29)

Still others have argued that "scaffolding" can support the construction of an otherwise inaccessible structure, such as an arch; when the arch is completed, the scaffolding atrophies, leaving a structure that is IC. Yet others claim that irreducible complexity is an illusion, because in any system that appears to be IC, there is some hidden form of redundancy. For example, John McDonald (web site, 2002) claims that candidate IC systems are actually reducibly complex: provided the reduced set of parts is reconfigured, the same function can be performed. Similarly, Shanks and Joplin (1999) claim that candidate IC systems are in fact redundantly complex.

Proponents of ID have responded to all of these proposals in detail (Behe 2000b, 2001; Dembski 2002). Most fundamentally, they have argued that demonstrating the conceivability of a scenario falls short of establishing its realistic probability. At issue here are rival hermeneutics for the assessment of probability. If Darwinian evolution is accorded a high degree of initial probability based on the many successes that (it is claimed) it has had in other areas, then it does not take much more than a plausible narrative to convince one that it probably works in a difficult case. On the other hand, if Darwinism is given a lower degree of initial probability, because one doubts the standard case for it, then only strong evidence that a causally specific Darwinian pathway actually exists is going to convince.

More generally, Dembski (1999, 2002) has argued that irreducible complexity is only a special case of complex specified information (CSI), that is, information that has a very low probability (hence high content) and that is specified by an independent pattern. Dembski argues that chance and necessity are unable to explain the appearance of CSI. Darwinists concede that neither chance alone nor necessity alone is capable of generating CSI, but they argue that the Darwinian interaction of chance and necessity is sufficient. However, Dembski has recently argued that the "No Free Lunch" theorems show that even Darwinian resources cannot account for the generation of CSI, only for its relocation and recombination (Dembski 2002, Chapter 4). Obviously, this claim will be much debated.

Much work remains to be done responding to the many critical reactions that ID proposals have provoked. Defenders of ID hope that this work will reveal that ID is a robust and fruitful paradigm, capable of significant refinements and precise enough to generate specific experiments designed to test the powers and alleged limitations of undirected causes.

2.3. Good and Evil

Even in Darwin's day, opinion was divided between those who praised the theory as licensing a progressive world order (or free market economics) and those who feared that it would rationalize racism, eugenics, and the abolition of human dignity. Today, the debate is at least as polarized, with those who defend the Darwinian contribution to ethics (Arnhart 1998; Ruse 2001) and those who denounce it as something positively pernicious (Wiker 2002). In the middle are some – including Darwinists such as the late Stephen Jay Gould, Richard Lewontin, and Kenneth Miller – who argue that there is no important connection between biology and ethics. Although these cultural debates are not directly relevant to the scientific issues, there is no question that they contribute to the very strong feelings on either side. Some see traditional values slipping away; others say "good riddance"; and yet others vie for a nuanced synthesis that holds the best of tradition and science in balance. These debates are far from settled, and much work remains to be done if ID is to convince people that its philosophy is required to support sound ethics.

Another long-standing debate is the theological problem of evil. From Hume (1779) until the present, many have argued against design in science on the grounds that it makes the designer responsible for natural evils such as parasitism. This suggests that either the designer lacks some of the traditional attributes of God or does not exist at all. Rather than be forced to this conclusion, would it not be wiser to suppose that the designer grants his creation a degree of autonomy, thereby avoiding direct responsibility for all that goes on in it?

In response, proponents of ID would agree that there are theological difficulties in understanding how the existence of evil can be reconciled with the existence of God. But, they would insist, these are not valid scientific objections against an empirical method for detecting design, such as Dembski's filter (Dembski 1998a). Defenders of ID have questioned both the correctness of Darwinian theology and the legitimacy of using it to exclude design as a scientific category (Hunter 2001; Nelson 1996). Nonetheless, a great deal of work remains to be done to show that ID does not have the unintended consequence of making the problem of evil even harder for the theologian to resolve.

3. CONCLUSION

The ID movement did not begin with Phillip Johnson. It is inaccurate to describe it as stealth creationism, both because of its clear public expression and because its philosophy is significantly different from that of traditional creationists. In fact, the ID movement began when some scientists encountered what they believed was a crisis in normal science that forced a reevaluation of the assumption that science must observe methodological naturalism. As the movement gained structure and numbers, its public voice became unavoidable. The rigor of the challenges to Darwinism in particular and to naturalism in general compelled a response, leading to the energetic and fruitful controversy that continues today. While critics may see design as a reactionary throwback to an outmoded model of science or as a confusion of science and religion, defenders of ID see themselves as revolutionaries who can build bridges between science and theology. The exchanges have not always been pretty. And much work remains to be done by ID and also (I suggest) by its critics. But perhaps no one has done more to move the debate forward than my colleagues and friends Bill Dembski and Michael Ruse.

Notes

1. My thanks to William Dembski, Stephen Meyer, and Michael Ruse for their comments on earlier versions of this essay.
2. Recently this center has simplified its name. It is now called the Center for Science and Culture. See <www.discovery.org>.
3. See the NOSHA "Who's Who" web page at <http://nosha.secularhumanism. net/whoswho.html>.
4. Stephen Meyer pursues this positive case for design in his chapter in this volume.
5. See Dr. Schaefer's biography at <http://www.leaderu.com/offices/schaefer/ docs/biosketch.html>.
6. The term "universal acid" derives from Daniel Dennett (1995), himself an enthusiastic supporter of naturalistic evolution.
7. Kenyon was interviewed about his experiences by Mars Hill Audio in 1994. See audiocassette volume 7, available from <http://www.marshillaudio.org>.
8. The ISCID home page is at <http://www.iscid.org>.
9. Information on MESA is available at the ISCID web site, <http://www.iscid.org/ mesa>.

References

Arnhart, L. 1998. *Darwinian Natural Right: The Biological Ethics of Human Nature.* Albany: State University of New York Press.

Behe, M. 1996. *Darwin's Black Box: The Biochemical Challenge to Evolution.* New York: The Free Press.

2000a. Correspondence with science journals: Response to critics concerning peer-review. Available at <http://www.discovery.org/viewDB/index.php3? program=CRSC%20 Responses&command=view&id=450>.

2000b. Self-organization and irreducibly complex systems: A reply to Shanks and Joplin. *Philosophy of Science* 67:1: 155–62.

2001. The modern intelligent design hypothesis. *Philosophia Christi* 3(1): 165–79.

Behe, M., W. Dembski, and S. Meyer. (eds.) 2000. *Science and Evidence for Design in the Universe.* San Francisco: Ignatius Press.

Bradley, W., R. Olsen, and C. Thaxton. 1984. *The Mystery of Life's Origin: Reassessing Current Theories.* New York: Philosophical Library.

Brooke, J., and M. Osler. (eds.) 2001. Science in theistic contexts: Cognitive dimensions. *Osiris* 16 (special edition).

Cairns-Smith, A. G. 1971. *The Life Puzzle.* Edinburgh: Oliver and Boyd.

Caplan, A. 1991. Creationist belief precludes credibility on science issues. *The Scientist,* February 18, available at <htttp://www.the-scientist.com/yr1991/feb/ opin3_910218.html>.

Coyne, J. 2001. Creationism by stealth. *Nature* 410 (April 12): 745–6.

Darwin, C. 1859. *On the Origin of Species.* London: John Murray.

Dawkins, R. 1996. *The Blind Watchmaker: Why the Evidence of Evolution Reveals a Universe without Design,* 2nd ed. New York: Norton.

Dembski, W. 1998a. *The Design Inference: Eliminating Chance through Small Probabilities.* Cambridge: Cambridge University Press.

   (ed.) 1998b. *Mere Creation: Science, Faith and Intelligent Design.* Downers Grove, IL: InterVarsity Press.

   1999. *Intelligent Design: The Bridge between Science and Theology.* Downers Grove, IL: InterVarsity Press.

   2002. *No Free Lunch: Why Specified Complexity Cannot Be Purchased without Intelligence.* Lanham, MD: Rowman and Littlefield.

Dennett, D. 1995. *Darwin's Dangerous Idea.* New York: Simon and Schuster.

Denton, M. 1986. *Evolution: A Theory in Crisis.* Chevy Chase, MD: Adler and Adler.

   1998. *Nature's Destiny: How the Laws of Physics Reveal Purpose in the Universe.* New York: The Free Press.

Dobzhansky, T. 1965. Discussion of a paper by G. Schramm. In *The Origins of Prebiological Systems and of Their Molecular Matrices,* ed. S. W. Fox. New York: Academic Press, p. 310.

Easterbrook, G. 1997. Science and God: A warming trend? *Science* 277: 890–3.

Forrest, B. 2001. The Wedge at work: How intelligent design creationism is wedging its way into the cultural and academic mainstream. In *Intelligent Design Creationism and Its Critics,* ed. R. Pennock. Cambridge, MA: MIT Press, pp. 5–53.

Forrest, B., and P. Gross. 2003. *Creationism's Trojan Horse: The Wedge of Intelligent Design.* New York: Oxford University Press.

Harrison, P. 1998. *The Bible, Protestantism, and the Rise of Natural Science.* Cambridge: Cambridge University Press.

Hume, D. 1779. *Dialogues Concerning Natural Religion.* Oxford: Clarendon Press.

Hunter, C. 2001. *Darwin's God: Evolution and the Problem of Evil.* Grand Rapids, MI: Brazos Press.

Jaki, S. 2000. *The Savior of Science.* Grand Rapids, MI: Eerdmans.

Johnson, Phillip. 1991. *Darwin on Trial.* Downers Grove, IL: InterVarsity Press.

   1995. *Reason in the Balance.* Downers Grove, IL: InterVarsity Press.

   2000. *The Wedge of Truth: Splitting the Foundations of Naturalism.* Downers Grove, IL: InterVarsity Press.

Kenyon, D., and G. Steinman. 1969. *Biochemical Predestination.* New York: McGraw Hill.

Koons, R. 2000. The incompatibility of naturalism and scientific realism. In *Naturalism: A Critical Analysis,* ed. W. L. Craig and J. P. Moreland. New York: Routledge, pp. 49–63.

McDonald, J. 2002. A reducibly complex mousetrap. Available at <http://udel.edu/~mcdonald/oldmousetrap.html>.

Menuge, A. 2001. Few signs of intelligence: The saga of Bill Dembski at Baylor. *Touchstone* (May): 54–5.

Meyer, S. 1993. "Danger: Indoctrination. A Scopes Trial for the '90s." *The Wall Street Journal,* December 6, p. A14. Available online from the "Article Database" at <www.discovery.org>.

2000. Evidence for design in physics and biology: From the origin of the universe to the origin of life. In *Science and Evidence for Design in the Universe*, ed. M. Behe, W. Dembski, and S. Meyer. San Francisco: Ignatius Press, pp. 53–111.

2001. Darwin in the dock. *Touchstone* (April): 57–9. Available on-line from the articles database at <www.touchstonemag.com>.

Miller, K. 1999. *Finding Darwin's God: A Scientist's Search for Common Ground between God and Evolution.* New York: HarperCollins.

Nelson, P. 1996. The role of theology in current evolutionary reasoning. *Biology and Philosophy* 11(4): 493–517. Reprinted in Pennock 2001, pp. 677–704.

Orr, H. A. 1996/97. Darwin v. intelligent design (again). *Boston Review* (December/January): 28–31. Available at <http://bostonreview.mit.edu/br21.6/orr.html>.

Osler, M. (ed.) *Rethinking the Scientific Revolution.* New York: Cambridge University Press, 2000.

Pearcey, N. and C. Thaxton. 1994. *The Soul of Science: Christian Faith and Natural Philosophy.* Wheaton, IL: Crossway Books.

Pennock, R. 1999. *Tower of Babel: The Evidence against the New Creationism.* Cambridge, MA: MIT Press.

(ed.) 2001. *Intelligent Design Creationism and Its Critics: Philosophical, Theological, and Scientific Perspectives.* Cambridge, MA: MIT Press.

Plantinga, A. 1993. *Warrant and Proper Function.* New York: Oxford University Press.

2000. *Warranted Christian Belief.* New York: Oxford University Press.

Ratzsch, D. 2001. *Nature, Design, and Science: The Status of Design in Natural Science.* Albany, NY: State University of New York Press.

Rea, M. 2000. Naturalism and material objects. In *Naturalism: A Critical Analysis*, ed. W. L. Craig and J. P. Moreland. New York: Routledge, pp. 110–32.

Ruse, M. 2001. *Can a Darwinian Be a Christian? The Relationship between Science and Religion.* New York: Cambridge University Press.

Scott, E. 2001. Fatally flawed iconoclasm. *Science* 292 (June 22): 2257–8.

Shanks, N. and K. Joplin. 1999. Redundant complexity: A critical analysis of intelligent design in biochemistry. *Philosophy of Science* 66: 268–82.

Stanley, S. 1979. *Macroevolution.* San Francisco: W. H. Freeman.

Ussery, D. 2001. The stealth creationists. *Skeptic* 8(4): 72–4.

Wells, J. 2000. *Icons of Evolution: Science or Myth? Why Much of What We Teach about Evolution Is Wrong.* Washington, DC: Regnery.

2002. Catch 23. Available at <http://www.discovery.org/viewDB/index.php3?program=CRSC&command=view&id=1212>.

Wiker, B. 2002. *Moral Darwinism: How We Became Hedonists.* Downers Grove, IL: InterVarsity Press.

PART I

DARWINISM

4

Design without Designer

Darwin's Greatest Discovery

Francisco J. Ayala[1]

It is also frequently asked what our belief must be about the form and shape of heaven according to Sacred Scripture. Many scholars engage in lengthy discussions on these matters.... Such subjects are of no profit for those who seek beatitude, and, what is worse, they take up very precious time that ought to be given to what is spiritually beneficial. What concern is it of mine whether heaven is like a sphere and the earth is enclosed by it and suspended in the middle of the universe?... In the matter of the shape of heaven the sacred writers ... did not wish to teach men these facts that would be of no avail for their salvation.

Saint Augustine, *The Literal Meaning of Genesis*, Book II, Chapter 9[2]

New knowledge has led us to realize that the theory of evolution is no longer a mere hypothesis. It is indeed remarkable that this theory has been progressively accepted by researchers, following a series of discoveries in various fields of knowledge. The convergence, neither sought nor fabricated, of the results of work that was conducted independently is in itself a significant argument in favor of this theory.

Pope John Paul II, *Address to the Pontifical Academy of Sciences*, October 22, 1996[3]

SYNOPSIS

I advance three propositions and conclude with two additional arguments. The first proposition is that Darwin's most significant intellectual contribution is that he brought the origin and diversity of organisms into the realm of science. The Copernican revolution consisted in a commitment to the postulate that the universe is governed by natural laws that account for natural phenomena. Darwin completed the Copernican revolution by extending that commitment to the living world.

The second proposition is that natural selection is a creative process that can account for the appearance of genuine novelty. How natural selection creates is shown by using a simple example and then clarified using two analogies – artistic creation and the "typing monkeys" – with which it shares important similarities and differences. The creative power of natural selection arises from a distinctive interaction between chance and necessity, or between random and deterministic processes.

The third proposition is that teleological explanations are necessary in order to give a full account of the attributes of living organisms, whereas they are neither necessary nor appropriate in the explanation of natural inanimate phenomena. I give a definition of "teleology" and clarify the matter by distinguishing between internal and external teleology, and between bounded and unbounded teleology. The human eye, so obviously constituted for seeing but resulting from a natural process, is an example of internal (or natural) teleology. A knife has external (or artificial) teleology, because it has been purposefully designed by an external agent. The development of an egg into a chicken is an example of bounded (or necessary) teleology, whereas the evolutionary origin of the mammals is a case of unbounded (or contingent) teleology, because there was nothing in the make-up of the first living cells that necessitated the eventual appearance of mammals.

An argument follows that the "design" of organisms is not "intelligent," but rather quite incompatible with the design that we would expect of an intelligent designer or even of a human engineer, and so full of dysfunctions, wastes, and cruelties as to unwarrant its attribution to any being endowed with superior intelligence, wisdom, and benevolence.

My second argument simply asserts that as successful and encompassing as science is as a way of knowing, it is not the only way.

DARWIN'S REVOLUTION

The publication in 1859 of *On the Origin of Species* by Charles Darwin ushered in a new era in the intellectual history of mankind. Darwin is deservedly given credit for the theory of biological evolution: he accumulated evidence demonstrating that organisms evolve and discovered the process – natural selection – by which they evolve. But the import of Darwin's achievement is that it completed the Copernican revolution initiated three centuries earlier, and that it thereby radically changed our conception of the universe and the place of mankind in it.

The discoveries of Copernicus, Kepler, Galileo, and Newton during the sixteenth and seventeenth centuries had gradually ushered in the notion that the workings of the universe could be explained by human reason. It was shown that the Earth is not the center of the universe, but a small planet rotating around an average star; that the universe is immense in space and in time; and that the motions of the planets around the sun can be explained by the same simple laws that account for the motion of physical objects on our planet. These and other discoveries greatly expanded human knowledge, but the intellectual revolution these scientists brought about was more fundamental: a commitment to the postulate that the universe obeys immanent laws that account for natural phenomena. The workings of the universe were brought into the realm of science: explanation through natural laws. Physical phenomena could be accounted for whenever the causes were adequately known.

Darwin completed the Copernican revolution by drawing out for biology the notion of nature as a lawful system of matter in motion. The adaptations and diversity of organisms, the origin of novel and highly organized forms, even the origin of mankind itself – all could now be explained by an orderly process of change governed by natural laws.

The origin of organisms and their marvelous adaptations were, however, either left unexplained or attributed to the design of an omniscient Creator. God had created the birds and bees, the fish and corals, the trees in the forest, and, best of all, man. God had given us eyes so that we might see, and He had provided fish with gills with which to breathe in water. Philosophers and theologians argued that the functional design of organisms manifests the existence of an all-wise Creator. Wherever there is design, there is a designer; the existence of a watch evinces the existence of a watchmaker.

The English theologian William Paley, in his *Natural Theology* (1802), elaborated the argument from design as a forceful demonstration of the existence of the Creator. The functional design of the human eye, argued Paley, provided conclusive evidence of an all-wise Creator. It would be absurd to suppose, he wrote, that by mere chance the human eye "should have consisted, first, of a series of transparent lenses . . . secondly of a black cloth or canvas spread out behind these lenses so as to receive the image formed by pencils of light transmitted through them, and placed at the precise geometrical distance at which, and at which alone, a distinct image could be formed . . . thirdly of a large nerve communicating between this membrane and the brain." The Bridgewater Treatises, published between 1833 and 1840, were written by eminent scientists and philosophers to set forth "the Power, Wisdom, and Goodness of God as manifested in the Creation." The structure and mechanisms of the human hand, for example, were cited as incontrovertible evidence that the hand had been designed by the same omniscient Power that had created the world.[4]

The advances of physical science had thus driven mankind's conception of the universe to a split-personality state of affairs, which persisted well into the mid nineteenth century. Scientific explanations, derived from natural laws, dominated the world of nonliving matter, on the Earth as well as in the heavens. Supernatural explanations, depending on the unfathomable deeds of the Creator, accounted for the origin and configuration of living creatures – the most diversified, complex, and interesting realities of the world. It was Darwin's genius to resolve this conceptual schizophrenia.

DARWIN'S DISCOVERY: DESIGN WITHOUT DESIGNER

The conundrum faced by Darwin can hardly be overestimated. The strength of the argument from design for demonstrating the role of the Creator is easily set forth. Wherever there is function or design we look for its author. A knife is made for cutting, and a clock is made to tell time; their functional designs have been contrived by a knife maker and a watchmaker. The exquisite design of Leonardo da Vinci's *Mona Lisa* proclaims that it was created by a gifted artist following a preconceived plan. Similarly, the structures, organs, and behaviors of living beings are directly organized to serve certain functions. The functional design of organisms and their features would therefore seem to argue for the existence of a designer. It was Darwin's greatest accomplishment to show that the directive organization of living beings can be explained as the result of a natural process – natural selection – without any need to resort to a Creator or other external agent. The origin and adaptation of organisms in all of their profusion and wondrous variation were thus brought into the realm of science.

Darwin accepted that organisms are "designed" for certain purposes, that is, that they are functionally organized. Organisms are adapted to certain ways of life, and their parts are adapted to perform certain functions. Fish are adapted to live in water; kidneys are designed to regulate the composition of blood; the human hand is made for grasping. But Darwin went on to provide a natural explanation of the design. He thereby brought the seemingly purposeful aspects of living beings into the realm of science.

Darwin's revolutionary achievement is that he extended the Copernican revolution to the world of living things. The origin and adaptive nature of organisms could now be explained, like the phenomena of the inanimate world, as the result of natural laws manifested in natural processes. Darwin's theory encountered opposition in some religious circles, not so much because he proposed the evolutionary origin of living things (which had been proposed before, and had been accepted even by Christian theologians), but because the causal mechanism – natural selection – excluded God as the explanation for the obvious design of organisms.[5] The configuration of the universe was no longer perceived as the result of God's design, but simply as the outcome of immanent, blind processes. There were, however, many theologians, philosophers, and scientists who saw no contradiction then – and many who see none today – between the evolution of species and Christian faith. Some see evolution as the "method of divine intelligence," in the words of the nineteenth-century theologian A. H. Strong. Others, such as Henry Ward Beecher (1818–1887), an American contemporary of Darwin, made evolution the cornerstone of their theology. These two traditions have persisted to the present. As cited at the beginning of this chapter, Pope John Paul II has stated that "the theory of evolution is no longer a mere hypothesis. It is . . . accepted by researchers, following a series of discoveries in various fields of knowledge." "Process" theologians perceive evolutionary dynamics as a pervasive element of a Christian view of the world.[6]

NATURAL SELECTION AS A NONCHANCE PROCESS

The central argument of the theory of natural selection is summarized by Darwin in his *Origin of Species* as follows:

As more individuals are produced than can possibly survive, there must in every case be a struggle for existence, either one individual with another of the same species, or with the individuals of distinct species, or with the physical conditions of life.... Can it, then, be thought improbable, seeing that variations useful to man have undoubtedly occurred, that other variations useful in some way to each being in the great and complex battle of life, should sometimes occur in the course of thousands of generations? If such do occur, can we doubt (remembering that more individuals are born than can possibly survive) that individuals having any advantage, however slight, over others, would have the best chance of surviving and of procreating their kind? On the other hand, we may feel sure that any variation in the least degree injurious would be rigidly destroyed. This preservation of favorable variation and the rejection of injurious variations, I call Natural Selection.[7]

Darwin's argument addresses the problem of explaining the adaptive character of organisms. Darwin argues that adaptive variations ("variations useful in some way to each being") occasionally appear, and that these are likely to increase the reproductive chances of their carriers. Over the generations, favorable variations will be preserved, and injurious ones will be eliminated. In one place, Darwin adds: "I can see no limit to this power [natural selection] in slowly and beautifully *adapting* each form to the most complex relations of life." Natural selection was proposed by Darwin primarily in order to account for the adaptive organization, or "design," of living beings; it is a process that promotes or maintains adaptation. Evolutionary change through time and evolutionary diversification (multiplication of species) are not directly promoted by natural selection (hence the so-called evolutionary stasis – the numerous examples of organisms with morphology that has changed little, if at all, for millions of years, as pointed out by the proponents of the theory of punctuated equilibria). But change and diversification often ensue as by-products of natural selection fostering adaptation.

Darwin formulated natural selection primarily as differential survival. The modern understanding of the principle of natural selection is formulated in genetic and statistical terms as differential reproduction. Natural selection implies that, on the average, some genes and genetic combinations are transmitted to the following generations more frequently than their alternative genetic units. Such genetic units will become more common in every subsequent generation, and the alternative units less common. Natural selection is a statistical bias in the relative rate of reproduction of alternative genetic units.

Natural selection has been compared to a sieve that retains the rarely arising useful genes and lets go the more frequently arising harmful mutants. Natural selection acts in that way, but it is much more than a purely negative process, for it is also able to generate novelty by increasing the probability of otherwise extremely improbable genetic combinations. Natural selection is thus in a way creative. It does not "create" the entities upon which it operates, but it produces adaptive genetic combinations that would not have existed otherwise.

The creative role of natural selection must not be understood in the sense of the "absolute" creation that traditional Christian theology predicates of the Divine act by which the universe was brought into being ex nihilo. Natural selection may instead be compared to a painter who creates a picture by mixing and distributing pigments in various ways over the canvas. The canvas and the pigments are not created by the artist, but the painting is. It is conceivable that a random combination of the pigments might result in the orderly whole that is the final work of art. But the probability of Leonardo's *Mona Lisa* resulting from a random combination of pigments, or of Saint Peter's Basilica resulting from a random association of marble, bricks, and other materials, is infinitely small. In the same way, the combination of genetic units that carries the hereditary information responsible for the formation of the vertebrate eye could never have been produced by a random process such as mutation – not even allowing for the more than three billion years during which life has existed on Earth. The complicated anatomy of the eye, like the exact functioning of the kidney, is the result of a nonrandom process – natural selection.

Critics have sometimes alleged as evidence against Darwin's theory of evolution examples showing that random processes cannot yield meaningful, organized outcomes. It is thus pointed out that a series of monkeys randomly striking letters on a typewriter would never write *On the Origin of Species*, even if we allowed for millions of years and many generations of monkeys pounding on typewriters.

This criticism would be valid if evolution depended only on random processes. But natural selection is a nonrandom process that promotes adaptation by selecting combinations that "make sense" – that is, that are useful to the organisms. The analogy of the monkeys would be more appropriate if a process existed by which, first, meaningful words would be chosen every time they appeared on the typewriter; and then we would also have typewriters with previously selected words rather than just letters as the keys; and again there would be a process that selected meaningful sentences every time they appeared in this second typewriter. If every time words such as "the," "origin," "species," and so on appeared in the first kind of typewriter, they each became a key in the second kind of typewriter, meaningful sentences would occasionally be produced in this second typewriter. If such sentences became incorporated into the keys of a third kind of typewriter, in which meaningful paragraphs were selected whenever they appeared, it is clear that pages and even chapters "making sense" would eventually be produced.

We need not carry the analogy too far, since the analogy is not fully satisfactory; but the point is clear. Evolution is not the outcome of purely random processes; rather, there is a "selecting" process, which picks up adaptive combinations because these reproduce more effectively and thus become established in populations. These adaptive combinations constitute, in turn, new levels of organization upon which the mutation (random) plus selection (nonrandom or directional) process again operates.

The manner in which natural selection can generate novelty in the form of accumulated hereditary information may be illustrated by the following example. In order to be able to reproduce in a culture medium, some strains of the colon bacterium *Escherichia coli* require that a certain substance, the amino acid histidine, be provided in the medium. When a few such bacteria are added to a cubic centimeter of liquid culture medium, they multiply rapidly and produce between two and three billion bacteria in a few hours. Spontaneous mutations to streptomycin resistance occur in normal (i.e., sensitive) bacteria at rates of the order of one in one hundred million (1×10^{-8}) cells. In our bacterial culture, we would expect between twenty and thirty bacteria to be resistant to streptomycin due to spontaneous mutation. If a proper concentration of the antibiotic is added to the culture, only the resistant cells survive. The twenty or thirty surviving bacteria will start reproducing, however, and – allowing a few hours for the necessary number of cell divisions – several billion bacteria will then be produced, all resistant to streptomycin. Among cells requiring histidine as a growth factor, spontaneous mutations able to reproduce in the absence of histidine arise at a rate of about four in one hundred million (4×10^{-8}) bacteria. The streptomycin-resistant cells may now be transferred to a culture with streptomycin but with no histidine. Most of them will not be able to reproduce, but about a hundred will start reproducing until the available medium is saturated.

Natural selection has produced, in two steps, bacterial cells resistant to streptomycin and not requiring histidine for growth. The probability of the two mutational events happening in the same bacterium is of about four in ten million billion $(1 \times 10^{-8} \times 4 \times 10^{-8} = 4 \times 10^{-16})$ cells. An event of such low probability is unlikely to occur even in a large laboratory culture of bacterial cells. With natural selection, cells having both properties are the common result.

As illustrated by the bacterial example, natural selection produces combinations of genes that would otherwise be highly improbable, because natural selection proceeds stepwise. The vertebrate eye did not appear suddenly in all its present perfection. Its formation requires the appropriate integration of many genetic units, and thus the eye could not have resulted from random processes alone. For more than half a billion years, the ancestors of today's vertebrates had some kind of organ sensitive to light. Perception of light, and later vision, were important for these organisms' survival and reproductive success. Accordingly, natural selection favored genes and gene combinations that increased the functional efficiency of the eye. Such genetic units gradually accumulated, eventually leading to the highly complex and efficient vertebrate eye. Natural selection can account for the rise and spread of genetic constitutions, and therefore of types of organisms, that would never have resulted from the uncontrolled action of random mutation. In this sense, natural selection is a creative process, although it does not create the raw materials – the genes – upon which it acts.[8]

CHANCE AND NECESSITY

There is an important respect in which artistic creation makes a poor analogy to the process of natural selection. A painter usually has a preconception of what he wants to paint and will consciously modify the painting so that it represents what he wants. Natural selection has no foresight, nor does it operate according to some preconceived plan. Rather, it is a purely natural process resulting from the interacting properties of physico-chemical and biological entities. Natural selection is simply a consequence of the differential multiplication of living beings. It has some appearance of purposefulness, because it is conditioned by the environment: which organisms reproduce more effectively depends on which variations they possess that are useful in the organism's environment. But natural selection does not anticipate the environments of the future; drastic environmental changes may be insuperable obstacles to organisms that were previously thriving.

The team of typing monkeys is also a bad analogy to evolution by natural selection, because it assumes that there is "somebody" who selects letter combinations and word combinations that make sense. In evolution, there is no one selecting adaptive combinations. These select themselves, because they multiply more effectively than less adaptive ones.

There is a sense in which the analogy of the typing monkeys is better than the analogy of the artist, at least if we assume that no particular statement was to be obtained from the monkeys' typing endeavors, just any statement making sense. Natural selection strives to produce not predetermined kinds of organisms, but only organisms that are adapted to their present environments. Which characteristics will be selected depends on which variations happen to be present at a given time and in a given place. This, in turn, depends on the random process of mutation, as well as on the previous history of the organism (i.e., on its genetic make-up as a consequence of previous evolution). Natural selection is an "opportunistic" process. The variables determining in which direction it will go are the environment, the preexisting constitution of the organisms, and the randomly arising mutations.

Thus, adaptation to a given environment may occur in a variety of different ways. An example may be taken from the adaptations of plant life to a desert climate. The fundamental adaptation is to the condition of dryness, which involves the danger of desiccation. During a major part of the year – sometimes for several years in succession – there is no rain. Plants have accomplished the urgent necessity of saving water in different ways. Cacti have transformed their leaves into spines, having made their stems into barrels containing a reserve of water; photosynthesis is performed in the surface of the stem instead of in the leaves. Other plants have no leaves during the dry season, but after it rains they burst into leaves and flowers and produce seeds. Ephemeral plants germinate from seeds, grow, flower, and produce seeds all within the space of a few weeks, when rainwater is available; during the rest of the year the seeds lie quiescent in the soil.

The opportunistic character of natural selection is also well evidenced by the phenomenon of adaptive radiation. The evolution of *Drosophila* fruitflies in Hawaii is a relatively recent adaptive radiation. There are about 1,500 *Drosophila* species in the world. Approximately 500 of them have evolved in the Hawaiian archipelago, although this island group has a small area, about one twenty-fifth the size of California. Moreover, the morphological, ecological, and behavioral diversity of Hawaiian *Drosophila* exceeds that of *Drosophila* in the rest of the world.

Why should have such "explosive" evolution have occurred in Hawaii? The overabundance of *Drosophila* fruitflies there contrasts with the absence of many other insects. The ancestors of Hawaiian *Drosophila* reached the archipelago before other groups of insects did, and thus they found a multitude of unexploited opportunities for living. They responded by a rapid adaptive radiation; although they are all probably derived from a single colonizing species, they adapted to the diversity of opportunities available in diverse places and at different times by developing appropriate adaptations, which varied widely from one to another species.

The process of natural selection can explain the adaptive organization of organisms, as well as their diversity and evolution as a consequence of their adaptation to the multifarious and ever-changing conditions of life. The fossil record shows that life has evolved in a haphazard fashion. The radiations, expansions, relays of one form by another, occasional but irregular trends, and the ever-present extinctions are best explained by natural selection of organisms subject to the vagaries of genetic mutation and environmental challenge. The scientific account of these events does not necessitate recourse to a preordained plan, whether imprinted from without by an omniscient and all-powerful Designer, or resulting from some immanent force driving the process towards definite outcomes. Biological evolution differs from a painting or an artifact in that it is not the outcome of a design preconceived by an artist or artisan.

Natural selection accounts for the "design" of organisms, because adaptive variations tend to increase the probability of survival and reproduction of their carriers at the expense of maladaptive, or less adaptive, variations. The arguments of Paley and the authors of the Bridgewater Treatises against the incredible improbability of chance accounts of the origin of organisms and their adaptations are well taken, as far as they go. But neither these scholars, nor any other writers before Darwin, were able to discern that there is a natural process (namely, natural selection) that is not random, but rather oriented and able to generate order, or to "create."[9] The traits that organisms acquire in their evolutionary histories are not fortuitous but determined by their functional utility to the organisms, and they come about in small steps that accumulate over time, each step providing some reproductive advantage over the previous condition.

Chance is, nevertheless, an integral part of the evolutionary process. The mutations that yield the hereditary variations available to natural selection arise at random, independent of whether they are beneficial or harmful to their carriers. But this random process (as well as others that come to play in the great theater of life) is counteracted by natural selection, which preserves what is useful and eliminates the harmful. Without mutation, evolution could not happen, because there would be no variations that could be differentially conveyed from one to another generation. But without natural selection, the mutation process would yield disorganization and extinction, because most mutations are disadvantageous. Mutation and selection have jointly driven the marvelous process that, starting from microscopic organisms, has produced orchids, birds, and humans.

The theory of evolution manifests chance and necessity jointly intertwined in the stuff of life; randomness and determinism interlocked in a natural process that has spurted the most complex, diverse, and beautiful entities in the universe: the organisms that populate the Earth, including humans, who think and love, who are endowed with free will and creative powers, and who are able to analyze the very process of evolution that brought them into existence. This is Darwin's fundamental discovery, that there is a process that is creative though not conscious. And this is the conceptual revolution that Darwin completed: that everything in nature, including the origin of living organisms, can be accounted for as a result of natural processes governed by natural laws. This is nothing if not a fundamental vision that has forever changed how human beings perceive themselves and their place in the universe.

TELEOLOGY AND TELEOLOGICAL EXPLANATIONS

Explanation by design, or teleology, is, according to a dictionary definition, "the use of design, purpose, or utility as an explanation of any natural phenomenon."[10] An object or a behavior is said to be teleological when it gives evidence of design or appears to be directed toward certain ends. For example, the behavior of human beings is often teleological. A person who buys an airplane ticket, reads a book, or cultivates the earth is trying to achieve a certain end: getting to a given city, acquiring knowledge, or getting food. Objects and machines made by people are also usually teleological: a knife is made for cutting; a clock is made for telling time; a thermostat is made to regulate temperature. Similarly, many features of organisms are teleological: a bird's wings are *for* flying; eyes are *for* seeing; kidneys are constituted *for* regulating the composition of the blood. The features of organisms that may be said to be teleological are those that can be identified as adaptations, whether they are structures such as a wing or a hand, organs such as a kidney, or behaviors such as the courtship displays of a peacock. Adaptations are features of organisms that have come about by natural selection because they serve certain functions and thus increase the reproductive success of their carriers.

Inanimate objects and processes (other than those created by people) are not teleological in the sense just explained, because we gain no additional scientific understanding by perceiving them as directed toward specific ends or as serving certain purposes. The configuration of a sodium chloride molecule (table salt) depends on the structure of sodium and chlorine, but it makes no sense to say that that structure is made in order to serve a certain purpose, such as tasting salty. Similarly, the shape of a mountain is the result of certain geological processes, but it did not come about in order to serve a certain purpose, such as providing slopes suitable for skiing. The motion of the Earth around the sun results from the laws of gravity, but it does not exist in order that the seasons may occur. We may use sodium chloride as food, a mountain for skiing, and we may take advantage of the seasons, but the use that we make of these objects or phenomena is not the reason why they came into existence or why they have certain configurations. On the other hand, a knife and a car exist and have particular configurations precisely in order to serve the purposes of cutting and transportation. Similarly, the wings of birds came about precisely because they permitted flying, which was reproductively advantageous. The mating displays of peacocks came about because they increased the chances of mating and thus of leaving progeny.

The previous comments point out the essential characteristics of teleological phenomena, which may be encompassed in the following definition: "Teleological explanations account for the *existence* of a certain feature in a system by demonstrating the feature's contribution to a specific property

of state of the system." Teleological explanations require that the feature
or behavior contribute to the persistence of a certain state or property of
the system: wings serve for flying; the sharpness of a knife serves for cutting.
Moreover – and this is the essential component of the concept – this contribution must be the reason why the feature or behavior exists at all: the
reason why wings came into existence is because they serve for flying; the
reason why a knife is sharp is that it is intended for cutting.

The configuration of a molecule of sodium chloride contributes to its
property of tasting salty and therefore to its use as food, not vice versa; the
potential use of sodium chloride as food is not the reason why it has a particular molecular configuration or tastes salty. The motion of the Earth around
the sun is the reason why seasons exist; the existence of the seasons is not
the reason why the Earth moves about the sun. On the other hand, the
sharpness of a knife can be explained teleologically, because the knife has
been created precisely to serve the purpose of cutting. Motorcars and their
particular configurations exist because they serve the purpose of transportation, and thus can be explained teleologically. Many features and behaviors
of organisms meet the requirements of teleological explanation.[11] The human hand, the wings of birds, the structure and behavior of kidneys, and
the mating displays of peacocks are examples already given.[12]

It is useful to distinguish different kinds of design or teleological phenomena. Actions or objects are *purposeful* when the end state or goal is
consciously intended by an agent. Thus, a man mowing his lawn is acting
teleologically in the purposeful sense; a lion hunting deer and a bird building a nest have at least the appearance of purposeful behavior. Objects
resulting from purposeful behavior exhibit *artificial* (or *external*) teleology.
A knife, a table, a car, and a thermostat are examples of systems exhibiting
artificial teleology: their teleological features were consciously intended by
some agent.

Systems with teleological features that result not from the purposeful action of an agent but from some natural process exhibit *natural* (or *internal*)
teleology. The wings of birds have a natural teleology; they serve an end –
flying – but their configuration is not due to the conscious design of any
agent. We may distinguish two kinds of natural teleology: *bounded*, or *determinate*, or *necessary* teleology; and *unbounded*, or *indeterminate*, or *contingent*
teleology.

Bounded natural teleology exists when a specific end state is reached
in spite of environmental fluctuations. The development of an egg into a
chicken is an example of bounded natural teleological process. The regulation of body temperature in a mammal is another example. In general,
the homeostatic processes of organisms are instances of bounded natural
teleology.[13]

Unbounded design, or contingent teleology, occurs when the end state
is not specifically predetermined but rather is the result of selection of one

from among several available alternatives. The adaptations of organisms are designed, or teleological, in this indeterminate sense. The wings of birds call for teleological explanation: the genetic constitutions responsible for their configuration came about because wings serve to fly and because flying contributes to the reproductive success of birds. But there was nothing in the constitution of the remote ancestors of birds that would necessitate the appearance of wings in their descendants. Wings came about as the consequence of a long sequence of events. At each stage, the most advantageous alternative was selected among those that happened to be available; but which alternatives were available at any one time depended, at least in part, on chance events.[14]

Teleological explanations are fully compatible with (efficient) causal explanations.[15] It is possible, at least in principle, to give a causal account of the various physical and chemical processes in the development of an egg into a chicken, or of the physico-chemical, neural, and muscular interactions involved in the functioning of the eye. (I use "in principle" in order to imply that any component of the process can be elucidated as a causal process if it is investigated in sufficient detail and in depth; but not all steps in almost any developmental process have been so investigated, with the possible exception of the flatworm *Caenorhabditis elegans*. The development of *Drosophila* fruitflies has also become known in much detail, even if not yet completely.) It is also possible, in principle, to describe the causal processes by which one genetic variant becomes eventually established in a population by natural selection. But these causal explanations do not make it unnecessary to provide teleological explanations where appropriate. Both teleological and causal explanations are called for in such cases.

Paley's claim that the design of living beings evinces the existence of a Designer was shown to be erroneous by Darwin's discovery of the process of natural selection, just as the pre-Copernican explanation for the motions of celestial bodies (and the argument for the existence of God based on the unmoved mover) was shown to be erroneous by the discoveries of Copernicus, Galileo, and Newton. There is no more reason to consider Darwin's theory of evolution and explanation of design anti-Christian than to consider Newton's laws of motion anti-Christian. Divine action in the universe must be sought in ways other than those that postulate it as the means to account for gaps in the scientific account of the workings of the universe.

Since the Copernican and Darwinian revolutions, all natural objects and processes have become subjects of scientific investigation. Is there any important missing link in the scientific account of natural phenomena? I believe there is – namely, the origin of the universe. The creation or origin of the universe involves a transition from nothing into being. But a transition can only be scientifically investigated if we have some knowledge about the states or entities on both sides of the boundary. Nothingness, however, is not a subject for scientific investigation or understanding. Therefore, as far as science is concerned, the origin of the universe will remain forever a mystery.

UNINTELLIGENT DESIGN

William Paley, in the much-cited first paragraph of *Natural Theology*, set the argument against chance as an explanation of the organized complexity of organisms and their parts:

In crossing a heath, suppose I pitched my foot against a *stone*, and were asked how the stone came to be there, I might possibly answer, that for any thing I knew to the contrary it had lain there for ever; nor would it, perhaps, be very easy to show the absurdity of this answer. But suppose I had found a *watch* upon the ground, and it should be inquired how the watch happened to be in that place, I should hardly think of the answer which I had before given, that for any thing I knew the watch might have always been there. Yet why should not this answer serve for the watch as well as for the stone; why is it not as admissible in the second case as in the first? For this reason, and for no other, namely, that when we come to inspect the watch, we perceive – what we could not discover in the stone – that its several parts are framed and put together for a purpose, *e.g.* that they are so formed and adjusted as to produce motion, and that motion so regulated as to point out the hour of the day; that if the different parts had been differently shaped from what they are, or placed after any other manner or in any other order than that in which they are placed, either no motion at all would have been carried on in the machine, or none which would have answered the use that is now served by it.[16]

The strength of the argument against chance derives, Paley tells us, from what he calls "relation," a notion akin to what contemporary authors have called "irreducible complexity."

When several different parts contribute to one effect, or, which is the same thing, when an effect is produced by the joint action of different instruments, the fitness of such parts or instruments to one another for the purpose of producing, by their united action, the effect, is what I call *relation*; and wherever this is observed in the works of nature or of man, it appears to me to carry along with it decisive evidence of understanding, intention, art ... all depending upon the motions within, all upon the system of intermediate actions.[17]

A remarkable example of complex "parts," fit together so that they cannot function one without the other, is provided by the two sexes, "manifestly made for each other ..., subsisting, like the clearest relations of art, in different individuals, unequivocal, inexplicable without design."[18]

The outcomes of chance do not exhibit relation among the parts or, as we might say, organized complexity:

[T]he question is, whether a useful or imitative conformation be the produce of chance.... Universal experience is against it. What does chance ever do for us? In the human body, for instance, chance, that is, the operation of causes without design, may produce a wen, a wart, a mole, a pimple, but never an eye. Among inanimate substances, a clod, a pebble, a liquid drop might be; but never was a watch, a telescope, an organized body of any kind, answering a valuable purpose by a complicated mechanism, the effect of chance. In no assignable instance has such a thing existed without intention somewhere.[19]

I am filled with amazement and respect for Paley's extensive and profound biological knowledge. He discusses the air bladder of fish, the fang of vipers, the claw of herons, the camel's stomach, the woodpecker's tongue, the elephant's proboscis, the hook in the bat's wing, the spider's web, the compound eyes of insects and their metamorphosis, the glowworm, univalve and bivalve mollusks, seed dispersal, and on and on, with accuracy and as much detail as known to the best biologists of his time. Paley, moreover, takes notice of the imperfections, defects, pain, and cruelty of nature and seeks to account for them in a chapter entitled "Of the Personality of the Deity," which strikes me by its well-meaning, if naïve, arrogance, as Paley seems convinced that he can determine God's "personality."

Contrivance, if established, appears to me to prove...the *personality* [Paley's emphasis] of the Deity, as distinguished from what is sometimes called nature, sometimes called a principle.... Now, that which can contrive, which can design, must be a person. These capacities constitute personality, for they imply consciousness and thought.... The acts of a mind prove the existence of a mind; and in whatever a mind resides, is a person. The seat of intellect is a person.[20]

One recent author who has reformulated Paley's argument-from-design responds to the critics who point out the imperfections of organisms in the following way.

The most basic problem is that the argument [against intelligent design] demands perfection at all. Clearly, designers who have the ability to make better designs do not necessarily do so.... I do not give my children the best, fanciest toys because I don't want to spoil them, and because I want them to learn the value of a dollar. The argument from imperfection overlooks the possibility that the designer might have multiple motives, with engineering excellence oftentimes relegated to a secondary role.... Another problem with the argument from imperfection is that it critically depends on psychoanalysis of the unidentified designer. Yet the reasons that a designer would or would not do anything are virtually impossible to know unless the designer tells you specifically what those reasons are.[21]

So, God may have had his reasons for not designing organisms to be as perfect as they could have been.

A problem with this explanation is that it destroys Intelligent Design as a scientific hypothesis, because it provides it with an empirically impenetrable shield.[22] If we cannot reject Intelligent Design because the designer may have reasons that we could not possibly ascertain, there would seem to be no way to test Intelligent Design by drawing out predictions, logically derived from the hypothesis, that are expected to be observed in the world of experience. Intelligent Design as an explanation for the adaptations of organisms could be a form of (natural) theology, as Paley would have it; but, whatever it is, it is not a scientific hypothesis.

I would argue, moreover, that it is not good theology either, because it leads to attributing to the designer qualities quite different from the omniscience, omnipotence, and omnibenevolence of the Creator.[23] It is not only that organisms and their parts are less than perfect, but also that deficiencies and dysfunctions are pervasive, evidencing defective "design." Consider the human jaw. We have too many teeth for the jaw's size, so that wisdom teeth need to be removed and orthodontists make a decent living straightening the others. Would we want to blame God for such defective design? A human engineer could have done better. Evolution gives a good account of this imperfection. Brain size increased over time in our ancestors, and the remodeling of the skull in order to fit the larger brain entailed a reduction of the jaw. Evolution responds to the organism's needs through natural selection, not by optimal design but by "tinkering," as it were, by slowly modifying existing structures. Consider now the birth canal of women, much too narrow for easy passage of the infant's head, so that thousands upon thousands of babies die during delivery. Surely we don't want to blame God for this defective design or for the children's deaths. Science makes it understandable, a consequence of the evolutionary enlargement of our brain. Females of other animals do not experience this difficulty. Theologians in the past struggled with the issue of dysfunction, because they thought it had to be attributed to God's design. Science, much to the relief of many theologians, provides an explanation that convincingly attributes defects, deformities, and dysfunctions to natural causes.

One more example: why are our arms and our legs, which are used for such different functions, made of the same materials – the same bones, muscles, and nerves, all arranged in the same overall pattern? Evolution makes sense of the anomaly. Our remote ancestors' forelimbs were legs. After our ancestors became bipedal and started using their forelimbs for functions other than walking, these were gradually modified, but they retained their original composition and arrangement. Engineers start with raw materials and a design suited for a particular purpose; evolution can modify only what is already there. An engineer who would design cars and airplanes, or wings and wheels, using the same materials arranged in a similar pattern, would surely be fired.

Examples of deficiencies and dysfunctions in all sorts of organisms can be endlessly multiplied, reflecting the opportunistic, tinkererlike character of natural selection, rather than Intelligent Design. The world of organisms also abounds in characteristics that might be called "oddities," as well as those that have been characterized as "cruelties," an apposite qualifier if the cruel behaviors were designed outcomes of a being holding to human (or higher) standards of morality. But the "cruelties" of biological nature are only metaphoric "cruelties" when applied to the outcomes of natural selection. Instances of "cruelty" not only involve the familiar predators (say, a chimpanzee) tearing apart their prey (say, a small monkey held alive by a chimpanzee biting large flesh morsels from the screaming monkey), or parasites destroying the functional organs of their hosts, but also involve, and very abundantly, organisms of the same species, even individuals of different sexes in association with their mating. A well-known example is the female praying mantis that devours the male after coitus is completed. Less familiar is the fact that, if she gets the opportunity, the praying mantis female will eat the head of the male *before* mating, which thrashes the headless male mantis into spasms of "sexual frenzy" that allow the female to connect his genitalia to hers.[24] In some midges (tiny flies), the female captures the male as if he were any other prey and with the tip of her proboscis injects into his head her spittle, which starts digesting the male's innards, which are then sucked by the female; partly protected from digestion are the relatively intact male organs, which break off inside the female and fertilize her.[25] Male cannibalism is known in dozens of species, particularly spiders and scorpions. Diverse sorts of oddities associated with mating behavior are described in the delightful, but accurate and documented, book by Olivia Judson, *Dr. Tatiana's Sex Advice to All Creation.*[26]

The defective design of organisms could be attributed to the gods of the ancient Greeks, Romans, and Egyptians, who fought with one another, made blunders, and were clumsy in their endeavors. But, in my view, it is not compatible with special action by the omniscient and omnipotent God of Judaism, Christianity, and Islam.[27]

SCIENCE AS A WAY OF KNOWING: POWER AND LIMITS

Science is a wondrously successful way of knowing. Science seeks explanations of the natural world by formulating hypotheses that are subject to the possibility of empirical falsification or corroboration. A scientific hypothesis is tested by ascertaining whether or not predictions about the world of experience derived as logical consequences from the hypothesis agree with what is actually observed.[28] Science as a mode of inquiry into the nature of the universe has been successful and of great consequence. Witness the proliferation of academic departments of science in universities and other research institutions, the enormous budgets that the body politic and the private sector willingly commit to scientific research, and its economic impact. The Office of Management and the Budget (OMB) of the U.S. government has estimated that fifty percent of all economic growth in the United States since the Second World War can directly be attributed to scientific knowledge and technical advances. The technology derived from scientific knowledge pervades our lives: the high-rise buildings of our cities, expressways and long span-bridges, rockets that take men to the moon,

telephones that provide instant communication across continents, computers that perform complex calculations in millionths of a second, vaccines
and drugs that keep bacterial parasites at bay, gene therapies that replace
DNA in defective cells. All these remarkable achievements bear witness to
the validity of the scientific knowledge from which they originated.

Scientific knowledge is also remarkable in the way in which it emerges by
way of consensus and agreement among scientists, and in the way in which
new knowledge builds upon past accomplishment rather than starting anew
with each generation or with each new practitioner. Surely scientists disagree
with each other on many matters; but these are issues not yet settled, and
the points of disagreement generally do not bring into question previous
knowledge. Modern scientists do not challenge the notions that atoms exist,
that there is a universe with a myriad stars, or that heredity is encased in the
DNA.

Science is a way of knowing, but it is not the only way. Knowledge also
derives from other sources, such as common sense, artistic and religious
experience, and philosophical reflection. In *The Myth of Sisyphus*, the great
French writer Albert Camus asserted that we learn more about ourselves and
the world from a relaxed evening's perception of the starry heavens and the
scent of grass than from science's reductionistic ways.[29] The validity of the
knowledge acquired by nonscientific modes of inquiry can be established
simply by pointing out that science (in the modern sense of empirically
tested laws and theories) dawned in the sixteenth century, but that mankind
had for centuries built cities and roads, brought forth political institutions
and sophisticated codes of law, advanced profound philosophies and value
systems, and created magnificent plastic art as well as music and literature.
We thus learn about ourselves and about the world in which we live, and
we also benefit from products of this nonscientific knowledge. The crops
we harvest and the animals we husband emerged millennia before science's
dawn, from practices set down by farmers in the Middle East, the Andean
sierras, and the Mayan plateaus.

It is not my intention here to belabor the extraordinary fruits of nonscientific modes of inquiry. But I have set forth the view that nothing in the
world of nature escapes the scientific mode of knowledge, and that we owe
this universality to Darwin's revolution. Now I wish simply to state something
that is obvious, but at times clouded by the hubris of some scientists. Successful as it is, and universally encompassing as its subject is, a scientific view of
the world is hopelessly incomplete. There are matters of value and meaning
that are outside science's scope. Even when we have a satisfying scientific
understanding of a natural object or process, we are still missing knowledge
concerning matters that may well be thought by many to be of equal or
greater import. Scientific knowledge may enrich aesthetic and moral perceptions, and illuminate the significance of life and the world, but these are
matters outside science's realm.

On April 28, 1937, early in the Spanish Civil War, Nazi airplanes bombed the small Basque town of Guernica, the first time that a civilian population had been determinedly destroyed from the air. The Spanish painter Pablo Picasso had recently been commissioned by the Spanish republican government to paint a large composition for the Spanish Pavilion at the Paris World Exhibition of 1937. In a frenzy of manic energy, the enraged Picasso sketched in two days and fully outlined in ten more days his famous *Guernica*, an immense painting measuring 25 feet, 8 inches by 11 feet, 6 inches. Suppose that I now were to describe the images represented in the painting, their sizes and positions, as well as the pigments used and the quality of the canvas. This description would be of interest, but it would hardly be satisfying if I had completely omitted aesthetic analysis and considerations of meaning, the dramatic message of man's inhumanity to man conveyed by the outstretched figure of the mother pulling her killed baby, the bellowing faces, the wounded horse, or the satanic image of the bull.

Let *Guernica* be a metaphor for the final point I wish to make. Scientific knowledge, like the description of the size, materials, and geometry of *Guernica*, is satisfying and useful. But once science has had its say, there remains much about reality that is of interest, questions of value and meaning that are forever beyond science's scope.

Notes

1. This paper incorporates most of my "Darwin's Devolution: Design without Designer" in *Evolutionary and Molecular Biology: Scientific Perspectives on Divine Action*, ed. R. J. Russell, W. R. Stoeger, and F. J. Ayala (Vatican City: Vatican Observatory and Berkeley, CA: Center for Theology and the Natural Sciences, 1998), pp. 101–16. The text has been updated and modified, and a new section ("Unintelligent Design") has been added. Accordingly, the original title has been changed.

2. Pope John Paul II, addressing the Pontifical Academy of Sciences on October 3, 1981 (*L'Osservatore Romano*, October 4, 1981), said: "The Bible speaks to us of the origins of the universe and its makeup, not in order to provide us with a scientific treatise, but in order to state the correct relationship of man with God and the universe. Sacred Scripture wishes simply to declare that the world was created by God, and in order to teach this truth, it expresses itself in the terms of the cosmology in use at the time of the writer. The sacred book likewise wishes to tell men that the world was... created for the service of man and the glory of God. Any other teaching about the origin and makeup of the universe is alien to the intentions of the Bible, which does not wish to teach how heaven was made but how one goes to heaven." The point made by Saint Augustine and the Pope is that it is a blunder to mistake the Bible for an elementary textbook of astronomy, geology, and biology.

   In Book III, Chapter 14, Augustine writes, "As for the other small creatures... there was present from the beginning in all living bodies a natural power, and, I might say, there were interwoven with these bodies the seminal principles of animals later to appear . . . each according to its kind and with its special properties." One may surmise that Augustine would have found no conflict between the theory of evolution and the teachings of Genesis, which are the subject of his commentary.

3. The Pope's full address has been published in the original French in *L'Osservatore Romano*, October 23, 1996, and in English translation in *L'Osservatore Romano*, October 30, 1996. The French and English texts can also be found in *Evolutionary and Molecular Biology*, ed. Russell, Stoeger, and Ayala, pp. 2–9.

4. About Paley's *Natural Theology* and the argument from design, see my "Intelligent Design: The Original Version" in *Theology and Science* 1 (2003): pp. 9–32.

5. The Roman Catholic Church's opposition to Galileo in the seventeenth century had been similarly motivated not only by the apparent contradiction between the heliocentric theory and a literal interpretation of the Bible, but also by the unseemly attempt to comprehend the workings of the universe, the "mind of God." It may be worth noting that Darwin's theory of evolution by natural selection also encountered vehement opposition among scientists, but for different reasons.

There are many thoughtful discussions of the dialogue between Darwinism and Christianity; see for example, *God and Nature*, ed. David C. Lindberg and Ronald L. Numbers (Berkeley: University of California Press, 1986), Chapters 13–16. In 1874, Charles Hodge (1793–1878), an influential Protestant theologian, published *What Is Darwinism?*, one of the most articulate attacks on evolutionism. Hodge perceived Darwin's theory as "the most thoroughly naturalistic that can be imagined and far more atheistic than that of his predecessor Lamarck." He concluded that "the denial of design in nature is actually the denial of God." However, a principle of solution was seen by other Protestant theologians in the notion that God operates through intermediate causes. The origin and motion of the planets could be explained by the law of gravity and other natural processes without denying God's creation and Providence. Similarly, evolution could be seen as the natural process through which God brought living beings into existence. Thus, A. H. Strong, president of Rochester Theological Seminary, wrote in his *Systematic Theology*: "We grant the principle of evolution, but we regard it as only the method of divine intelligence." The brute ancestry of man was not incompatible with his exalted status as a creature in the image of God. Strong drew an analogy to Christ's miraculous conversion of water into wine: "The wine in the miracle was not water because water had been used in the making of it, nor is man a brute because the brute has made some contributions to its creation."

Arguments for and against Darwin's theory came from Catholic theologians as well. Gradually, well into the twentieth century, evolution by natural selection came to be accepted by the enlightened majority of Christian writers. Pius XII accepted, in his encyclical *Humani Generis*, (1950) that biological evolution is compatible with the Christian faith, although he argued that God's intervention would be necessary for the creation of the human soul. Pope John Paul II's address to the Pontifical Academy of Sciences of October 3, 1981 (see note 2), is an argument against the biblical literalism of fundamentalists and shares with most

Protestant theologians a view of Christian belief that is not incompatible with evolution and, more generally, with science. The Pope's address of October 22, 1996, cited at the beginning of this chapter, expresses definite acceptance of the theory of evolution on the basis of accumulated evidence.

6. See, e.g., John Haught's, chapter in this volume, "Darwin, Design, and Divine Providence," and his "Darwin's Gift to Theology," in *Evolutionary and Molecular Biology*, ed. Russell, Stoeger, and Ayala, pp. 393–418.

7. Charles Darwin, *On The Origin of Species*, facsimile of the first edition (New York: Atheneum, 1967), Chapter 3, p. 63, and Chapter 4, pp. 80–1.

8. A common objection posed to the account I have sketched of how natural selection gives rise to otherwise improbable features – is that some postulated transitions – for example, from a leg to a wing – cannot be adaptive. The answer to this kind of objection is well known to evolutionists. For example, there are rodents, primates, and other living animals that exhibit modified legs used for both running and gliding. The fossil record famously includes the reptile *Archaeopterix* and many other intermediate showing limbs incipiently transformed into wings endowed with feathers. One challenging transition involves the bones that make up the lower jaw of reptiles but have evolved into bones now found in the mammalian ear. What possible function could a bone have, either in the mandible or in the ear, during the intermediate stages? However, two transitional forms of therapsids (mammal-like reptiles) with a double jaw joint – one joint consisting of the bones that persist in the mammalian jaw, the other composed of the quadrate and articular bones, which eventually became the hammer and anvil of the mammalian ear – are known from the fossil record. Recent discoveries of fossil ancestors (and their relatives) of birds have provided unexpected details about the evolution of flight. Some diminutive dinosaurs, called microraptors, evolved feathers on both the forelimbs and the hind limbs. *Microraptor gui* had a trunk 15 cm long covered with downy feathers; the tail had a tuft of longer ones. The front and hind limbs had "hands" and feet but were partly covered by about a dozen primary, flightlike feathers plus about eighteen shorter, secondary feathers. *Microraptor gui* is dated as 126 million years old, some 25 million years younger than *Archaeopteryx*, and lived mostly on trees, using the rear feathers for gliding (*Nature* 421 (2003): 323–4 and 335–40; *Science* 299 (2003): 491. Functional morphology research suggests that the first protobirds used their primitive wings to help them to scale inclined objects and trees (*Science* 299 (2003): 329 and 402–4).

9. See Ayala, "Intelligent Design: The Original Version."

10. *Webster's Third New International Dictionary* (1966).

11. Not all features of a car contribute to efficient transportation – some features are added for esthetic or other reasons. But as long as a feature is added because it exhibits certain properties – such as its appeal to the esthetic preferences of potential customers – it may be explained teleologically. Nevertheless, there may be features in a car, a knife, or any other man-made object that need not be explained teleologically. That knives have handles may be explained teleologically, but the fact that a particular handle is made of pine rather than oak might be explained simply by the availability of material. Similarly, not all features of organisms have teleological explanations.

12. In general, as pointed out earlier, those features and behaviors that are considered adaptations are explained teleologically. This is simply because adaptations are features that come about by natural selection. Among alternative genetic variants that may arise by mutation or recombination, the ones that become established in a population are those that contribute most to the reproductive success of their carriers. "Fitness" is the measure used by evolutionists to quantify reproductive success. But reproductive success is usually mediated by some function or property. Wings and hands acquired their present configuration through the long-term accumulation of genetic variants adaptive to their carriers. How natural selection yields adaptive features may be explained by citing examples in which the adaptation arises as a consequence of a single gene mutation. One example is the presence of normal hemoglobin rather than hemoglobin S in humans. One amino acid substitution in the beta chain in humans results in hemoglobin molecules less efficient for oxygen transport. The general occurrence in human populations of normal rather than S hemoglobin is explained teleologically by the contribution of hemoglobin to effective oxygen transport and thus to reproductive success. A second example, the difference between peppered gray moths and melanic moths, is also a consequence of differences in one or only a few genes. The replacement of gray moths by melanics in polluted regions is explained teleologically by the fact that in such regions melanism decreases the probability that a moth will be eaten by a bird. The predominance of peppered forms in nonpolluted regions is similarly explained.

   Not all features of organisms need to be explained teleologically, because not all come about as a direct result of natural selection. Some features may become established by random genetic drift, by chance association with adaptive traits, or in general, by processes other than natural selection. Proponents of the neutrality theory of protein evolution argue that many alternative protein variants are adaptively equivalent. At least in certain cases, most evolutionists would admit that the selective differences between alternative nucleotides at a certain site in the DNA of a gene coding for a particular protein must be virtually nil, particularly when population size is very small. In such cases, the presence in a population of one nucleotide sequence rather than a slightly different one, adaptively equivalent to the first, would not be explained teleologically. Needless to say, in such cases there would be nucleotide changes in the DNA that would not be adaptive. The presence of an adaptive DNA sequence rather than a nonadaptive one would be explained teleologically; but among those adaptively equivalent, the presence of one DNA sequence rather than another would not require a teleological explanation.

13. Two types of homeostasis are usually distinguished – physiological and developmental – although intermediate conditions exist. Physiological homeostatic reactions enable organisms to maintain a certain physiological steady state in spite of environmental fluctuations. The regulation of the concentration of salt in blood by the kidneys, or the hypertrophy of muscle owing to strenuous use, are examples of physiological homeostasis. Developmental homeostasis refers to the regulation of the different paths that an organism may follow in the progression from fertilized egg to adult. The process can be influenced by the environment in various ways, but the characteristics of the adult

individual, at least within a certain range, are largely predetermined in the fertilized egg.

14. In spite of the role played by stochastic events in the phylogenetic history of birds, it would be mistaken to say that wings are not teleological features. As pointed out earlier, there are differences between the teleology of an organism's adaptations and the nonteleological potential uses of natural inanimate objects. A mountain may have features appropriate for skiing, but those features did not come about in order to provide skiing slopes. On the other hand, the wings of birds came about precisely because they serve for flying, although they may have derived from features that originally served other purposes, such as climbing or gliding (see note 8). The explanatory reason for the existence of wings and their configuration is the end they serve – flying – which in turn contributes to the reproductive success of birds. If wings did not serve an adaptive function, they would never have come about or would gradually disappear over the generations.

The indeterminate character of the outcome of natural selection over time is due to a variety of nondeterministic factors. The outcome of natural selection depends, first, on what alternative genetic variants happen to be available at any one time. This, in turn, depends on the stochastic processes of mutation and recombination, and also on the past history of any given population. (What new genes may arise by mutation and what new genetic constitutions may arise by recombination depend on what genes happen to be present – which depends on previous history.) The outcome of natural selection also depends on the conditions of the physical and biotic environment. Which alternatives among available genetic variants may be favored by selection depends on the particular set of environmental conditions to which a population is exposed.

It is important, for historical reasons, to reiterate that the process of evolution by natural selection is not teleological in the purposeful sense. The natural theologians of the nineteenth century erroneously claimed that the directive organization of living beings evinces the existence of a Designer. The adaptations of organisms can be explained as the result of natural processes without recourse to consciously intended end products. There is purposeful activity in the world, at least in man; but the existence and particular structures of organisms, including humans, need not be explained as the result of purposeful behavior.

Some scientists and philosophers who held that evolution is a natural process erred, nevertheless, in seeing evolution as a determinate, or bounded, process. Lamarck (1809) thought that evolutionary change necessarily proceeded along determined paths, from simpler to more complex organisms. Similarly, the evolutionary philosophies of Bergson (1907), Teilhard de Chardin (1959), and the theories of nomogenesis (Berg 1926), aristogenesis (Osborn 1934), orthogenesis, and the like are erroneous because they all claim that evolutionary change necessarily proceeds along determined paths. These theories mistakenly take embryological development as the model of evolutionary change, regarding the teleology of evolution as determinate. Although there are teleologically determinate processes in the living world, such as embryological development and physiological homeostasis, the evolutionary origin of living beings is teleological only in the indeterminate sense. Natural selection does not in any way direct evolution toward any particular kind of organism or toward any particular properties.

15. See F. J. Ayala "Adaptation and Novelty: Teleological Explanations in Evolutionary Biology" *History and Philosophy of Life Sciences* 21 (1999): 3–33. See also F. J. Ayala, "Teleological Explanations in Evolutionary Biology," *Philosophy of Science* 37 (1970): 1–15; and F. J. Ayala, "The Distinctness of Biology," in *Laws of Nature: Essays on the Philosophical, Scientific, and Historical Dimensions*, ed. Friedel Weinert (Berlin and New York: Walter de Gruyter, 1995), pp. 268–85.

16. William Paley, *Natural Theology* (New York: American Tract Society, n.d.), p. 1. I will cite pages from this American edition, which is undated but seems to have been printed in the late nineteenth century.

17. Ibid., pp. 175–6.

18. Ibid., p. 180.

19. Ibid., p. 49.

20. Ibid., p. 265.

21. Michael J. Behe, *Darwin's Black Box: The Biochemical Challenge to Evolution* (New York: Touchstone/Simon and Schuster, 1996), p. 223.

22. Robert T. Pennock, ed., *Intelligent Design Creationism and Its Critics: Philosophical, Theological, and Scientific Perspectives* (Cambridge, MA: MIT Press, 2001), p. 249.

23. Paley, *Natural Theology*, Chapter 23.

24. S. E. Lawrence, "Sexual Cannibalism in the Praying Mantis, *Mantis religiosa:* A Field Study," *Animal Behaviour* 43 (1992): 569–83; see also M. A. Elgar, "Sexual Cannibalism in Spiders and Other Invertebrates," in *Cannibalism: Ecology and Evolution among Diverse Taxa*, ed. M. A. Elgar and B. J. Crespi (Oxford: Oxford University Press, 1992).

25. J. A. Downes, "Feeding and Mating in the Insectivorous Ceratopogoninae (Diptera)," *Memoirs of the Entomological Society of Canada* 104 (1978): 1–62.

26. Olivia Judson, *Dr. Tatiana's Sex Advice to All Creation* (New York: Holt, 2002).

27. David Hull, a distinguished American philosopher of biology, has made the same point in a somewhat more strident tone: "What kind of God can one infer from the sort of phenomena epitomized by the species on Darwin's Galapagos Islands? The evolutionary process is rife with happenstance, contingency, incredible waste, death, pain and horror.... Whatever the God implied by evolutionary theory and the data of natural selection may be like, he is not the Protestant God of waste not, want not. He is also not the loving God who cares about his productions. He is not even the awful God pictured in the Book of Job. The God of the Galapagos is careless, wasteful, indifferent, almost diabolical. He is certainly not the sort of God to whom anyone would be inclined to pray." David L. Hull, "God of the Galapagos," *Nature* 352 (1992): 485–6.

28. Testing a scientific hypothesis involves at least four different activities. First, the hypothesis must be examined for internal consistency. A hypothesis that is self-contradictory or logically ill-formed in some other way should be rejected. Second, the logical structure of the hypothesis must be examined in order to ascertain whether it has explanatory value, i.e., whether it makes the observed phenomena intelligible in some sense, whether it provides an understanding of why the phenomena do in fact occur as observed. A hypothesis that is purely

tautological should be rejected because it has no explanatory value. A scientific hypothesis identifies the conditions, processes, or mechanisms that account for the phenomena it purports to explain. Thus, hypotheses establish general relationships between certain conditions and their consequences or between certain causes and their effects. For example, the motions of the planets around the sun are explained as a consequence of gravity, and respiration as an effect of red blood cells that carry oxygen from the lungs to various parts of the body.

Third, the hypothesis must be examined for its consistency with hypotheses and theories commonly accepted in the particular field of science, or to see whether it represents any advance with respect to well-established alternative hypotheses. Lack of consistency with other theories is not always ground for rejection of a hypothesis, although it often will be. Some of the greatest scientific advances occur precisely when it is shown that a widely held and well supported hypothesis can be replaced by a new one that accounts for the same phenomena that were explained by the preexisting hypothesis as well as other phenomena it could not account for. One example is the replacement of Newtonian mechanics by the theory of relativity, which rejects the conservation of matter and the simultaneity of events that occur at a distance – two fundamental tenets of Newton's theory.

Examples of this kind are common in rapidly advancing disciplines, such as molecular biology at present. The so-called central dogma holds that molecular information flows only in one direction, from DNA to RNA to protein. The DNA contains the genetic information that determines what the organism is, but that information has to be expressed in enzymes (a particular class of proteins) that guide all chemical processes in cells. The information contained in the DNA molecules is conveyed to proteins by means of intermediate molecules, called messenger RNA. David Baltimore and Howard Temin were awarded the Nobel Prize for discovering that information could flow in the opposite direction, from RNA to DNA, by means of the enzyme reverse transcriptase. They showed that some viruses, as they infect cells, are able to copy their RNA into DNA, which then becomes integrated into the DNA of the infected cell, where it is used as if it were the cell's own DNA.

Other examples are the following. Until very recently, it was universally thought that only the proteins known as enzymes could mediate (technically, "catalyze") the chemical reactions in cells. However, Thomas Cech and Sidney Altman received a Nobel Prize in 1989 for showing that certain RNA molecules act as enzymes and catalyze their own reactions. One more example concerns the so-called colinearity between DNA and protein. It was generally thought that the sequence of nucleotides in the DNA of a gene is expressed consecutively in the sequence of amino acids in the protein. This conception was shaken by the discovery that genes come in pieces, separated by intervening DNA segments that do not carry genetic information; Richard Roberts and Philip Sharp received a Nobel Prize in 1993 for this discovery.

The fourth and most distinctive test is the one that I have identified, which consists of putting on trial an empirically scientific hypothesis by ascertaining whether or not predictions about the world of experience derived as logical consequences from the hypothesis agree with what is actually observed. This is the critical element that distinguishes the empirical sciences from other forms of knowledge: the requirement that scientific hypotheses be empirically falsifiable. Scientific hypotheses cannot be consistent with all possible states of affairs in the empirical world. A hypothesis is scientific only if it is consistent with some but not other possible states of affairs not yet observed in the world, so that it may be subject to the possibility of falsification by observation. The predictions derived from a scientific hypothesis must be sufficiently precise that they limit the range of possible observations with which they are compatible. If the results of an empirical test agree with the predictions derived from a hypothesis, the hypothesis is said to be provisionally corroborated; otherwise, it is falsified.

The requirement that a scientific hypothesis be falsifiable has been called by Karl Popper the *criterion of demarcation* of the empirical sciences, because it sets apart the empirical sciences from other forms of knowledge. A hypothesis that is not subject to the possibility of empirical falsification does not belong in the realm of science. See F. J. Ayala, "On the Scientific Method, Its Practice and Pitfalls," *History and Philosophy of Life Sciences* 16 (1994): 205–40.

29. This point has been made in sensuous prose by John Updike in his recent *Toward the End of Time* (New York: Knopf, 1997), p. 214: "It makes sense: all those blazing suns, red and swollen or white and shrunken or yellow like our moderate own, blue and new or black and collapsed, madly spinning neutron stars or else all-swallowing black holes denser yet, not to mention planets and cinderlike planetoids and picturesque clouds of glowing gas and dark matter hypothetical or real and titanic streaming soups of neutrinos, could scarcely be expected to converge exactly upon a singularity smaller, by many orders of magnitude, than a pinhead. The Weyl curvature, in other words, was very very *very* near zero at the Big Bang, but will be much larger at the Big Crunch. But, I ignorantly wonder, how does time's arrow know this, in our trifling immediate vicinity? What keeps it from spinning about like the arrow of a compass, jumping broken cups back on the table intact and restoring me, if not to a childhood self, to the suburban buck I was when still married." The mystery of aging and ultimate personal demise receive, in Updike's view, but little help from considerations of the immensity and endurance of the physicist's universe.

I sometimes make the same point by rephrasing a witticism that I once heard from a friend: "In the matter of the value and meaning of the universe, science has all the answers, except the interesting ones."

5

The Flagellum Unspun

The Collapse of "Irreducible Complexity"

Kenneth R. Miller

Almost from the moment *On the Origin of Species* was published in 1859, the opponents of evolution have fought a long, losing battle against their Darwinian foes. Today, like a prizefighter in the late rounds losing badly on points, they've placed their hopes on one big punch – a single claim that might smash through the overwhelming weight of scientific evidence to bring Darwin to the canvas once and for all. Their name for this virtual roundhouse right is "Intelligent Design."

In the last several years, the Intelligent Design (ID) movement has attempted to move against the standards of science education in several American states, most famously in Kansas and Ohio (Holden 1999; Gura 2002). The principal claim made by adherents of this view is that they can detect the presence of "Intelligent Design" in complex biological systems. As evidence, they cite a number of specific examples, including the vertebrate blood clotting cascade, the eukaryotic cilium, and most notably, the eubacterial flagellum (Behe 1996a; Behe 2002).

Of all these examples, the flagellum has been presented so often as a counterexample to evolution that it might well be considered the "poster child" of the modern anti-evolution movement. Variations of its image (Figure 5.1) now appear on web pages of anti-evolution groups such as the Discovery Institute, and on the covers of "Intelligent Design" books such as William Dembski's *No Free Lunch* (Dembski 2002a). To anti-evolutionists, the high status of the flagellum reflects the supposed fact that it could not possibly have been produced by an evolutionary pathway.

There is, to be sure, nothing new or novel in an anti-evolutionist pointing to a complex or intricate natural structure and professing skepticism that it could have been produced by the "random" processes of mutation and natural selection. Nonetheless, the "argument from personal incredulity," as such sentiments have been appropriately described, has been a weapon of little value in the anti-evolution movement. Anyone can state at any time

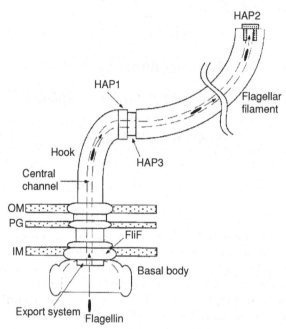

FIGURE 5.1. The eubacterial flagellum. The flagellum is an ion-powered rotary motor, anchored in the membranes surrounding the bacterial cell. This schematic diagram highlights the assembly process of the bacterial flagellar filament and the cap–filament complex. OM = outer membrane; PG = peptidoglycan layer; IM = cytoplasmic membrane. (From Yonekura et al. 2000.)

that he or she cannot imagine how evolutionary mechanisms might have produced a certain species, organ, or structure. Such statements, obviously, are personal – and they say more about the limitations of those who make them than they do about the limitations of Darwinian mechanisms.

The hallmark of the Intelligent Design movement, however, is that it purports to rise above the level of personal skepticism. It claims to have found a *reason* why evolution could not have produced a structure like the bacterial flagellum – a reason based on sound, solid scientific evidence.

Why does the intelligent design movement regard the flagellum as unevolvable? Because it is said to possesses a quality known as "irreducible complexity." Irreducibly complex structures, we are told, could not have been produced by evolution – or, for that matter, by any natural process. They do exist, however, and therefore they must have been produced by something. That something could only be an outside intelligent agency operating beyond the laws of nature – an intelligent designer. That, simply stated, is the core of the new argument from design, and the intellectual basis of the Intelligent Design movement.

The great irony of the flagellum's increasing acceptance as an icon of the anti-evolutionist movement is that fact that research had demolished its status as an example of irreducible complexity almost at the very moment it was first proclaimed. The purpose of this chapter is to explore the arguments by which the flagellum's notoriety has been achieved, and to review the research developments that have now undermined the very foundations of those arguments.

THE ARGUMENT'S ORIGINS

The flagellum owes its status principally to *Darwin's Black Box* (Behe 1996a), a book by Michael Behe that employed it in a carefully crafted anti-evolution argument. Building upon William Paley's well-known "argument from design," Behe sought to bring the argument two centuries forward into the realm of biochemistry. Like Paley, Behe appealed to his readers to appreciate the intricate complexity of living organisms as evidence for the work of a designer. Unlike Paley, however, he raised the argument to a new level, claiming to have discovered a scientific principle that could be used to prove that certain structures could not have been produced by evolution. That principle goes by the name of "irreducible complexity."

An irreducibly complex structure is defined as "a single system composed of several well-matched, interacting parts that contribute to the basic function, wherein the removal of any one of the parts causes the system to effectively cease functioning" (Behe 1996a, 39). Why would such systems present difficulties for Darwinism? Because they could not possibly have been produced by the process of evolution:

An irreducibly complex system cannot be produced directly by numerous, successive, slight modifications of a precursor system, because any precursor to an irreducibly complex system that is missing a part is by definition nonfunctional. . . . Since natural selection can only choose systems that are already working, then if a biological system cannot be produced gradually it would have to arise as an integrated unit, in one fell swoop, for natural selection to have anything to act on. (Behe 1996b)

The phrase "numerous, successive, slight modifications" is not accidental. The very same words were used by Charles Darwin in the *Origin of Species* in describing the conditions that had to be met for his theory to be true. As Darwin wrote, if one could find an organ or structure that could not have been formed by "numerous, successive, slight modifications," his "theory would absolutely break down" (Darwin 1859, 191). To anti-evolutionists, the bacterial flagellum is now regarded as exactly such a case – an "irreducibly complex system" that "cannot be produced directly by numerous successive, slight modifications." A system that could not have evolved – a desperation punch that just might win the fight in the final round, a tool with which the theory of evolution might be brought down.

THE LOGIC OF IRREDUCIBLE COMPLEXITY

Living cells are filled, of course, with complex structures whose detailed evolutionary origins are not known. Therefore, in fashioning an argument against evolution one might pick nearly any cellular structure – the ribosome, for example – and claim, correctly, that its origin has not been explained in detail by evolution.

Such arguments are easy to make, of course, but the nature of scientific progress renders them far from compelling. The lack of a detailed current explanation for a structure, organ, or process does not mean that science will never come up with one. As an example, one might consider the question of how left-right asymmetry arises in vertebrate development, a question that was beyond explanation until the 1990s (Belmonte 1999). In 1990, one might have argued that the body's left-right asymmetry could just as well be explained by the intervention of a designer as by an unknown molecular mechanism. Only a decade later, the actual molecular mechanism was identified (Stern 2002), and any claim one might have made for the intervention of a designer would have been discarded. The same point can be made, of course, regarding any structure or mechanism whose origins are not yet understood.

The utility of the bacterial flagellum is that it seems to rise above this "argument from ignorance." By asserting that it is a structure "in which the removal of an element would cause the whole system to cease functioning" (Behe 2002), the flagellum is presented as a "molecular machine" whose individual parts must have been specifically crafted to work as a unified assembly. The existence of such a multipart machine therefore provides genuine scientific proof of the actions of an intelligent designer.

In the case of the flagellum, the assertion of irreducible complexity means that a minimum number of protein components, perhaps thirty, are required to produce a working biological function. By the logic of irreducible complexity, these individual components should have no function until all thirty are put into place, at which point the function of motility appears. What this means, of course, is that evolution could not have fashioned those components a few at a time, since they do not have functions that could be favored by natural selection. As Behe wrote, "natural selection can only choose among systems that are already working" (Behe 2002), and an irreducibly complex system does not work unless all of its parts are in place. The flagellum is irreducibly complex, and therefore, it must have been designed. Case closed.

ANSWERING THE ARGUMENT

The assertion that cellular machines are irreducibly complex, and therefore provide proof of design, has not gone unnoticed by the scientific community.

A number of detailed rebuttals have appeared in the literature, and many have pointed out the poor reasoning of recasting the classic argument from design in the modern language of biochemistry (Coyne 1996; Miller 1996; Depew 1998; Thornhill and Ussery 2000). I have suggested elsewhere that the scientific literature contains counterexamples to any assertion that evolution cannot explain biochemical complexity (Miller 1999, 147), and other workers have addressed the issue of how evolutionary mechanisms allow biological systems to increase in information content (Adami, Ofria, and Collier 2000; Schneider 2000).

The most powerful rebuttals to the flagellum story, however, have not come from direct attempts to answer the critics of evolution. Rather, they have emerged from the steady progress of scientific work on the genes and proteins associated with the flagellum and other cellular structures. Such studies have now established that the entire premise by which this molecular machine has been advanced as an argument against evolution is wrong – **the bacterial flagellum is not irreducibly complex**. As we will see, the flagellum – the supreme example of the power of this new "science of design" – has failed its most basic scientific test. Remember the claim that "any precursor to an irreducibly complex system that is missing a part is by definition nonfunctional"? As the evidence has shown, nature is filled with examples of "precursors" to the flagellum that are indeed "missing a part," and yet are fully functional – functional enough, in some cases, to pose a serious threat to human life.

THE TYPE III SECRETORY APPARATUS

In the popular imagination, bacteria are "germs" – tiny microscopic bugs that make us sick. Microbiologists smile at that generalization, knowing that most bacteria are perfectly benign, and that many are beneficial – even essential – to human life. Nonetheless, there are indeed bacteria that produce diseases, ranging from the mildly unpleasant to the truly dangerous. Pathogenic, or disease-causing, bacteria threaten the organisms they infect in a variety of ways, one of which is by producing poisons and injecting them directly into the cells of the body. Once inside, these toxins break down and destroy the host cells, producing illness, tissue damage, and sometimes even death.

In order to carry out this diabolical work, bacteria not only must produce the protein toxins that bring about the demise of their hosts, but also must efficiently inject them across the cell membranes and into the cells of their hosts. They do this by means of any number of specialized protein secretory systems. One, known as the type III secretory system (TTSS), allows gram-negative bacteria to translocate proteins directly into the cytoplasm of a host cell (Heuck 1998). The proteins transferred through the TTSS include a variety of truly dangerous molecules, some of which are known as "virulence

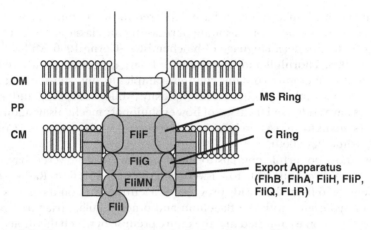

FIGURE 5.2. There are extensive homologies between type III secretory proteins and proteins involved in export in the basal region of the bacterial flagellum. These homologies demonstrate that the bacterial flagellum is not "irreducibly complex." In this diagram (redrawn from Heuck 1998), the shaded portions of the basal region indicate proteins in the *E. coli* flagellum homologous to the Type III secretory structure of *Yersinia.* OM = outer membrane; PP = periplasmic space; CM = cytoplasmic membrane.

factors," that are directly responsible for the pathogenic activity of some of the most deadly bacteria in existence (Heuck 1998; Büttner and Bonas 2002).

At first glance, the existence of the TTSS, a nasty little device that allows bacteria to inject these toxins through the cell membranes of their unsuspecting hosts, would seem to have little to do with the flagellum. However, molecular studies of proteins in the TTSS have revealed a surprising fact: the proteins of the TTSS are directly homologous to the proteins in the basal portion of the bacterial flagellum. As figure 5.2 (Heuck 1998) shows, these homologies extend to a cluster of closely associated proteins found in both of these molecular "machines." On the basis of these homologies, McNab (1999) has argued that the flagellum itself should be regarded as a type III secretory system. Extending such studies with a detailed comparison of the proteins associated with both systems, Aizawa has seconded this suggestion, noting that the two systems "consist of homologous component proteins with common physico-chemical properties" (Aizawa 2001, 163). It is now clear, therefore, that a smaller subset of the full complement of proteins in the flagellum makes up the functional transmembrane portion of the TTSS.

Stated directly, the TTSS does its dirty work using a handful of proteins from the base of the flagellum. From the evolutionary point of view, this relationship is hardly surprising. In fact, it is to be expected that the opportunism of evolutionary processes would mix and match proteins in order to produce new and novel functions. According to the doctrine of irreducible complexity, however, this should not be possible. If the flagellum is indeed irreducibly complex, then removing just one part, let alone ten or fifteen, should render what remains "by definition nonfunctional." Yet the TTSS is indeed fully functional, even though it is missing most of the parts of the flagellum. The TTSS may be bad news for us, but for the bacteria that possess it, it is a truly valuable biochemical machine.

The existence of the TTSS in a wide variety of bacteria demonstrates that a small portion of the "irreducibly complex" flagellum can indeed carry out an important biological function. Because such a function is clearly favored by natural selection, the contention that the flagellum must be fully assembled before any of its component parts can be useful is obviously incorrect. What this means is that the argument for intelligent design of the flagellum has failed.

COUNTERATTACK

Classically, one of the most widely repeated charges made by anti-evolutionists is that the fossil record contains wide "gaps" for which transitional fossils have never been found. Therefore, the intervention of a creative agency – an intelligent designer – must be invoked to account for each gap. Such gaps, of course, have been filled with increasing frequency by paleontologists – the increasingly rich fossil sequences demonstrating the origins of whales are a useful example (Thewissen, Hussain, and Arif 1994; Thewissen et al. 2001). Ironically, the response of anti-evolutionists to such discoveries is frequently to claim that things have only gotten worse for evolution. Where previously there had been just one gap, as a result of the transitional fossil there are now two (one on either side of the newly discovered specimen).

As word of the relationship between the eubacterial flagellum and the TTSS has begun to spread among the "design" community, the first hints of a remarkably similar reaction have emerged. The TTSS only makes problems worse for evolution, according to this response, because now there are two irreducibly complex systems to deal with. The flagellum is still irreducibly complex – but so is the TTSS. So now there are two systems for evolutionists to explain instead of just one.

Unfortunately for this line of argument, the claim that one irreducibly complex system might contain another is self-contradictory. To understand this, we need to remember that the entire point of the design argument, as exemplified by the flagellum, is that only the entire biochemical machine, with all of its parts, is functional. For the Intelligent Design argument to stand, this must be the case, since it provides the basis for their claim that only the complete flagellum can be favored by natural selection, not any of its component parts.

However, if the flagellum contains within it a smaller functional set of components such as the TTSS, then the flagellum itself cannot be irreducibly complex – by definition. Since we now know that this is indeed the case, it is obviously true that the flagellum is not irreducibly complex.

A second reaction, which I have heard directly after describing the relationship between the secretory apparatus and the flagellum, is the objection that the TTSS does not tell us how either it or the flagellum evolved. This is certainly true, although Aizawa has suggested that the TTSS may indeed be an evolutionary precursor of the flagellum (Aizawa 2001). Nonetheless, until we have produced a step-by-step account of the evolutionary derivation of the flagellum, one may indeed invoke the argument from ignorance for this and every other complex biochemical machine.

However, in agreeing to this, one must keep in mind that the doctrine of irreducible complexity was intended to go one step beyond the claim of ignorance. It was fashioned in order to provide a rationale for claiming that the bacterial flagellum could not have evolved, even in principle, because it is irreducibly complex. Now that a simpler, functional system (the TTSS) has been discovered among the protein components of the flagellum, the claim of irreducible complexity has collapsed, and with it any "evidence" that the flagellum was designed.

THE COMBINATORIAL ARGUMENT

At first glance, William Dembski's case for Intelligent Design seems to follow a distinctly different strategy in dealing with biological complexity. His recent book, *No Free Lunch* (Dembski 2002a), lays out this case, using information theory and mathematics to show that life is the result of Intelligent Design. Dembski makes the assertion that living organisms contain what he calls "complex specified information" (CSI), and he claims to have shown that the evolutionary mechanism of natural selection cannot produce CSI. Therefore, any instance of CSI in a living organism must be the result of intelligent design. And living organisms, according to Dembski, are chock-full of CSI.

Dembski's arguments, couched in the language of information theory, are highly technical and are defended, almost exclusively, by reference to their utility in detecting information produced by human beings. These include phone and credit card numbers, symphonies, and artistic woodcuts, to name just a few. One might then expect that Dembski, having shown how the presence of CSI can be demonstrated in man-made objects, would then turn to a variety of biological objects. Instead, he turns to just one such object, the bacterial flagellum.

Dembski then offers his readers a calculation showing that the flagellum could not possibly have have evolved. Significantly, he begins that calculation by linking his arguments to those of Behe, writing: "I want therefore in this section to show how irreducible complexity is a special case of specified complexity, and in particular I want to sketch how one calculates the relevant probabilities needed to eliminate chance and infer design for such systems" (Dembski 2002a, 289). Dembski then tells us that an irreducibly complex system, like the flagellum, is a "discrete combinatorial object." What this means, as he explains, is that the probability of assembling such an object can be calculated by determining the probabilities that each of its components might have originated by chance, that they might have been localized to the same region of the cell, and that they would have been assembled in precisely the right order. Dembski refers to these three probabilities as Porig, Plocal, and Pconfig, and he regards each of them as separate and independent (Dembski 2002a, 291).

This approach overlooks the fact that the last two probabilities are actually contained within the first. Localization and self-assembly of complex protein structures in prokaryotic cells are properties generally determined by signals built into the primary structures of the proteins themselves. The same is probably true for the amino acid sequences of the thirty or so protein components of the flagellum and the approximately twenty proteins involved in the flagellum's assembly (McNab 1999; Yonekura et al. 2000). Therefore, if one gets the sequences of all the proteins right, localization and assembly will take care of themselves.

To the ID enthusiast, however, this is a point of little concern. According to Dembski, evolution still could not construct the thirty proteins needed for the flagellum. His reason is that the probability of their assembly falls below what he terms the "universal probability bound." According to Dembski, the probability bound is a sensible allowance for the fact that highly improbable events do occur from time to time in nature. To allow for such events, he argues that given enough time, any event with a probability larger than 10^{-150} might well take place. Therefore, if a sequence of events, such as a presumed evolutionary pathway, has a calculated probability less than 10^{-150}, we may conclude that the pathway is impossible. If the calculated probability is greater than 10^{-150}, it is possible (even if unlikely).

When Dembski turns his attention to the chances of evolving the thirty proteins of the bacterial flagellum, he makes what he regards as a generous assumption. Guessing that each of the proteins of the flagellum have about 300 amino acids, one might calculate that the chance of getting just one such protein to assemble from "random" evolutionary processes would be 20^{-300}, since there are 20 amino acids specified by the genetic code. Dembski, however, concedes that proteins need not get the *exact* amino acid sequence right in order to be functional, so he cuts the odds to just 20^{-30}, which he tells his readers is "on the order of 10^{-39} (Dembski 2002a, 301). Since the flagellum requires thirty such proteins, he explains that thirty such probabilities "will all need to be multiplied to form the origination probability" (Dembski 2002a, 301). That would give us an origination probability for the flagellum of 10^{-1170}, far below the universal probability bound. The flagellum could not have evolved, and now we have the numbers to prove it. Right?

ASSUMING IMPOSSIBILITY

I have no doubt that to the casual reader, a quick glance over the pages of numbers and symbols in Dembski's books is impressive, if not downright intimidating. Nonetheless, the way in which he calculates the probability of an evolutionary origin for the flagellum shows how little biology actually stands behind those numbers. His computation calculates only the probability of spontaneous, random assembly for each of the proteins of the flagellum. Having come up with a probability value on the order of 10^{-1170}, he assures us that he has shown the flagellum to be unevolvable. This conclusion, of course, fits comfortably with his view that "[t]he Darwinian mechanism is powerless to produce irreducibly complex systems" (Dembski 2002a, 289).

However complex Dembski's analysis, the scientific problem with his calculations is almost too easy to spot. By treating the flagellum as a "discrete combinatorial object" he has shown only that it is unlikely that the parts of the flagellum could assemble spontaneously. Unfortunately for his argument, no scientist has ever proposed that the flagellum, or any other complex object, evolved in that way. Dembski, therefore, has constructed a classic "straw man" and blown it away with an irrelevant calculation.

By treating the flagellum as a discrete combinatorial object, he has assumed in his calculation that no subset of the thirty or so proteins of the flagellum could have biological activity. As we have already seen, this is wrong. Nearly a third of those proteins are closely related to components of the TTSS, which does indeed have biological activity. A calculation that ignores that fact has no scientific validity.

More importantly, Dembski's willingness to ignore the TTSS lays bare the underlying assumption of his entire approach to the calculation of probabilities and the detection of "design." *He assumes what he is trying to prove.*

According to Dembski, the detection of "design" requires that an object display complexity that could not be produced by what he calls "natural causes." In order to do that, one must first examine all of the possibilities by which an object, such as the flagellum, might have been generated naturally. Dembski and Behe, of course, come to the conclusion that there are no such natural causes. But how did they determine that? What is the scientific method used to support such a conclusion? Could it be that their assertions of the lack of natural causes simply amount to an unsupported personal belief? Could it be that there are such causes but that they simply happened not to think of them? Dembski actually seems to realize that this is a serious problem. He writes: "Now it can happen that we may not know enough to determine all the relevant chance hypotheses. Alternatively, we might think we know the relevant chance hypotheses, but later discover that we missed a crucial one. In the one case a design inference could not even get going; in the other, it would be mistaken" (Dembski 2002a, 123, note 80).

What Dembski is telling us is that in order to "detect" design in a biological object, one must first come to the conclusion that the object could not have been produced by any "relevant chance hypotheses" (meaning, naturally, evolution). Then, and only then, are Dembski's calculations brought into play. Stated more bluntly, what this really means is that the "method" first involves *assuming the absence* of an evolutionary pathway leading to the object, followed by a calculation "proving" the impossibility of spontaneous assembly. Incredibly, this a priori reasoning is exactly the sort of logic upon which the new "science of design" has been constructed.

Not surprisingly, scientific reviewers have not missed this point – Dembski's arguments have been repeatedly criticized on this issue and on many others (Charlesworth 2002; Orr 2002; Padian 2002).

DESIGNING THE CYCLE

In assessing the design argument, therefore, it only *seems* as though two distinct arguments have been raised for the unevolvability of the flagellum. In reality, those two arguments, one invoking irreducible complexity and the other specified complex information, both depend upon a single scientifically insupportable position – namely, that we can look at a complex biological object and determine with absolute certainty that none of its component parts could have been first selected to perform other functions. Now the discovery of extensive homologies between the type III secretory system and the flagellum has shown just how wrong that position was.

When anti-evolutionary arguments featuring the bacterial flagellum rose to prominence, beginning with the 1996 publication of *Darwin's Black Box* (Behe 1996a), they were predicated upon the assertion that each of the protein components of the flagellum was crafted, in a single act of design, to fit the specific purpose of the flagellum. The flagellum was said to be unevolvable because the entire complex system had to be assembled first in order to produce any selectable biological function. This claim was broadened to include all complex biological systems, and to assert further that science would never find an evolutionary pathway to any of these systems. After all, it hadn't so far, at least according to one of the principal advocates of "design":

There is no publication in the scientific literature – in prestigious journals, specialty journals, or books – that describes how molecular evolution of any real, complex, biochemical system either did occur or even might have occurred. (Behe 1996a, 185)

As many critics of intelligent design have pointed out, that statement is simply false. Consider, as just one example, the Krebs cycle, an intricate biochemical pathway consisting of nine enzymes and a number of cofactors that occupies center stage in the pathways of cellular metabolism. The Krebs cycle is "real," "complex," and "biochemical." Does it also present a problem for evolution? Apparently it does, according to the authors of a 1996 paper in the *Journal of Molecular Evolution*, who wrote:

"The Krebs cycle has been frequently quoted as a key problem in the evolution of living cells, hard to explain by Darwin's natural selection: How could natural selection explain the building of a complicated structure in toto, when the intermediate stages have no obvious fitness functionality? (Melendez-Hevia, Wadell, and Cascante 1996)

Where Intelligent Design theorists threw up their hands and declared defeat for evolution, however, these researchers decided to do the hard scientific work of analyzing the components of the cycle and seeing if any of them might have been selected for other biochemical tasks. What they found should be a lesson to anyone who asserts that evolution can act only by direct selection for a final function. In fact, nearly all of the proteins of the complex cycle can serve different biochemical purposes within the cell, making it possible to explain in detail how they evolved:

In the Krebs cycle problem the intermediary stages were also useful, but for different purposes, and, therefore, its complete design was a very clear case of opportunism. . . . . the Krebs cycle was built through the process that Jacob (1977) called "evolution by molecular tinkering," stating that evolution does not produce novelties from scratch: It works on what already exists. The most novel result of our analysis is seeing how, with minimal new material, evolution created the most important pathway of metabolism, achieving the best chemically possible design. In this case, a chemical engineer who was looking for the best design of the process could not have found a better design than the cycle which works in living cells. (Melendez-Hevia, Wadell, and Cascante 1996)

Since this paper appeared, a study based on genomic DNA sequences has confirmed the validity of its approach (Huynen, Dandekar, and Bork 1999). By contrast, how would Intelligent Design have approached the Krebs cycle? Using Dembski's calculations as our guide, we would first determine the amino acid sequences of each of the proteins of the cycle, and then calculate the probability of their spontaneous assembly. When this is done, an origination probability of less than 10^{-400} is the result. Therefore, applying "design" as a predictive science would have told both groups of researchers that their ultimately successful studies would have been fruitless, since the probability of spontaneous assembly falls below the "universal probability bound."

We already know, however, the reason that such calculations fail. They carry a built-in assumption that the component parts of a complex biochemical system have no possible function beyond serving the completely assembled system itself. As we have seen, this assumption is false. The Krebs cycle researchers knew better, of course, and were able to produce two important studies describing how a real, complex, biochemical system might have evolved – the very thing that design theorists once claimed did not exist in the scientific literature.

THE FAILURE OF DESIGN

It is no secret that concepts like "irreducible complexity" and "Intelligent Design" have failed to take the scientific community by storm (Forrest 2002). Design has not prompted new research studies, new breakthroughs, or novel insights on so much as a single scientific question. Design advocates acknowledge this from time to time, but they often claim that this is because the scientific deck is stacked against them. The Darwinist establishment, they say, prevents them from getting a foot in the laboratory door.

I would suggest that the real reason for the cold shoulder given "design" by the scientific community – particularly by life science researchers – is that time and time again its principal scientific claims have turned out to be wrong. Science is a pragmatic activity, and if your hypothesis doesn't work, it is quickly discarded.

The claim of irreducible complexity for the bacterial flagellum is an obvious example of this, but there are many others. Consider, for example, the intricate cascade of proteins involved in the clotting of vertebrate blood. This has been cited as one of the principal examples of the kind of complexity that evolution cannot generate, despite the elegant work of Russell Doolittle (Doolittle and Feng 1987; Doolittle 1993) to the contrary. A number of proteins are involved in this complex pathway, as described by Behe:

When an animal is cut, a protein called Hagemann factor (XII) sticks to the surface of cells near the wound. Bound Hagemann factor is then cleaved by a protein called HMK to yield activated Hagemann factor. Immediately the activated Hagemann factor converts another protein, called prekallikrein, to its active form, kallikrein. (Behe 1996a, 84)

How important are each of these proteins? In line with the dogma of irreducible complexity, Behe argues that each and every component must be in place before the system will work, and he is perfectly clear on this point:

[N]one of the cascade proteins are used for anything except controlling the formation of a clot. Yet in the absence of any of the components, blood does not clot, and the system fails. (Behe 1996a, 86)

As we have seen, the claim that every one of the components must be present in order for clotting to work is central to the "evidence" for design. One of those components, as these quotations indicate, is factor XII, which initiates the cascade. Once again, however, a nasty little fact gets in the way of Intelligent Design theory. Dolphins lack factor XII (Robinson, Kasting, and Aggeler 1969), yet their blood clots perfectly well. How can this be, if the clotting cascade is indeed irreducibly complex? It cannot, of course, and therefore the claim of irreducible complexity is wrong for this system as well. I would suggest, therefore, that the real reason for the rejection of "design" by the scientific community is remarkably simple – the claims of the Intelligent Design movement are contradicted time and time again by the scientific evidence.

THE FLAGELLUM UNSPUN

In any discussion of the question of "Intelligent Design," it is absolutely essential to determine what is meant by the term itself. If, for example, the advocates of design wish to suggest that the intricacies of nature, life, and the universe reveal a world of meaning and purpose consistent with an overarching, possibly Divine, intelligence, then their point is philosophical, not scientific. It is a philosophical point of view, incidentally, that I share, along with many scientists. As H. Allen Orr pointed out in a recent review:

Plenty of scientists have, after all, been attracted to the notion that natural laws reflect (in some way that's necessarily poorly articulated) an intelligence or aesthetic sensibility. This is the religion of Einstein, who spoke of "the grandeur of reason incarnate in existence" and of the scientist's religious feeling [that] takes the form of a rapturous amazement at the harmony of natural law. (Orr 2002).

This, however, is not what is meant by "Intelligent Design" in the parlance of the new anti-evolutionists. Their views demand not a universe in which the beauty and harmony of natural law has brought a world of vibrant and fruitful life into existence, but rather a universe in which the emergence and evolution of life is expressly made impossible by the very same rules. Their view requires that behind each and every novelty of life we find the direct and active involvement of an outside Designer whose work violates the very laws of nature that He had fashioned. The world of Intelligent Design is not the bright and innovative world of life that we have come to know through science. Rather, it is a brittle and unchanging landscape, frozen in form and unable to adapt except at the whims of its designer.

Certainly, the issue of design and purpose in nature is a philosophical one that scientists can and should discuss with great vigor. However, the notion at the heart of today's Intelligent Design movement is that the direct intervention of an outside designer can be demonstrated by the very existence of complex biochemical systems. What even they acknowledge is that their entire scientific position rests upon a single assertion – that the living cell contains biochemical machines that are irreducibly complex. And, they assert, the bacterial flagellum is the prime example of such a machine.

Such an assertion, as we have seen, can be put to the test in a very direct way. If we are able to find contained within the flagellum an example of a machine with fewer protein parts that serves a purpose distinct from motility, the claim of irreducible complexity is refuted. As we have also seen, the flagellum does indeed contain such a machine, a protein-secreting apparatus that carries out an important function even in species that lack the flagellum altogether. A scientific idea rises or falls on the weight of the evidence, and the evidence in the case of the bacterial flagellum is abundantly clear.

As an icon of the anti-evolutionist movement, the flagellum has fallen.

The very existence of the type III secretory system shows that the bacterial flagellum is not irreducibly complex. It also demonstrates, more generally, that the claim of "irreducible complexity" is scientifically meaningless, constructed as it is upon the flimsiest of foundations – the assertion that because science has not yet found selectable functions for the components of a certain structure, it never will. In the final analysis, as the claims of Intelligent Design fall by the wayside, its advocates are left with a single remaining tool with which to battle against the rising tide of scientific evidence. That tool may be effective in some circles, of course, but the scientific community will be quick to recognize it for what it really is – the classic argument from ignorance, dressed up in the shiny cloth of biochemistry and information theory.

When three leading advocates of Intelligent Design were recently given a chance to make their case in an issue of *Natural History* magazine, they each concluded their articles with a plea for design. One wrote that we should recognize "the design inherent in life and the universe" (Behe 2002), another that "design remains a possibility" (Wells 2002), and the third "that the natural sciences need to leave room for design" (Dembski 2002b). Yes, it is true. Design does remain a possibility, but not the type of "Intelligent Design" of which they speak.

As Darwin wrote, there is grandeur in an evolutionary view of life, a grandeur that is there for all to see, regardless of their philosophical views on the meaning and purpose of life. I do not believe, even for an instant, that Darwin's vision has weakened or diminished the sense of wonder and awe that one should feel in confronting the magnificence and diversity of the living world. Rather, to a person of faith it should enhance the sense of the Creator's majesty and wisdom (Miller 1999). Against such a backdrop, the struggles of the Intelligent Design movement are best understood as clamorous and disappointing double failures – rejected by science because they do not fit the facts, and having failed religion because they think too little of God.

References

Adami, C., C. Ofria, and T. C. Collier. 2000. Evolution of biological complexity. *Proceedings of the National Academy of Sciences* 97: 4463–8.

Aizawa, S.-I. 2001. Bacterial flagella and type III secretion systems. *FEMS Microbiology Letters* 202: 157–64.

Behe, M. 1996a. *Darwin's Black Box.* New York: The Free Press.

1996b. Evidence for intelligent design from biochemistry. Speech given at the Discovery Institute's God and Culture Conference, August 10, 1996, Seattle, Washington. Available at <http://www.arn.org/docs/behe/mb_idfrom biochemistry.htm>.

2002. The challenge of irreducible complexity. *Natural History* 111 (April): 74.

Büttner D., and U. Bonas. 2002. Port of entry – the Type III secretion translocon. *Trends in Microbiology* 10: 186–91.

Belmonte, J. C. I. 1999. How the body tells right from left. *Scientific American* 280 (June): 46–51.

Charlesworth, B. 2002. Evolution by design? *Nature* 418: 129.

Coyne, J. A. 1996. God in the details. *Nature* 383: 227–8.

Darwin, C. 1872. *On the Origin of Species,* 6th ed. London: Oxford University Press.

Dembski, W. 2002a. No free lunch: Why specified complexity cannot be purchased without intelligence. Lanham, MD: Rowman and Littlefield.

2002b. Detecting design in the natural sciences. *Natural History* 111 (April): 76.

Depew, D. J. 1998. Intelligent design and irreducible complexity: A rejoinder. *Rhetoric and Public Affairs* 1: 571–8.

Doolittle, R. F. 1993. The evolution of vertebrate blood coagulation: A case of yin and yang. *Thrombosis and Heamostasis* 70: 24–8.

Doolittle, R. F., and D. F. Feng. 1987. Reconstructing the evolution of vertebrate blood coagulation from a consideration of the amino acid sequences of clotting proteins. *Cold Spring Harbor Symposia on Quantitative Biology* 52: 869–74.

Forrest B. 2002. The newest evolution of creationism. *Natural History* 111 (April): 80.

Gura, T. 2002. Evolution critics seek role for unseen hand in education. *Nature* 416: 250.

Heuck, C. J. 1998. Type III protein secretion systems in bacterial pathogens of animals and plants. *Microbiology and Molecular Biology Reviews* 62: 379–433.

Holden, C. 1999. Kansas dumps Darwin, raises alarm across the United States. *Science* 285: 1186–7.

Huynen, M. A., T. Dandekar, and P. Bork. 1999. Variation and evolution of the citric-acid cycle: A genomic perspective. *Trends in Microbiology* 7: 281–91.

McNab, R. M. 1999. The bacterial flagellum: Reversible rotary propellor and type III export apparatus. *Journal of Bacteriology* 181: 7149–53.

Melendez-Hevia, E., T. G. Wadell, and M. Cascante. 1996. The puzzle of the Krebs citric acid cycle: Assembling the pieces of chemically feasible reactions, and opportunism in the design of metabolic pathways during evolution. *Journal of Molecular Evolution* 43: 293–303.

Miller, K. R. 1999. *Finding Darwin's God.* New York: HarperCollins.

1996. A review of *Darwin's Black Box. Creation/Evolution* 16: 36–40.

Orr, H. A. 2002. The return of intelligent design. *The Boston Review* (Summer 2002): 53–6.

Padian, K. 2002. Waiting for the watchmaker. *Science* 295: 2373–4.

Pennock, R. T. 2001. *Intelligent Design Creationism and Its Critics: Philosophical, Theological, and Scientific Perspectives.* Cambridge, MA: MIT Press.

Robinson, A. J., M. Kropatkin, and P. M. Aggeler. 1969. Hagemann factor (factor XII) deficiency in marine mammals. *Science* 166: 1420–2.

Schneider, T. D. 2000. Evolution of biological information. *Nucleic Acids Research* 28: 2794–9.

Stern, C. 2002. Embryology: Fluid flow and broken symmetry. *Nature* 418: 29–30.

Thewissen, J. G. M., S. T. Hussain, and M. Arif. 1994. Fossil evidence for the origin of aquatic locomotion in archaeocete whales. *Science* 286: 210–12.

Thewissen, J. G. M., E. M. Williams, L. J. Roe, and S. T. Hussain. 2001. Skeletons of terrestrial cetaceans and the relationship of whales to artiodactyls. *Nature* 413: 277–81.

Thornhill, R. H., and D. W. Ussery. 2000. A classification of possible routes of Darwinian evolution. *Journal of Theoretical Biology* 203: 111–16.

Wells, J, 2002. Elusive icons of evolution. *Natural History* 111 (April): 78.

Yonekura, K., S. Maki, D. G. Morgan, D. J. DeRosier, F. Vonderviszt, K. Imada, and K. Namba. 2000. The bacterial flagellar cap as the rotary promoter of flagellin self-assembly. *Science* 290: 2148–52.

6

The Design Argument

Elliott Sober[1]

The design argument is one of three main arguments for the existence of God; the others are the ontological argument and the cosmological argument. Unlike the ontological argument, the design argument and the cosmological argument are a posteriori. And whereas the cosmological argument can focus on any present event to get the ball rolling (arguing that it must trace back to a first cause, namely God), design theorists are usually more selective.

Design arguments have typically been of two types – *organismic* and *cosmic*. Organismic design arguments start with the observation that organisms have features that adapt them to the environments in which they live and that exhibit a kind of *delicacy*. Consider, for example, the vertebrate eye. This organ helps organisms to survive by permitting them to perceive objects in their environment. And were the parts of the eye even slightly different in their shape and assembly, the resulting organ would not allow us to see. Cosmic design arguments begin with an observation concerning features of the entire cosmos – the universe obeys simple laws; it has a kind of stability; its physical features permit life, and intelligent life, to exist. However, not all design arguments fit into these two neat compartments. Kepler, for example, thought that the face we see when we look at the moon requires explanation in terms of Intelligent Design. Still, the common thread is that design theorists describe some empirical feature of the world and argue that this feature points toward an explanation in terms of God's intentional planning and away from an explanation in terms of mindless natural processes.

The design argument raises epistemological questions that go beyond its traditional theological context. As William Paley (1802) observed, when we find a watch while walking across a heath, we unhesitatingly infer that it was produced by an intelligent designer. No such inference forces itself upon us when we observe a stone. Why is explanation in terms of Intelligent Design so compelling in the one case but not in the other? Similarly, when we observe the behavior of our fellow human beings, we find it irresistible

98

to think that they have minds that are filled with beliefs and desires. And when we observe nonhuman organisms, the impulse to invoke mentalistic explanations is often very strong, especially when they look a lot like us. When does the behavior of an organism – human or not – warrant this mentalistic interpretation? The same question can be posed about machines. Few of us feel tempted to attribute beliefs and desires to hand calculators. We use calculators to help us add, but they don't literally figure out sums; in this respect, calculators are like the pieces of paper on which we scribble calculations. There is an important difference between a device that we use to help *us* think and a device that *itself* thinks. However, when a computer plays a decent game of chess, we may find it useful to explain and predict its behavior by thinking of it as having goals and deploying strategies (Dennett 1987b). Is this merely a useful fiction, or does the machine really have a mind? And if we think that present day chess-playing computers are, strictly speaking, mindless, what would it take for a machine to pass the test? Surely, as Turing (1950) observed, it needn't look like us. In all of these contexts, we face *the problem of other minds* (Sober 2000a). If we understood the ground rules of this general epistemological problem, that would help us to think about the design argument for the existence of God. And conversely – if we could get clear on the theological design argument, that might throw light on epistemological problems that are not theological in character.

WHAT IS THE DESIGN ARGUMENT?

The design argument, like the ontological argument, raises subtle questions concerning what the logical structure of the argument really is. My main concern here will not be to describe how various thinkers have presented the design argument, but to find the soundest formulation that the argument can be given.

The best version of the design argument, in my opinion, uses an inferential idea that probabilists call *the Likelihood Principle.* This can be illustrated by way of Paley's (1802) example of the watch on the heath. Paley describes an observation that he claims discriminates between two hypotheses:

(W) O1: The watch has features G1 . . . Gn.
   W1: The watch was created by an intelligent designer.
   W2: The watch was produced by a mindless chance process.

Paley's idea is that O1 would be unsurprising if W1 were true, but would be very surprising if W2 were true. This is supposed to show that O1 *favors* W1 over W2; O1 supports W1 more than it supports W2. Surprise is a matter of degree; it can be captured by the concept of conditional probability. The probability of O given H – $\Pr(O \mid H)$ – represents how unsurprising O would be if H were true. The Likelihood Principle says that we can decide in which direction the evidence is pointing by comparing such conditional probabilities:

(LP) Observation O supports hypothesis H1 more than it supports hypothesis H2 if and only if $Pr(O \mid H1) > Pr(O \mid H2)$.

There is a lot to say on the question of why the Likelihood Principle should be accepted (Hacking 1965; Edwards 1972; Royall 1997; Forster and Sober 2003; Sober 2002); for the purposes of this essay, I will take it as a given.

We now can describe the likelihood version of the design argument for the existence of God, again taking our lead from one of Paley's favorite examples of a delicate adaptation. The basic format is to compare two hypotheses as possible explanations of a single observation:

(E) O2: The vertebrate eye has features F1 ... Fn.
    E1: The vertebrate eye was created by an intelligent designer.
    E2: The vertebrate eye was produced by a mindless chance process.

We do not hesitate to conclude that the observations strongly favor design over chance in the case of argument (W); Paley claims that precisely the same conclusion should be drawn in the case of the propositions assembled in (E).[2]

CLARIFICATIONS

Several points of clarification are needed here concerning likelihood in general, and the likelihood version of the design argument in particular. First, I use the term "likelihood" in a technical sense. Likelihood is not the same as probability. To say that H has a high likelihood, given observation O, is to comment on the value of $Pr(O \mid H)$, not on the value of $Pr(H \mid O)$; the latter is H's *posterior probability*. It is perfectly possible for a hypothesis to have a high likelihood and a low posterior probability. When you hear noises in your attic, this confers a high likelihood on the hypothesis that there are gremlins up there bowling, but few of us would conclude that this hypothesis is probably true.

Although the likelihood of H (given O) and the probability of H (given O) are different quantities, they are related. The relationship is given by Bayes' Theorem:

$$Pr(H \mid O) = Pr(O \mid H)Pr(H)/Pr(O).$$

$Pr(H)$ is the *prior probability* of the hypothesis – the probability that H has before we take the observation O into account. From Bayes' Theorem we can deduce the following:

$$Pr(H1 \mid O) > Pr(H2 \mid O) \text{ if and only if}$$
$$Pr(O \mid H1)Pr(H1) > Pr(O \mid H2)Pr(H2).$$

Which hypothesis has the higher posterior probability depends not only on how their likelihoods are related, but also on how their prior probabilities are related. This explains why the likelihood version of the design argument does not show that design is more probable than chance. To draw this further conclusion, we'd have to say something about the prior probabilities of the two hypotheses. It is here that I wish to demur (and this is what separates me from card-carrying Bayesians). Each of us perhaps has some subjective degree of belief, before we consider the design argument, in each of the two hypotheses (E1) and (E2). However, I see no way to understand the idea that the two hypotheses have *objective* prior probabilities. Since I would like to restrict the design argument as much as possible to matters that are objective, I will not represent it as an argument concerning which hypothesis is more probable.[3] However, those who have prior degrees of belief in (E1) and (E2) should use the likelihood argument to update their subjective probabilities. The likelihood version of the design argument says that the observation O2 should lead you to increase your degree of belief in (E1) and reduce your degree of belief in (E2).

My restriction of the design argument to an assessment of likelihoods, not probabilities, reflects a more general point of view. Scientific theories often have implications about which observations are probable (and which are improbable), but it rarely makes sense to describe them as having objective probabilities. Newton's law of gravitation (along with suitable background assumptions) says that the return of Haley's comet was to be expected, but what is the probability that Newton's law is true? Hypotheses have objective probabilities when they describe possible outcomes of a chance process. But as far as anyone knows, the laws that govern our universe are not the result of a chance process. Bayesians think that *all* hypotheses have probabilities; the position I am advocating sees this as a special feature of *some* hypotheses.[4]

Just as likelihood considerations leave open which probabilities one should assign to the competing hypotheses, they also don't tell you which hypothesis you should *believe*. I take it that belief is a dichotomous concept – you either believe a proposition or you do not. Consistent with this is the idea that there are three attitudes one might take to a statement – you can believe it true, believe it false, or withhold judgment. However, there is no simple connection between the matter-of-degree concept of probability and the dichotomous (or trichotomous) concept of belief. This is the lesson I extract from the lottery paradox (Kyburg 1961). Suppose 100,000 tickets are sold in a fair lottery; one ticket will win, and each has the same chance of winning. It follows that each ticket has a very high probability of not winning. If you adopt the policy of believing a proposition when it has a high probability, you will believe of each ticket that it will not win. However, this conclusion contradicts the assumption that the lottery is fair. What this shows is that high probability does not suffice for

belief (and low probability does not suffice for disbelief). It is for this reason that many Bayesians prefer to say that individuals have *degrees* of belief.
The rules for the dichotomous concept are unclear; the matter-of-degree
concept at least has the advantage of being anchored to the probability
calculus.

In summary, likelihood arguments have rather modest pretensions. They
don't tell you which hypotheses to believe; in fact, they don't even tell you
which hypotheses are probably true. Rather, they evaluate how the observations at hand discriminate among the hypotheses under consideration.

I now turn to some details concerning the likelihood version of the design
argument. The first concerns the meaning of the Intelligent Design hypothesis. This hypothesis occurs in (W1) in connection with the watch and in
(E1) in connection with the vertebrate eye. In the case of the watch, Paley
did not dream that he was offering an argument for the existence of *God*.
However, in the case of the eye, Paley thought that the intelligent designer
under discussion was God himself. Why are these cases different? The bare
bones of the likelihood arguments (W) and (E) do not say. What Paley had
in mind is that building the vertebrate eye and the other adaptive features
that organisms exhibit requires an intelligence far greater than anything
that human beings could muster. This is a point that we will revisit at the
end of this chapter.

It also is important to understand the nature of the hypothesis with which
the Intelligent Design hypothesis competes. I have used the term "chance" to
express this alternative hypothesis. In large measure, this is because design
theorists often think of chance as the alternative to design. Paley is again
exemplary. *Natural Theology* is filled with examples like that of the vertebrate
eye. Paley was not content to describe a few cases of delicate adaptations; he
wanted to make sure that even if he got a few details wrong, the weight of
the evidence would still be overwhelming. For example, in Chapter 15 he
considers the fact that our eyes point in the same direction as our feet; this
has the convenient consequence that we can see where we are going. The
obvious explanation, Paley (1802, p. 179) says, is Intelligent Design. This
is because the alternative explanation is that the direction of our eyes and
the direction of our gait were determined by chance, which would mean
that there was only a 1/4 probability that our eyes would be able to scan the
quadrant into which we are about to step.

I construe the idea of chance in a particular way. To say that an outcome
is the result of a *uniform chance process* means that it was one of a number
of *equally probable* outcomes. Examples in the real world that come close
to being uniform chance processes may be found in gambling devices –
spinning a roulette wheel, drawing from a deck of cards, tossing a coin.
The term "random" becomes more and more appropriate as real world
systems approximate uniform chance processes. However, as R. A. Fisher
once pointed out, it is not a "matter of chance" that casinos turn a profit

each year, nor should this be regarded as a "random" event. The financial bottom line at a casino is the result of a large number of chance events, but the rules of the game make it enormously probable (though not certain) that casinos end each year in the black. All uniform chance processes are probabilistic, but not all probabilistic outcomes are "due to chance."

It follows that the two hypotheses considered in my likelihood rendition of the design argument are not exhaustive. Mindless uniform chance is one alternative to Intelligent Design, but it is not the only one. This point has an important bearing on the dramatic change in fortunes that the design argument experienced with the advent of Darwin's (1859) theory of evolution. The process of evolution by natural selection is *not* a uniform chance process. The process has two parts. Novel traits arise in individual organisms "by chance"; however, whether they then disappear from the population or increase in frequency and eventually reach 100 percent representation is anything but a "matter of chance." The central idea of natural selection is that traits that help organisms to survive and reproduce have a better chance of becoming common than traits that hurt their prospects. The essence of natural selection is that evolutionary outcomes have *un*equal probabilities. Paley and other design theorists writing before Darwin did not and could not cover all possible mindless natural processes. Paley addressed the alternative of uniform chance, not the alternative of natural selection.[5]

Just to nail down this point, I want to describe a version of the design argument formulated by John Arbuthnot. Arbuthnot (1710) carefully tabulated birth records in London over eighty-two years and noticed that in each year, slightly more sons than daughters were born. Realizing that boys die in greater numbers than girls, he saw that this slight bias in the sex ratio at birth gradually subsides, until there are equal numbers of males and females at the age of marriage. Arbuthnot took this to be evidence of Intelligent Design; God, in his benevolence, wanted each man to have a wife and each woman to have a husband. To draw this conclusion, Arbuthnot considered what he took to be the relevant competing hypothesis – that the sex ratio at birth is determined by a uniform chance process. He was able to show that if the probability is 1/2 that a baby will be a boy and 1/2 that it will be a girl, then it is enormously improbable that the sex ratio should be skewed in favor of males in each and every year he surveyed (Stigler 1986, 225–6).

Arbuthnot could not have known that R. A. Fisher (1930) would bring sex ratio within the purview of the theory of natural selection. Fisher's insight was to see that a mother's mix of sons and daughters affects the number of *grand*-offspring she will have. Fisher demonstrated that when there is random mating in a large population, the sex ratio strategy that evolves is one in which a mother invests equally in sons and daughters (Sober 1993, 17). A mother will put half her reproductive resources into producing sons and half into producing daughters. This equal division means that she should have more sons than daughters, if sons tend to die sooner. Fisher's model therefore predicts the slightly uneven sex ratio at birth that Arbuthnot observed.[6]

My point in describing Fisher's idea is not to fault Arbuthnot for living in the eighteenth century. Rather, the thing to notice is that what Arbuthnot meant by "chance" was very different from what Fisher was talking about when he described how a selection process might shape the sex ratio found in a population. Arbuthnot was right that the probability of there being more males than females at birth in each of eighty-two years is extremely low, if each birth has the same chance of producing a male as it does of producing a female. However, if Fisher's hypothesized process is doing the work, a male-biased sex ratio in the population is extremely probable. Showing that design is more likely than chance leaves it open that some third, mindless process might still have a higher likelihood than design. This is not a defect in the design argument, so long as the conclusion of that argument is not overstated. Here the modesty of the likelihood version of the design argument is a point in its favor. To draw a stronger conclusion – that the design hypothesis is more likely than *any* hypothesis involving mindless natural processes – one would have to attend to more alternatives than just design and (uniform) chance.[7]

I now want to draw the reader's attention to some features of the likelihood version of the design argument (E) concerning how the observation and the competing hypotheses are formulated. First, notice that I have kept the observation (O2) conceptually separate from the two hypotheses (E1) and (E2). If the observation were simply that "the vertebrate eye exists," then, since (E1) and (E2) both entail this proposition, each would have a likelihood of unity. According to the Likelihood Principle, this observation does not favor design over chance. Better to formulate the question in terms of explaining the properties of the vertebrate eye, not in terms of explaining why the eye exists. Notice also that I have not formulated the design hypothesis as the claim that God exists; this existence claim says nothing about the putative Designer's involvement in the creation of the vertebrate eye. Finally, I should point out that it would do no harm to have the design hypothesis say that God created the vertebrate eye; this possible reformulation is something I'll return to later.

OTHER FORMULATIONS OF THE DESIGN ARGUMENT, AND THEIR DEFECTS

Given the various provisos that govern probability arguments, it would be nice if the design argument could be formulated deductively. For example, if the hypothesis of mindless chance processes entailed that it is *impossible* that organisms exhibit delicate adaptations, then a quick application of *modus tollens* would sweep that hypothesis from the field. However much design theorists might yearn for an argument of this kind, there apparently is none to be had. As the story about monkeys and typewriters illustrates, it is *not* impossible that mindless chance processes should produce delicate adaptations; it is merely very *improbable* that they should do so.

If *modus tollens* cannot be pressed into service, perhaps there is a probabilistic version of *modus tollens* that can achieve the same result. Is there a Law of Improbability that begins with the premise that $Pr(O \mid H)$ is very low and concludes that H should be rejected? There is no such principle (Royall 1997, Chapter 3). The fact that you won the lottery does not, by itself, show that there is something wrong with the conjunctive hypothesis that the lottery was fair and a million tickets were sold and you bought just one ticket. And if we randomly drop a very sharp pin onto a line that is 1,000 miles long, the probability of its landing where it does is negligible; however, that outcome does not falsify the hypothesis that the pin was dropped at random.

The fact that there is no probabilistic *modus tollens* has great significance for understanding the design argument. The logic of this problem is essentially comparative. In order to evaluate the design hypothesis, we must know what it predicts and compare this with the predictions made by other hypotheses. The design hypothesis cannot win by default. The fact that an observation would be very improbable if it arose by chance is not enough to refute the chance hypothesis. One must show that the design hypothesis confers on the observation a higher probability; and even then, the conclusion will merely be that the observation *favors* the design hypothesis, not that that hypothesis *must be true.*[8]

In the continuing conflict (in the United States) between evolutionary biology and creationism, creationists attack evolutionary theory, but they never take even the first step toward developing a positive theory of their own. The three-word slogan "God did it" seems to satisfy whatever craving for explanation they may have. Is the sterility of this intellectual tradition a mere accident? Could Intelligent Design theory be turned into a scientific research program? I am doubtful, but the present point concerns the logic of the design argument, not its future prospects. Creationists sometimes assert that evolutionary theory "cannot explain" this or that finding (e.g., Behe 1996). What they mean is that certain outcomes are *very improbable* according to the evolutionary hypothesis. Even this more modest claim needs to be scrutinized. However, if it were true, what would follow about the plausibility of creationism? In a word – *nothing.*

It isn't just defenders of the design hypothesis who have fallen into the trap of supposing that there is a probabilistic version of *modus tollens.* For example, the biologist Richard Dawkins (1986, 144–6) takes up the question of how one should evaluate hypotheses that attempt to explain the origin of life by appeal to strictly mindless natural processes. He says that an acceptable theory of this sort can say that the origin of life on Earth was somewhat improbable, but it must not go too far. If there are N planets in the universe that are "suitable" locales for life to originate, then an acceptable theory of the origin of life on Earth must say that that event had a probability of at least 1/N. Theories that say that terrestrial life was less probable than this should be rejected. How does Dawkins obtain this lower bound? Why is the number of planets relevant? Perhaps he is thinking that if α is the actual frequency of life-bearing planets among "suitable" planets (i.e., planets on which it is possible for life to evolve), then the true probability of life's evolving on Earth must also be α. There is a mistake here, which we can uncover by examining how actual frequency and probability are related. With small sample size, it is perfectly possible for these quantities to have very different values (consider a fair coin that is tossed three times and then destroyed). However, Dawkins is obviously thinking that the sample size is very large, and here he is right that the actual frequency provides a good estimate of the true probability. It is interesting that Dawkins tells us to reject a theory if the probability it assigns is too *low*. Why doesn't he also say that it should be rejected if the probability it assigns is too *high*? The reason, presumably, is that we cannot rule out the possibility that the Earth was not just *suitable* but *highly conducive* to the evolution of life. However, this point cuts both ways. Although α is the *average* probability of a suitable planet's having life evolve, it still is possible that different suitable planets might have different probabilities – some may have values greater than α while others have values that are lower. Dawkins's lower bound assumes that the Earth was above average; this is a mistake that might be termed the Lake Woebegone Fallacy.

Some of Hume's (1779) criticisms of the design argument in his *Dialogues Concerning Natural Religion* depend on formulating the argument as something other than a likelihood inference. For example, Hume at one point has Philo say that the design argument is an argument from analogy, and that the conclusion of the argument is supported only very weakly by its premises. His point can be formulated by thinking of the design argument as follows:

> Watches are produced by intelligent design.
> Organisms are similar to watches to degree p.
> p[===
> Organisms were produced by intelligent design.

Notice that the letter "p" appears twice in this argument. It represents the degree of similarity of organisms and watches, and it represents the probability that the premises confer on the conclusion. Think of similarity as the proportion of shared characteristics. Things that are 0 percent similar have no traits in common; things that are 100 percent similar have all traits in common. The analogy argument says that the more similar watches and organisms are, the more probable it is that organisms were produced by intelligent design.

Let us grant the Humean point that watches and organisms have relatively few characteristics in common. (It is doubtful that there is a well-defined totality consisting of all the traits of each, but let that pass.) After all, watches are made of metal and glass and go "tick tock"; organisms metabolize and reproduce and go "oink" and "bow wow." If the design argument is a likelihood inference, this is all true but entirely irrelevant. It doesn't matter how similar watches and organisms are. With respect to argument (W), what matters is how one should explain the fact that watches are well adapted for the task of telling time; with respect to (E), what matters is how one should explain the fact that organisms are well adapted to their environments. Paley's analogy between watches and organisms is merely heuristic. The likelihood argument about organisms stands on its own (Sober 1993).

Hume also has Philo construe the design argument as an inductive argument and then complain that the inductive evidence is weak. Philo suggests that for us to have good reason to think that our world was produced by an intelligent designer, we'd have to visit other worlds and observe that all or most of them were produced by Intelligent Design. But how many other worlds have we visited? The answer is – not even one. Apparently, the design argument is an inductive argument that could not be weaker; its sample size is zero. This objection dissolves once we move from the model of inductive sampling to that of likelihood. You don't have to observe the processes of Intelligent Design and chance at work in different worlds in order to maintain that the two hypotheses confer different probabilities on your observations.

THREE POSSIBLE OBJECTIONS TO THE LIKELIHOOD ARGUMENT

There is another objection that Hume makes to the design argument, one that apparently pertains to the likelihood version of the argument that I have formulated and that many philosophers think is devastating. Hume points out that the design argument does not establish the attributes of the designer. The argument does not show that the designer who made the universe, or who made organisms, is morally perfect, or all-knowing, or all-powerful, or that there is just one such being. Perhaps this undercuts some versions of the design argument, but it does not touch the likelihood argument we are considering. Paley, perhaps responding to this Humean point, makes it clear that his design argument aims to establish the *existence* of the designer, and that the question of the designer's *characteristics* must be addressed separately.[9] My own rendition of the argument follows Paley in this regard. Does this limitation of the argument render it trivial? Not at all – it is *not* trivial to claim that the adaptive contrivances of organisms are due to intelligent design, even when details about the designer are not supplied. This supposed "triviality" would be *big* news to evolutionary biologists.

The likelihood version of the design argument consists of two premises: $Pr(O \mid Chance)$ is very low, and $Pr(O \mid Design)$ is higher. Here O describes some observation of the features of organisms or some feature of the entire cosmos. The first of these claims is sometimes rejected by appeal to a theory that Hume describes under the heading of the Epicurean hypothesis. This is the monkeys-and-typewriters idea that if there are a finite number of particles that have a finite number of possible states, then, if they swarm about at random, they eventually will visit all possible configurations, including configurations of great order.[10] Thus, the order we see in our universe, and the delicate adaptations we observe in organisms, in fact had a high probability of eventually coming into being, according to the hypothesis of chance. Van Inwagen (1993, 144) gives voice to this objection and explains it by way of an analogy: suppose you toss a coin twenty times, and it lands heads every time. You should not be surprised at this outcome if you are one among millions of people who toss a fair coin twenty times. After all, with so many people tossing, it is all but inevitable that some people will get twenty heads. The outcome you obtained, therefore, was not improbable, according to the chance hypothesis.

There is a fallacy in this criticism of the design argument, which Hacking (1987) calls "the inverse gambler's fallacy." He illustrates his idea by describing a gambler who walks into a casino and immediately observes two dice being rolled that land double-six. The gambler considers whether this result favors the hypothesis that the dice had been rolled many times before the roll he just observed or the hypothesis that this was the first roll of the evening. The gambler reasons that the outcome of double-six would be more probable under the first hypothesis:

Pr(double-six on this roll | there were many rolls) >

Pr(double-six on this roll | there was just one roll).

In fact, the gambler's assessment of the likelihoods is erroneous. Rolls of dice have the *Markov property*; the probability of double-six on this roll is the same (1/36) regardless of what may have happened in the past. What is true is that the probability that a double-six will occur *at some time or other* increases as the number of trials is increased:

Pr(a double-six occurs sometime | there were many rolls) >

Pr(a double-six occurs sometime | there was just one roll).

However, the *principle of total evidence* says that we should assess hypotheses by considering *all* the evidence we have. This means that the relevant observation is that *this* roll landed double-six; we should not focus on the logically weaker proposition that a double-six occurred *at some time or other*. Relative to the stronger description of the observations, the hypotheses have identical likelihoods.

Applying this point to the criticism of the design argument that we are presently considering, we must conclude that the criticism is mistaken. It *is* highly probable (let us suppose), according to the chance hypothesis, that the universe will contain order and adaptation somewhere and at some time. However, the relevant observation is more specific – *our* corner of the universe is orderly, and the organisms now on Earth are well adapted. These events *do* have very low probability, according to the chance hypothesis, and the fact that a weaker description of the observations has high probability on the chance hypothesis is not relevant (see also White 2000).[11]

If the first premise in the likelihood formulation of the design argument – that $Pr(O \mid Chance)$ is very low – is correct, then the only question that remains is whether $Pr(O \mid Design)$ is higher. This, I believe, is the Achilles heel of the design argument. The problem is to say how probable it is, for example, that the vertebrate eye would have features F1 . . . Fn if the eye were produced by an intelligent designer. What is required is not the specification of a single probability value, or even of a precisely delimited range of values. All that is needed is an argument that shows that this probability is indeed higher than the probability that chance confers on the observation.

The problem is that the design hypothesis confers a probability on the observation only when it is supplemented with further assumptions about what the Designer's goals and abilities would be if He existed. Perhaps the Designer would never build the vertebrate eye with features F1 . . . Fn, either because He would lack the goals or because He would lack the ability. If so, the likelihood of the design hypothesis is zero. On the other hand, perhaps the Designer would want above all to build the eye with features F1 . . . Fn and would be entirely competent to bring this plan to fruition. If so, the likelihood of the design hypothesis is unity. There are as many likelihoods as there are suppositions concerning the goals and abilities of the putative designer. Which of these, or which class of these, should we take seriously?

It is no good answering this question by assuming that the eye was built by an intelligent Designer and then inferring that the designer must have wanted to give the eye features F1 . . . Fn and must have had the ability to do so – since, after all, these are the features we observe. For one thing, this pattern of argument is question-begging. One needs *independent* evidence as to what the Designer's plans and abilities would be if He existed; one can't obtain this evidence by *assuming* that the design hypothesis is true (Sober 1999). Furthermore, even if we assume that the eye was built by an intelligent designer, we can't tell from this what the probability is that the eye would have the features we observe. Designers sometimes bring about outcomes that are not very probable, given the plans they had in mind.

This objection to the design argument is an old one; it was presented by Keynes (1921) and before him by Venn (1866). In fact, the basic idea was formulated by Hume. When we behold the watch on the heath, we know that the watch's features are not particularly improbable, on the hypothesis that the watch was produced by a Designer who has the sorts of *human* goals and abilities with which we are familiar. This is the deep disanalogy between the watchmaker and the putative maker of organisms and universes. We are invited, in the latter case, to imagine a Designer who is radically different from the human craftsmen with whom we are familiar. But if this Designer is so different, why are we so sure that this being would build the vertebrate eye in the form in which we find it?

This challenge is not turned back by pointing out that we often infer the existence of intelligent designers when we have no clue as to what they were trying to achieve. The biologist John Maynard Smith tells the story of a job he had during World War II inspecting a warehouse filled with German war materiel. He and his coworkers often came across machines whose functions were entirely opaque to them. Yet, they had no trouble seeing that these objects were built by intelligent designers. Similar stories can be told about archaeologists who work in museums; they often have objects in their collections that they know are artefacts, although they have no idea what the makers of these artefacts had in mind.

My claim is not that design theorists must have independent evidence that singles out the exact goals and abilities of the putative intelligent designer. They may be uncertain as to which of the goal/ability pairs GA-1, GA-2, . . . , GA-n is correct. However, since

$$\text{Pr}(\text{the eye has F1} \dots \text{Fn} \mid \text{Design}) =$$
$$\sum_i \text{Pr}(\text{the eye has F1} \dots \text{Fn} \mid \text{Design \& GA-i}) \text{Pr}(\text{GA-i} \mid \text{Design}),$$

they do have to show that

$$\sum_i \text{Pr}(\text{the eye has F1} \dots \text{Fn} \mid \text{Design \& GA-i}) \text{Pr}(\text{GA-i} \mid \text{Design}) >$$
$$\text{Pr}(\text{the eye has F1} \dots \text{Fn} \mid \text{Chance}).$$

I think that Maynard Smith in his warehouse and archaeologists in their museums are able to do this. They aren't sure exactly what the intelligent designer was trying to achieve (e.g., they aren't certain that GA-1 is true and that all the other GA pairs are false), but they are able to see that it is not terribly improbable that the object should have the features one observes if it were made by a human intelligent designer. After all, the items in Maynard Smith's warehouse were symmetrical and smooth metal containers that had what appeared to be switches, dials, and gauges on them. And the "arte-facts of unknown function" in anthropology museums likewise bear signs of human handiwork.

It is interesting in this connection to consider the epistemological problem of how one would go about detecting intelligent life elsewhere in the universe (if it exists). The SETI (Search for Extraterrestrial Intelligence) project, funded until 1993 by the U.S. National Aeronautics and Space Administration and now supported privately, dealt with this problem in two ways (Dick 1996). First, the scientists wanted to send a message into deep space that would allow any intelligent extraterrestrials who received it to figure out that it was produced by intelligent designers (namely, us). Second, they scanned the night sky hoping to detect signs of intelligent life elsewhere.

The message, transmitted in 1974 from the Arecibo Observatory, was a simple picture of our solar system, a representation of oxygen and carbon, a picture of a double helix representing DNA, a stick figure of a human being, and a picture of the Arecibo telescope. How sure are we that if intelligent aliens find these clues, that they will realize that they were produced by intelligent designers? The hope is that this message will strike the aliens who receive it as evidence favoring the hypothesis of intelligent design over the hypothesis that some mindless physical process (not necessarily one involving uniform chance) was responsible for it. It is hard to see how the SETI engineers could have done any better, but one still cannot dismiss the possibility that they will fail. If extraterrestrial minds are very different from our own – either because they have different beliefs and desires or because they process information in different ways – it may turn out that their interpretation of the evidence will differ profoundly from the interpretation that human beings would arrive at, were they on the receiving end. To say anything more precise about this, we'd have to be able provide specifics about the aliens' mental characteristics. If we are uncertain as to how the mind of an extraterrestrial will interpret this evidence, how can we be so sure that God, if he were to build the vertebrate eye, would endow it with the features that we find it to have?

When SETI engineers search for signs of intelligent life elsewhere in the universe, what are they looking for? The answer is surprisingly simple. They look for narrow-band radio emissions. This is because human beings have built machines that produce these signals, and as far as we know, such emissions are not produced by mindless natural processes. The SETI engineers search for this kind of signal not because it is "complex" or fulfills some a priori criterion that would make it a "sign of intelligence," but simply because they think they know what sorts of mechanisms are needed to produce it.[12] This strategy may not work, but it is hard to see how the scientists could do any better. Our judgments about what counts as a sign of intelligent design must be based on empirical information about what designers often do and what they rarely do. As of now, these judgments are based on our knowledge of *human* intelligence. The more our hypotheses about intelligent designers depart from the human case, the less sure we are about what the ground rules are for inferring intelligent design. It is imaginable that these limitations will subside as human beings learn more about the cosmos. But for now, we are rather limited.

I have been emphasizing the fallibility of two assumptions – that we know what counts as a sign of extraterrestrial intelligence and that we know how extraterrestrials will interpret the signals we send. My point has been to shake a complacent assumption that figures in the design argument. However, I suspect that SETI engineers are on much firmer ground than theologians. If extraterrestrials evolved by the same type of evolutionary process that produced human intelligence, that may provide useful constraints on conjectures about the minds they have. No theologian, to my knowledge, thinks that God is the result of biological processes. Indeed, God is usually thought of as a *super*natural Being who is radically different from the things we observe *in* nature. The problem of extraterrestrial intelligence is therefore an intermediate case; it lies between the watch found on the heath and the God who purportedly built the universe and shaped the vertebrate eye, but is much closer to the first. The upshot of this point for Paley's design argument is this: *Design arguments for the existence of human (and humanlike) watchmakers are often unproblematic; it is design arguments for the existence of God that leave us at sea.*

I began by formulating the design hypothesis in argument (E) as the claim that an intelligent designer made the vertebrate eye. Yet I have sometimes discussed the hypothesis as if it asserted that *God* is the designer in question. I don't think this difference makes a difference with respect to the objection I have described. To say that some designer or other made the eye is to state a disjunctive hypothesis. To figure out the likelihood of this disjunction, one needs to address the question of what each putative designer's goals and intentions would be.[13] The theological formulation shifts the problem from the evaluation of a disjunction to the evaluation of a disjunct, but the problem remains the same. Even supposing that God is omniscient, omnipotent, and perfectly benevolent, what is the probability that the eye would have features F1 . . . Fn if God set his hand to making it? He *could* have produced those results if he had wanted to. But why should we think that this is what he *would* have wanted to do? The assumption that God can do anything is part of the problem, not the solution. An engineer who is more limited would be more predictable.

There is another reply to my criticism of the design argument that should be considered. I have complained that we have no way to evaluate the likelihood of the design hypothesis, since we don't know which auxiliary assumptions about goal/ability pairs we should use. But why not change the subject? Instead of evaluating the likelihood of design, why not evaluate the likelihood of various conjunctions – (Design & GA-1), (Design & GA-2), and so on? Some of these will have high likelihoods, others will have low, but it will no longer be a mystery what likelihoods these hypotheses possess. There are two problems with this tactic. First, it is a game that two can play. Consider the hypothesis that the vertebrate eye was created by the mindless process of electricity. If I simply get to *invent* auxiliary hypotheses without having to *justify* them independently, I can just stipulate the following assumption: if electricity created the vertebrate eye, the eye must have features F1 . . . Fn. The electricity hypothesis now is a conjunct in a conjunction that has maximum likelihood, just like the design hypothesis.

This is a dead end. My second objection is that it is an important part of scientific practice that conjunctions be broken apart (when possible) and their conjuncts scrutinized (Sober 1999; Sober 2000). If your doctor runs a test to see whether you have tuberculosis, you will not be satisfied if she reports that the likelihood of the conjunction "you have tuberculosis & auxiliary assumption 1" is high, while the likelihood of the conjunction "you have tuberculosis & auxiliary assumption 2" is low. You want your doctor to address the first *conjunct*, not just the various *conjunctions*. And you want her to do this by using a test procedure that is *independently* known to have small error probabilities. Demand no less of your theologian.

My formulation of the design argument as a likelihood inference, and my criticism of it, have implications for the *problem of evil*. It is a mistake to try to *deduce* the nonexistence of God from the fact that so much evil exists. Even supposing that God is all-powerful, all-knowing, and entirely benevolent, there is no contradiction in the hypothesis that God allows various evils to exist because they are necessary correlates of greater goods, where we don't understand in any detail what these correlations are or why they must obtain (Plantinga 1974). A similar reply to the argument from evil can be made when the argument is formulated *nondeductively* (Madden and Hare 1968; Plantinga 1979; Rowe 1979). Suppose it is suggested that the kinds of evil we observe, and the amount of evil that there is, *favor* the hypothesis that there is no God. Within the framework of likelihood inference, there are two quantities that we must evaluate: What is the probability that there would be as much evil as there is, if the universe were produced by an all-powerful, all-knowing, and entirely benevolent God? And what is the probability of there being as much evil as there is, if the universe were produced by mindless natural processes? Once again, if the ways of God are enormously mysterious, we will have no way to evaluate the first of these likelihoods. Those who like this approach to the problem of evil should agree with my criticism of the argument from design.

THE RELATIONSHIP OF THE ORGANISMIC DESIGN ARGUMENT TO DARWINISM

Philosophers who criticize the organismic design argument often believe that the argument was dealt its death blow by Hume. True, Paley wrote after Hume, and the many Bridgewater Treatises elaborating the design argument appeared after Hume's *Dialogues* were published posthumously. Nonetheless, for these philosophers, the design argument after Hume was merely a corpse that could be propped up and paraded. Hume had taken the life out of it.

Biologists often take a different view. Dawkins (1986, 4) puts the point provocatively by saying that it was not until Darwin that it was possible to be an intellectually fulfilled atheist. The thought here is that Hume's skeptical attack was not the decisive moment; rather, it was Darwin's development and confirmation of a substantive scientific explanation of the adaptive features of organisms that really undermined the design argument (at least in its organismic formulation). Philosophers who believe that a theory can't be rejected until a better theory is developed to take its place often sympathize with this point of view.

My own interpretation coincides with neither of these. As indicated earlier, I think that Hume's criticisms largely derive from an empiricist epistemology that is too narrow. However, seeing the design argument's fatal flaw does not depend on seeing the merits of Darwinian theory. The Likelihood Principle, it is true, says that theories must be evaluated comparatively, not on their own. But in order for this to be possible, each theory must make predictions. It is at this fundamental level that I think the design argument is defective.

Biologists often present two criticisms of creationism. First, they argue that the design hypothesis is untestable. Second, they contend that there is plenty of evidence that the hypothesis is false. Obviously, these two lines of argument are in conflict.[14] I have already endorsed the first criticism; now I want to say a little about the second. A useful example is Stephen Jay Gould's (1980) widely read article about the Panda's thumb. Pandas are vegetarian bears who have a spur of bone (a "thumb") protruding from their wrists. They use this device to strip bamboo, which is the main thing they eat. Gould says that the hypothesis of intelligent design predicts that pandas should *not* have this inefficient device. A benevolent, powerful, and intelligent Engineer could and would have done a lot better. Evolutionary theory, on the other hand, says that the panda's thumb is what we should expect. The thumb is a modification of the wrist bones found in the common ancestor that pandas share with carnivorous bears. Evolution by natural selection is a tinkerer; it does not design adaptations from scratch, but rather modifies preexisting features, with the result that adaptations are often imperfect.

Gould's argument, I hope it is clear, is a likelihood argument. I agree with what he says about evolutionary theory, but I think his discussion of the design hypothesis leads him into the same trap that ensnared Paley. Gould thinks that he knows what God would do if he built pandas, just as Paley thought he knew what God would do if he built the vertebrate eye. But neither of them knows anything of the sort. Both help themselves to *assumptions* about God's goals and abilities. However, it is not enough to make assumptions about these matters; one needs independent evidence that these auxiliary assumptions are true. Paley's problem is also Gould's.

ANTHROPIC REASONING AND COSMIC DESIGN ARGUMENTS

Evolutionary theory seeks to explain the adaptive features of organisms; it has nothing to say about the origin of the universe as a whole. For this reason, evolutionary theory conflicts with the organismic design hypothesis, but not with the cosmic design hypothesis. Still, the main criticism I presented of the first type of design argument also applies to the second. I now want to examine a further problem that cosmic design arguments sometimes encounter.[15]

Suppose I catch fifty fish from a lake, and you want to use my observation O to test two hypotheses:

O: All the fish I caught were more than ten inches long.
F1: All the fish in the lake are more than ten inches long.
F2: Only half the fish in the lake are more than ten inches long.

You might think that the Likelihood Principle says that F1 is better supported, since

(1) $$\Pr(O \mid F1) > \Pr(O \mid F2).$$

However, you then discover how I caught my fish:

(A1) I caught the fish by using a net that (because of the size of its holes) can't catch fish smaller than ten inches, and I left the net in the lake until there were fifty fish in it.

This leads you to replace the analysis provided by (1) with the following:

(2) $$\Pr(O \mid F1 \& A1) = \Pr(O \mid F2 \& A1) = 1.0.$$

Furthermore, you now realize that your first assessment, (1), was based on the erroneous assumption that

(A0) The fish I caught were a random sample from the fish in the lake.

Instead of (1), you should have written

$$\Pr(O \mid F1 \& A0) > \Pr(O \mid F2 \& A0).$$

This inequality is true; the problem, however, is that (A0) is false.

This example, from Eddington (1938), illustrates the idea of an *observational selection effect* (an OSE). When a hypothesis is said to render a set of observations probable (or improbable), ask yourself what assumptions allow the hypothesis to have this implication. The point illustrated here is that the procedure you use to obtain your observations can be relevant to assessing likelihoods.[16]

One version of the cosmic design argument begins with the observation that our universe is "fine-tuned." That is, the values of various physical constants are such as to permit life to exist, but if they had been even slightly different, life would have been impossible. McMullin (1993, 378) summarizes some of the relevant facts as follows:

If the strong nuclear force were to have been as little as 2% stronger (relative to the other forces), all hydrogen would have been converted into helium. If it were 5% weaker, no helium at all would have formed and there would be nothing but hydrogen. If the weak nuclear force were a little stronger, supernovas could not occur, and heavy elements could not have formed. If it were slightly weaker, only helium might have formed. If the electromagnetic forces were stronger, all stars would be red dwarfs, and there would be no planets. If it were a little weaker, all stars would be very hot and short-lived. If the electron charge were ever so slightly different, there would be no chemistry as we know it. Carbon (^{12}C) only just managed to form in the primal nucleosynthesis.

I'll abbreviate the fact that the values of these physical constants fall within the narrow limits specified by saying that "the constants are right." A design argument can now be constructed, one that claims that the constants' being right should be explained by postulating the existence of an intelligent designer, one who wanted life to exist and who arranged the universe so that this could occur (Swinburne 1990a). As with Paley's organismic design argument, we can represent the reasoning in this cosmic design argument as the assertion of a likelihood inequality:

(3) Pr(Constants are right | Design) > Pr(Constants are right | Chance).

However, there is a problem with (3) that resembles the problem with (1). Consider the fact that

(A3) We exist, and if we exist the constants must be right.

We need to take (A3) into account; instead of (3), we should have said:

(4) Pr(Constants are right | Design & A3) = Pr(Constants are right | Chance & A3) = 1.0.

That is, given (A3), the constants must be right, regardless of whether the universe was produced by intelligent design or by chance.

Proposition (4) reflects the fact that our observation that the constants are right is subject to an OSE. Recognizing this OSE is in accordance with a *weak anthropic principle* – "what we can expect to observe must be restricted by the conditions necessary for our presence as observers" (Carter 1974). The argument involves no commitment to *strong anthropic principles*. For example, there is no assertion that the correct cosmology must entail that the existence of observers such as ourselves was inevitable, nor is it claimed that our existence *explains* why the physical constants are right (Barrow 1988, Earman 1987, McMullin 1993).[17]

Although this point about OSEs undermines the version of the design argument that cites the fact that the physical constants are right, it does not touch other versions. For example, when Paley concludes that the vertebrate eye was produced by an intelligent designer, his argument cannot be refuted by claiming that

(A4) We exist, and if we exist vertebrates must have eyes with features F1 . . . Fn.

If (A4) were true, the likelihood inequality that Paley asserted would have to be replaced by an equality, just as (1) had to be replaced by (2), and (3) had to be replaced by (4). But fortunately for Paley, (A4) is false. However, matters change if we think of Paley as seeking to explain the modest fact that organisms have at least one adaptive contrivance. If this were false, we would not be able to make observations; indeed, we would not exist. Paley was right to focus on the details; the more minimal description of what we observe does not sustain the argument he wanted to endorse.

The issue of OSEs can be raised in connection with other cosmic versions of the design argument. Swinburne (1990b, 191) writes that "the hypothesis of theism is that the universe exists because there is a God who keeps it in being and that laws of nature operate because there is a God who brings it about that they do." Let us separate the *explananda*. The fact that the universe exists does *not* favor design over chance; after all, if the universe did not exist, we would not exist and so would not be able to observe that it does.[18] The same point holds with respect to the fact that the universe is law-governed. Even supposing that lawlessness is possible, could we exist and make observations if there were no laws? If not, then the lawful character of the universe does not discriminate between design and chance. Finally, we may consider the fact that our universe is governed by one set of laws rather than by another. Swinburne (1968) argues that the fact that our universe obeys *simple* laws is better explained by the hypothesis of design than by the hypothesis of chance. Whether this observation also is subject to an OSE depends on whether we could exist in a universe obeying alternative laws.

Before taking up an objection to this analysis of the argument from fine-tuning, I want to summarize what it has in common with the fishing example. In the fishing example, the source of the OSE is obvious – it is located in a device outside of ourselves. The net with big holes ensures that the observer will make a certain observation, regardless of which of two hypotheses is true. But where is the device that induces an OSE in the fine-tuning example? There is none; rather, it is the observer's own existence that does the work. Nonetheless, the effect is the same. Owing to the fact that we exist, we are bound to observe that the constants are right, regardless of whether our universe was produced by chance or by design.[19]

This structural similarity between fishing and fine-tuning may seem to be undermined by a disanalogy. In the latter case, we know that proposition (3) is correct: the probability that the constants will be right if the universe was created by a powerful Deity bent on having life exist is greater than it would be if the values of the constants were set by a uniform chance process. This inequality seems to hold regardless of how or whether we make our observations. The fishing example looks different; here we know that proposition (1) is false. There is no saying whether a likelihood inequality obtains until we specify the procedure used to obtain the observations;

once we do this, there *is* no likelihood inequality. Thus, in fine-tuning, we have an inequality that is true because it reflects the metaphysical facts; in fishing, we have an inequality that is false for epistemic reasons. My response is that I agree that this point of difference exists, but contend that it does nothing to save the argument from fine-tuning. Although proposition (3) is true, we are bound to observe that the constants are right, regardless of whether our universe arose by chance or by design. My objection to proposition (3) is not that it is false, but that it should not be used to interpret the observations; (4) is the relevant proposition to which we should attend.

In order to visualize this point, imagine that a Deity creates a million universes and that a chance process does the same for another million. Let's assume that the proportion of universes in which the constants are right is greater in the former case. Doesn't it follow that if we observe that the constants are right in our universe, that this observation favors the hypothesis that our universe arose by design? In fact, this does not follow. It *would* follow if we had the same probability of observing any of the first million universes if the design hypothesis were true, and had the same probability of observing any of the second million universes if the chance hypothesis were true. But this is not the case – our probability of observing a universe in which the constants are right is unity in each case.

What this means is that full understanding of the workings of OSEs must acknowledge that there are two stages at which a bias can be introduced. There is first the process by which the system described by the hypotheses under test generates some state of the world that we are able to observe. Second, there is the process by which we come to observe that state of the world. This two-step process occurs in fishing and fine-tuning as follows:

Composition of the lake $\longrightarrow$ Contents of the net $\longrightarrow$
We observe the contents of the net.

Origin of the universe $\longrightarrow$ Constants are right $\longrightarrow$
We observe that the constants are right.

The OSE in the fishing example arises in step 1; the OSE in fine-tuning crops up in step 2.

Leslie (1989, 13–14, 107–8), Swinburne (1990a, 171), and Van Inwagen (1993, 135, 144) all defend the fine-tuning argument against the criticism I have just described. Each mounts his defense by describing an analogy with a mundane example. Here is Swinburne's rendition of an example that Leslie presents:

On a certain occasion the firing squad aim their rifles at the prisoner to be executed. There are twelve expert marksmen in the firing squad, and they fire twelve rounds each. However, on this occasion all 144 shots miss. The prisoner laughs and comments that the event is not something requiring any explanation because if the marksmen had not missed, he would not be here to observe them having done so. But of course, the prisoner's comment is absurd; the marksmen all having missed is indeed something requiring explanation; and so too is what goes with it – the prisoner's being alive to observe it. And the explanation will be either that it was an accident (a most unusual chance event) or that it was planned (e.g., all the marksmen had been bribed to miss). Any interpretation of the anthropic principle which suggests that the evolution of observers is something which requires no explanation in terms of boundary conditions and laws being a certain way (either inexplicably or through choice) is false.

First a preliminary clarification – the issue isn't whether the prisoner's survival "requires explanation" but whether this observation provides evidence as to whether the marksmen intended to spare the prisoner or shot at random.[20]

My response takes the form of a dilemma. I'll argue, first, that if the firing squad example is analyzed in terms of the Likelihood Principle, the prisoner is right and Swinburne is wrong – the prisoner's survival does not allow him to conclude that design is more likely than chance. However, there is a different analysis of the prisoner's situation, in terms of the *probabilities* of hypotheses, not their *likelihoods*. This second analysis says that the prisoner *is* mistaken; however, it has the consequence that the prisoner's inference differs fundamentally from the design argument that appeals to fine-tuning. Each horn of this dilemma supports the conclusion that the firing squad example does nothing to save this version of the design argument.

So let us begin. If we understand Swinburne's claim in terms of the Likelihood Principle, we should read him as saying that

(L1) Pr(The prisoner survived | The marksmen intended to miss) > Pr(The prisoner survived | The marksmen fired at random).

He thinks that the anthropic principle requires us to replace this claim with the following irrelevancy:

(L2) Pr(The prisoner survived | The marksmen intended to miss & The prisoner survived) = Pr(The prisoner survived | The marksmen fired at random & The prisoner survived) = 1.0.

This equality would lead us to conclude (mistakenly, Swinburne thinks) that the prisoner's survival does not discriminate between the hypotheses of design and chance.

To assess the claim that the prisoner has made a mistake, it is useful to compare the prisoner's reasoning with that of a bystander who witnesses the prisoner survive the firing squad. The prisoner reasons as follows: "Given that I now am able to make observations, I must be alive, whether my survival was due to intelligent design or to chance." The bystander says the following: "Given that I now am able to make observations, the fact that the prisoner is now alive is made more probable by the design hypothesis than it is by the chance hypothesis." The prisoner is claiming that he is subject to an OSE, while the bystander is saying that he, the bystander, is not. Both, I submit, are correct.[21]

I suggest that part of the intuitive attractiveness of the claim that the prisoner has made a mistake derives from a shift between the prisoner's point of view and the bystander's. The bystander is right to use (L1) to interpret his observations; however, the prisoner has no business using (L1) to interpret his observations, because he, the prisoner, is subject to an OSE. The prisoner needs to replace (L1) with (L2). My hunch is that Swinburne thinks the prisoner errs in his assessment of likelihoods because we bystanders would be making a mistake if we reasoned as he does.[22]

The basic idea of an OSE is that we must take account of the procedures used to obtain the observations when we assess the likelihoods of hypotheses. This much was clear from the fishing example. What may seem strange about my reading of the firing squad story is my claim that the prisoner and the bystander are in different epistemic situations, even though their observational reports differ by a mere pronoun. After the marksmen fire, the prisoner thinks "I exist," while the bystander thinks "He exists"; the bystander, but not the prisoner, is able to use his observation to say that design is more likely than chance, or so I say. If this seems odd, it may be useful to reflect on Sorensen's (1988) concept of *blind spots*. A proposition p is a blind spot for an individual S just in case, if p were true, S would not be able to know that p is true. Although some propositions (e.g., "Nothing exists," "The constants are wrong") are blind spots for everyone, other propositions are blind spots for some people but not for others. Blind spots give rise to OSEs; if p is a blindspot for S, then if S makes an observation in order to determine the truth value of p, the outcome must be that not-p is observed. The prisoner, but not the bystander, has "The prisoner does not exist" as a blind spot. This is why "The prisoner exists" has an evidential significance for the bystander that it cannot have for the prisoner.[23]

To bolster my claim that the prisoner is right to think that likelihood does not distinguish between chance and design, I want to describe a slightly different problem. Suppose that a firing squad always subjects its victims to the same probabilistic process, which has the result that the prisoner either survives or is killed. A thousand prisoners who have survived the firing squad one by one are assembled and are asked to pool their knowledge and estimate the value of an unknown probability. What is the probability that a prisoner will survive if the firing squad fires? The standard methodology here is *maximum likelihood estimation*; one finds the value of the parameter of interest that maximizes the probability of the observations. This is why, if a coin lands heads 512 times out of a thousand tosses, the "best" estimate of the probability that the coin will land heads when it is tossed is 0.512. Those who believe that the single prisoner has evidence about his firing squad's intentions are obliged to conclude that the best estimate in this new problem is that the probability is unity. However, those persuaded that the single prisoner is subject to an OSE will want to maintain that the thousand prisoners are in the same boat. These skeptics will deny that the observations provide a basis for estimation. Isn't it *obvious* that testimony limited to survivors provides no evidence on which to base an estimate of the probability that someone will survive the firing squad? And if this is true of a *thousand* survivors, how can a *single* survivor be said to know that design is more likely than chance?

I now turn to a different analysis of the prisoner's situation. The prisoner, like the rest of us, knows how firing squads work. They always, or almost always, follow the orders they receive – which are, almost always, to execute someone. Occasionally, they produce fake executions. They almost never fire at random. What is more, firing squads have firm control over outcomes; if they want to kill (or spare) someone, they always, or almost always, succeed. This and related items of background knowledge support the following *probability* claim:

(Pf) Pr(The marksmen intended to spare the prisoner | The prisoner survived) >
Pr(The marksmen intended to spare the prisoner).

Firing squads rarely intend to spare their victims, but the survival of the prisoner makes it very probable that his firing squad had precisely that intention. The likelihood analysis led to the conclusion that the prisoner and the bystander are in different epistemic situations; the bystander should evaluate the hypotheses by using (L1), but the prisoner is obliged to use (L2). However, from the point of view of probabilities, the prisoner and the bystander can say the same thing; both can cite (Pf).[24]

What does this tell us about the fine-tuning version of the design argument? I construed that argument as a claim about likelihoods. As such, it is subject to an OSE; given that we exist, the constants must be right, regardless of whether our universe was produced by chance or by design. However, we now need to consider whether the fine-tuning argument can be formulated as a claim about probabilities. Can we assert that

(Pu) Pr(The universe was created by an intelligent designer | The constants are right) >
Pr(The universe was created by an intelligent designer)?

I don't think so. In the case of firing squads, we have frequency data and our general knowledge of human behavior on which to ground the probability statement (Pf). But we have neither data nor theory on which to ground (Pu). And we cannot defend (Pu) by saying that an intelligent designer would ensure that the constants are right, because this takes us back to the likelihood considerations we have already discussed. The prisoner's conclusion that he can say nothing about chance and design *is* mistaken if he is making a claim about probabilities. But the argument from fine-tuning can't be defended as a claim about probabilities.

The rabbit/duck quality of this problem merits review. I have discussed three examples – fishing, fine-tuning, and the firing squad. If we compare fine-tuning to fishing, they seem similar. This makes it intuitive to conclude that the design argument based on fine-tuning is wrong. However, if we compare fine-tuning to the firing squad, *they* seem similar. Since the prisoner apparently has evidence that favors design over chance, we are led to the conclusion that the fine-tuning argument must be right. This shifting gestalt can be stabilized by imposing a formalism. The first point is that OSEs are to be understood by comparing the *likelihoods* of hypotheses, not their *probabilities*. The second is that it is perfectly true that the prisoner can assert the *probability* claim (Pf). The question, then, is whether the design argument from fine-tuning is a likelihood argument or a probability argument. If the former, it is flawed because it fails to take account of the fact that there is an OSE. If the latter, it is also flawed, but for a different reason – it makes claims about probabilities that we have no reason to accept; indeed, we cannot even *understand* them as objective claims.[25]

A PREDICTION

It was obvious to Paley and to other purveyors of the organismic design argument that if an intelligent designer built organisms, that designer would have to be far more intelligent than any human being could ever be. This is why the organismic design argument was for them an argument for the existence of *God*. I predict that it will eventually become clear that the organismic design argument should never have been understood in this way. This is because I expect that human beings will eventually build organisms from nonliving materials. This achievement will not close down the question of whether the organisms we observe were created by intelligent design or by mindless natural processes; on the contrary, it will give that question a practical meaning, since the organisms we will see around us will be of both kinds.[26] However, it will be abundantly clear that the fact of organismic adaptation has nothing to do with whether God exists. When the Spanish conquistadors arrived in the New World, several indigenous peoples thought these intruders were gods, so powerful was the technology that the intruders possessed. Alas, the locals were mistaken; they did not realize that these beings with guns and horses were merely *human* beings. The organismic design argument for the existence of God embodies the same mistake. Human beings in the future will be the conquistadors, and Paley will be our Montezuma.

Notes

1. I am grateful to Martin Barrett, Nick Bostrom, David Christensen, Ellery Eells, Branden Fitelson, Malcolm Forster, Alan Hajek, Daniel Hausman, Stephen Leeds, Williams Mann, Lydia McGrew, Derk Pereboom, Roy Sorensen, and Richard Swinburne for useful comments. I have used portions of this chapter in seminars I've given in a large number of philosophy departments, too numerous to list here. My thanks to participants for their stimulating and productive discussion.

2. Does this construal of the design argument conflict with the idea that the argument is an *inference to the best explanation*? Not if one's theory of inference to the best explanation says that observations influence the assessment of explanations in this instance via the vehicle of likelihoods.

3. Another reason to restrict the design argument to likelihood considerations is that it is supposed to be an *empirical* argument. To invoke prior probabilities is to bring in considerations other than the observations at hand.

4. In light of the fact that it is possible for a hypothesis to have an objective likelihood without also having an objective probability, one should understand Bayes' Theorem as specifying how the quantities it mentions are related to each other, *if all are well defined*. And just as hypotheses can have likelihoods without having (objective) probabilities, it also is possible for the reverse situation to obtain. Suppose that I draw a card from a deck of unknown composition. I observe (O) that the card is the four of diamonds. I now consider the hypothesis (H) that the card is a four. The value of $Pr(H \mid O)$ is well defined, but the value of $Pr(O \mid H)$ is not.

5. Actually, Paley (1802) *does* consider a "selective retention" process, but only very briefly. In Chapter 5, pp. 49–51, he explores the hypothesis that a random process once generated a huge range of variation and that this variation was then culled, with only stable configurations surviving. Paley argues against this hypothesis by saying that we should see unicorns and mermaids if it were true. He also says that it mistakenly predicts that organisms should fail to form a taxonomic hierarchy. It is ironic that Darwin claimed that his own theory *predicts* hierarchy. In fact, Paley and Darwin are both right. Darwin's theory includes the idea that all living things have common ancestors, while the selection hypothesis that Paley considers does not.

6. More precisely, Fisher said that a mother should have a son with probability p and a daughter with probability $(1-p)$, where the effect of this is that the expected expenditures on the two sexes are the same; the argument is not undermined by the fact that some mothers have all sons while others have all daughters.

7. Dawkins (1986) makes the point that evolution by natural selection is not a uniform chance process by way of an analogy with a combination lock. This is discussed by Sober (1993, 36–9).

8. Dembski (1998) construes design inference as allowing one to argue in favor of the design hypothesis and "sweep from the field" all alternatives, without the

design hypothesis' ever having to make a prediction. For criticisms of Dembski's framework, see Fitelson, Stephens, and Sober (1999).

9. Paley (1802) argues in Chapter 16 that the benevolence of the Deity is demonstrated by the fact that organisms experience more pleasure than they need to (295). He also argues that pain is useful (320) and that few diseases are fatal; he defends the latter conclusion by citing statistics on the cure rate at a London hospital (321).

10. For it to be certain that all configurations will be visited, there must be infinite time. The shorter the time frame, the lower the probability that a given configuration will occur. This means that the estimated age of the universe may entail that it is very *im*probable that a given configuration will occur. I set this objection aside in what follows.

11. It is a standard feature of likelihood comparisons that O_s sometimes fails to discriminate between a pair of hypotheses even though O_w is able to do so, when O_s entails O_w. You are the cook in a restaurant. The waiter brings an order into the kitchen – someone ordered bacon and eggs. You wonder whether this information discriminates between the hypothesis that your friend Smith ordered the meal and the hypothesis that your friend Jones did. You know the eating habits of each. Here is the probability of the order's being for ±bacon and ±eggs, conditional on the order's coming from Smith and conditional on the order's coming from Jones:

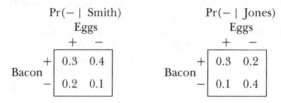

The fact that the customer ordered bacon and eggs does not discriminate between the two hypotheses (since $0.3 = 0.3$). However, the fact that the customer ordered bacon favors Smith over Jones (since $0.7 > 0.5$), and so does the fact that the customer ordered eggs (since $0.5 > 0.4$).

12. The example of the SETI project throws light on Paley's question as to why we think that watches must be the result of Intelligent Design but don't think this when we observe a stone. It is tempting to answer this question by saying that watches are "complicated" while stones are not. However, there are many complicated natural processes (like the turbulent flow of water coming from a faucet) that don't cry out for explanation in terms of intelligent design. Similarly, narrow-band radio emissions may be physically "simple," but that doesn't mean that the SETI engineers were wrong to search for them.

13. Assessing the likelihood of a disjunction involves an additional problem. Even if the values of $\Pr(O \mid D1)$ and $\Pr(O \mid D2)$ are known, what is the value of $\Pr(O \mid D1$ or $D2)$? The answer is that it must be somewhere in between. But exactly where depends on further considerations, since $\Pr(O \mid D1$ or $D2) = \Pr(O \mid D1)\Pr(D1 \mid D1$ or $D2) + \Pr(O \mid D2)\Pr(D2 \mid D1$ or $D2)$. If either God or a superintelligent extraterrestrial built the vertebrate eye, what is the probability that it was God who did so?

14. The statement "p is both false and untestable" is logically consistent (assuming that the verificationist theory of meaning is mistaken). However, the *assertion* of this conjunction is paradoxical, something akin to Moore's paradoxical statement "p is true but I don't believe it." Both conjunctions embody pragmatic, not semantic, paradoxes.

15. In order to isolate this new problem from the one already identified, I will assume in what follows that the design hypothesis and the chance hypothesis with which it competes have built into them auxiliary assumptions that suffice for their likelihoods to be well defined.

16. This general point surfaces in simple inference problems such as the ravens paradox (Hempel 1965). Does the fact that the object before you is a black raven confirm the generalization that all ravens are black? That depends on how you gathered your data. Perhaps you sampled at random from the set of *ravens;* alternatively, you may have sampled at random from the set of *black ravens.* In the first case, your observation confirms the generalization, but in the second case it does not. In the second case, notice that you were bound to observe that the object before you is a black raven, regardless of whether all ravens are black.

17. Although weak and strong anthropic principles differ, they have something in common. For example, the causal structure implicitly assumed in the weak anthropic principle is that of two effects of a common cause:

We exist now.

(WAP)   origin of universe

Constants now are right.

In contrast, one of the strong anthropic principles assumes the following causal arrangement:

(SAP) We exist now. $\longrightarrow$ origin of the universe $\longrightarrow$ Constants now are right.

Even though (WAP) is true and (SAP) is false, both entail a *correlation* between our existence and the constants' now having the values they do. In order to deal with the resulting OSEs, we must decide how to take these correlations into account in assessing likelihoods.

18. Similarly, the fact that there is something rather than nothing does not discriminate between chance and design.

19. The fishing and fine-tuning examples involve *extreme* OSEs. More modest OSEs are possible. If C describes the circumstances in which we make our observational determination as to whether proposition O is true, and we use the outcome of this determination to decide whether H1 or H2 is more likely, then a *quantitative* OSE is present precisely when

$$Pr(O \mid H1 \,\&\, C) \neq Pr(O \mid H1) \text{ or}$$
$$Pr(O \mid H2 \,\&\, C) \neq Pr(O \mid H2).$$

A *qualitative* OSE occurs when taking account of C alters the likelihood ordering:

$$\Pr(O \mid H1 \,\&\, C) > \Pr(O \mid H2 \,\&\, C) \text{ and } \Pr(O \mid H1) \not> \Pr(O \mid H2) \text{ or}$$
$$\Pr(O \mid H1 \,\&\, C) = \Pr(O \mid H2 \,\&\, C) \text{ and } \Pr(O \mid H1) \neq \Pr(O \mid H2).$$

Understood in this way, an OSE is just an example of *sampling bias*.

20. There is a third possibility – that the marksmen intended to kill the prisoner – but for the sake of simplicity (and also to make the firing squad argument more parallel with the argument from fine-tuning), I will ignore this possibility.

21. The issue, thus, is not whether (L1) or (L2) is true (both are), but which an agent should use in interpreting the bearing of observations on the likelihoods of hypotheses. In this respect, the injunction of the weak anthropic principle is like the principle of total evidence – it is a pragmatic principle, concerning which statements should be used for which purposes.

22. In order to replicate in the fine-tuning argument the difference between the prisoner's and the bystander's points of view, imagine that we observe through a telescope another universe in which the constants are right. We bystanders can use this observation in a way that the inhabitants of that universe cannot.

23. Notice that "I exist" when thought by the prisoner is a priori, whereas "the prisoner exists" when thought by the bystander is a posteriori. Is it so surprising that an a priori statement should have a different evidential significance than an a posteriori statement?

    I also should note that my claim is that the proposition "I am alive" does not permit the prisoner to conclude that design is more likely than chance. I do not say that there is no proposition he can cite after the marksmen fire that discriminates between the two hypotheses. Consider, for example, the observation that "no bullets hit me." This favors design over chance, even after the prisoner conditionalizes on the fact that he is alive. Notice also that if the prisoner were alive but riddled with bullets, it is not so clear that design would be more likely than chance.

24. I have argued that the prisoner should assign the same likelihoods to chance and design, but that he is entitled to think that his survival lowers the probability of chance and raises the probability of design. On its face, this contradicts the following consequence of Bayes' Theorem:

$$\frac{\Pr(\text{Chance} \mid \text{I survive})}{\Pr(\text{Design} \mid \text{I survive})} = \frac{\Pr(\text{I survive} \mid \text{Chance})}{\Pr(\text{I survive} \mid \text{Design})} \times \frac{\Pr(\text{Chance})}{\Pr(\text{Design})}.$$

If the ratio of posterior probabilities is greater than the ratio of priors, this must be because the two likelihoods have different values.

The reason my argument implies no such contradiction is that I have argued, first, that the relevant likelihoods are *not* the ones just displayed, but rather are ones that take account of the presence of an OSE. I have further imagined that the prisoner possesses knowledge (inferred from frequencies) that the two posterior probabilities displayed here are, respectively, low and high. This inference might be called "direct," since it proceeds without the prisoner's having to assign values to likelihoods. Bayes' Theorem describes how various quantities are related when each is well defined; it does not entail that all of them are well defined in every situation (Sober 2002). It is a familiar point made by critics of Bayesianism that likelihoods can be well defined even when prior and posterior probabilities are not. This severing of the connection between likelihoods and probabilities, or something like it, arises in the firing squad problem. Given his observation, after the firing squad fires, that he exists, the prisoner can know that chance is improbable and that design is highly probable, even though his evaluation of likelihoods should focus on likelihoods that are identical in value.

25. The hypothesis that our universe is one among many has been introduced as a possible explanation of the fact that the constants (in our universe) are right. A universe is here understood to be a region of space-time that is causally closed. See Leslie (1989) for discussion. If the point of the multiverse hypothesis is to challenge the design hypothesis, on the assumption that the design hypothesis has already vanquished the hypothesis of chance, then the multiverse hypothesis is not needed. Furthermore, in comparing the multiverse hypothesis and the design hypothesis, one needs to attend to the inverse gambler's fallacy discussed earlier. This is not to deny that there may be other evidence for the multiverse hypothesis; however, the mere fact that the constants are right in our universe does not favor that hypothesis.

26. As Dennett (1987a, 284–5) observes, human beings have been modifying the characteristics of animals and plants by *artificial selection* for thousands of years. However, the organisms thus modified were not *created* by human beings. If the design argument endorses a hypothesis about how organisms were brought into being, then the work of plant and animal breeders, per se, does not show that the design argument should be stripped of its theological trappings.

References

Arbuthnot, J. 1710. An argument for Divine Providence, taken from the constant regularity observ'd in the births of both sexes. *Philosophical Transactions of the Royal Society of London* 27: 186–90.

Barrow, J. 1988. *The World within the World*. Oxford: Clarendon Press.

Behe, M. 1996. *Darwin's Black Box*. New York: The Free Press.

Carter, B. 1974. Large number coincidences and the anthropic principle in cosmology. In *Confrontation of Cosmological Theories with Observational Data*, ed. M. S. Longair. Dordrecht: Reidel, pp. 291–8.

Darwin, C. [1859] 1964. *On the Origin of Species*. Cambridge, MA: Harvard University Press.

Dawkins, R. 1986. *The Blind Watchmaker*. New York: Norton.

Dembski, W. 1998. *The Design Inference*. Cambridge: Cambridge University Press.

Dennett, D. 1987a. Intentional systems in cognitive ethology: The "Panglossian Paradigm" defended. In his *The Intentional Stance*. Cambridge, MA: MIT Press, pp. 237–86.

1987b. True believers. In his *The Intentional Stance*. Cambridge, MA: MIT Press, pp. 13–42.

Dick, S. 1996. *The Biological Universe: The Twentieth-Century Extraterrestrial Life Debate and the Limits of Science*. Cambridge: Cambridge University Press.

Earman, J. 1987. The SAP also rises: A critical examination of the anthropic principle. *American Philosophical Quarterly* 24: 307–17.

Eddington, A. 1939. *The Philosophy of Physical Science*. Cambridge: Cambridge University Press.

Edwards, A. 1972. *Likelihood*. Cambridge: Cambridge University Press.

Fisher, R. 1930. *The Genetical Theory of Natural Selection*. New York: Dover.

Fitelson, B., C. Stephens, and E. Sober. 1999. How not to detect design: A review of W. Dembski's *The Design Inference*. *Philosophy of Science* 66: 472–488. Also available at <http://philosophy.wisc.edu/sober>.

Forster, M. and E. Sober. 2003. Why likelihood? In *The Nature of Scientific Evidence*, ed. M. Taper and S. Lee. Chicago: University of Chicago Press. Also available at <http://philosophy.wisc.edu/forster>.

Gould, S. 1980. *The Panda's Thumb*. New York: Norton.

Hacking, I. 1965. *The Logic of Statistical Inference*. Cambridge: Cambridge University Press.

  1987. The inverse gambler's fallacy: The argument from design. The anthropic principle applied to Wheeler universes. *Mind* 96: 331–40.

Hempel, C. 1965. Studies in the logic of confirmation. In his *Aspects of Scientific Explanation and Other Essays in the Philosophy of Science*. New York: The Free Press, pp. 3–46.

Hume, D. [1779] 1990. *Dialogues Concerning Natural Religion*. London: Penguin.

Keynes, J. 1921. *A Treatise on Probability*. London: Macmillan.

Kyburg, H. 1961. *Probability and the Logic of Rational Belief*. Middletown, CT: Wesleyan University Press.

Leslie, J. 1989. *Universes*. London: Routledge.

Madden, E. and P. Hare. 1968. *Evil and the Concept of God*. Springfield, MA: Charles Thomas.

McMullin, E. 1993. Indifference principle and anthropic principle in cosmology. *Studies in the History and Philosophy of Science* 24: 359–89.

Paley, W. 1802. *Natural Theology, or, Evidences of the Existence and Attributes of the Deity, Collected from the Appearances of Nature*. London: Rivington.

Plantinga, A. 1974. *The Nature of Necessity*. New York: Oxford.

  1979. The probabilistic argument from evil. *Philosophical Studies* 35: 1–53.

Rowe, W. 1979. The problem of evil and some varieties of atheism. *American Philosophical Quarterly* 16: 335–41.

Royall, R. 1997. *Statistical Evidence: A Likelihood Paradigm*. London: Chapman and Hall.

Sober, E. 1993. *Philosophy of Biology*. Boulder, CO.: Westview Press.

  1999. Testability. *Proceedings and Addresses of the American Philosophical Association* 73: 47–76. Also available at <http://philosophy.wisc.edu/sober>.

  2000a. Evolution and the problem of other minds. *Journal of Philosophyf* 97: 365–86.

  2000b. Quine's two dogmas. *Proceedings of the Aristotlean Society* supp. vol. 74: 237–80.

  2002. Bayesianism: Its scope and limits. In *Bayesianism*, ed. R. Swinburne. London: British Academy, pp. 21–38. Also available at <http://philosophy.wisc.edu/sober>.

Sorensen, R. 1988. *Blindspots*. Oxford: Oxford University Press.

Stigler, S. 1986. *The History of Statistics*. Cambridge: Harvard University Press.

Swinburne, R. 1968. The argument from design. *Philosophy* 43: 199–212.

    1990a. Argument from the fine-tuning of the universe. In *Physical Cosmology and Philosophy*, ed. J. Leslie. New York: Macmillan, pp. 160–79.

    1990b. The limits of explanation. In *Explanation and Its Limits*, ed. D. Knowles. Cambridge: Cambridge University Press, pp. 177–93.

Turing, A. 1950. Computing machinery and intelligence. *Mind* 59: 433–60.

Van Inwagen, P. 1993. *Metaphysics*. Boulder, CO.: Westview Press.

Venn, J. 1866. *The Logic of Chance*. New York: Chelsea.

White, R. 2000. Fine-tuning and multiple universes. *Nous* 34: 260–76.

7

DNA by Design?

Stephen Meyer and the Return of the God Hypothesis

Robert T. Pennock

In his keynote address at a recent Intelligent Design (ID) conference at Biola University, ID leader William Dembski began by quoting "a well-known ID sympathizer" whom he had asked to assess the current state of the ID movement. Dembski explained that he had asked because, "after some initial enthusiasm on his part three years ago, his interest seemed to have flagged" (Dembski 2002). The sympathizer replied that

[t]oo much stuff from the ID camp is repetitive, imprecise and immodest in its claims, and otherwise very unsatisfactory. The 'debate' is mostly going around in circles. (Dembski 2002)

Those of us who have been following the ID or "Wedge" movement since it coalesced around point man Philip Johnson during the early 1990s reached much the same assessment of its arguments years ago. In something of an understatement, Dembski told his supporters (the conference was closed to critical observers) that "the scientific research part of ID" was "lagging behind" its cultural penetration. He noted that there are only "a handful of academics and independent researchers" currently doing any work on the scholarly side of ID, and offered some suggestions to try to rally his troops[1] (Dembski 2002). We will have to wait to see if anything comes of this call, but judging from ID's track record, it seems unlikely. This chapter is a look back at nearly a decade and a half of repetitious, imprecise, immodest, and unsatisfactory arguments. So that our review does not entirely circle over old ground, I propose that we look at the ID arguments through the writings of Stephen C. Meyer. Meyer is certainly one of the core workers Dembski had in mind, but his work has so far received little critical attention.

Meyer is the longtime director of the Discovery Institute's Center for the Renewal of Science and Culture,[2] which is the de facto headquarters of the Wedge movement. With publications going back to the mid-1980s that helped to lay the groundwork for the Wedge arguments, Meyer was one of the earliest leaders of the movement and has continued to play a central role. As we will see, he had already published an attack on evolution that charged it is based upon naturalistic assumptions – the centerpiece of the ID attack that Philip Johnson would begin to press in 1991 – even before he was introduced to Johnson in 1987 (Meyer 2001). Meyer was coauthor of a special philosophy section in *Of Pandas and People*, the ID textbook supplement for junior high and high school biology courses that tries to make the case that ID is legitimate science (Hartwig and Meyer [1989] 1993). He also takes a leading role in the movement's persistent lobbying to get ID into the public schools, testifying at congressional and other hearings.[3]

As part of ID's lobbying efforts, Meyer is an active writer of op-ed pieces. In two editorials written in May and July of 1996 – "Limits of Natural Selection a Reason to Teach All Theories," in the *Tacoma News Tribune* (Meyer 1996b) and "'Don't Ask, Don't Tell' in Biology Instruction," in the *Washington Times* (Meyer 1996a) – Meyer (in the first article) defended teaching anti-evolutionism in the Sultan, Washington schools and (in the second) attacked the way California's science guidelines recommended that teachers help students with religious objections to science. Reading these and other op-ed pieces gives a clear picture of the points that the Wedge wants to hammer home.

In both articles, Meyer faults biology textbooks for presenting only "half of the picture," leaving out information about the Cambrian explosion that, he says, confirms a pattern of abrupt appearance rather than an evolutionary process. These texts purportedly failed to define "evolution" adequately – it can refer, he claims, to anything from "trivial" microevolutionary change to "the creation of life by strictly mindless, material forces" – and they failed to mention scientists who reject evolution in favor of "alternative theories," such as Intelligent Design. He cites ID theorist Michael Behe and his idea that the "irreducibly complex" bacterial flagellum provides evidence against the "superstitions" of the self-assembly of life. He criticizes biologists (mentioning Douglas Futuyma and Kenneth Miller) who, he says, make no attempt to hide the anti-theistic implications of Darwinism.

Meyer does not just make the same points in both articles; the paragraphs discussing these main ideas, comprising over two of the three pages of the July article, are actually copied word for word from the May article. We will reply to Meyer's other points along the way, but here let us just note that Darwinian evolution has "anti-theistic" implications only for those who think they already know, rather specifically, what God did and did not do. Meyer's misrepresentation of Miller makes sense only given ID's own narrow view, since Miller is a Christian theist who explicitly rejects the contention that Darwinian evolution is anti-theistic (Miller 1999).

In a 1998 op-ed piece in Spokane's *Spokesman-Review* – "Let Schools Provide Full Disclosure" (Meyer 1998) – Meyer gave advice to school board

members in Post Falls, Washington, some of whom wanted to accommo-
date parents who were pressing to teach creationism. Biblically based Young
Earth creationism would be "legally problematic," he said, but the Intelligent
Design approach could probably escape a legal challenge. Again, following
this introduction, most of the paragraphs are repeated verbatim from the
earlier articles, without citation.

Nor has Meyer changed his cut-and-paste approach in subsequent years.[4]
A few weeks ago (as I write this), most of the same core paragraphs were
copied into yet another article by Meyer, titled "Darwin Would Love This
Debate" (Chapman and Meyer 2002) in the June 10, 2002, issue of the
Seattle Times. This time Meyer addresses the controversy over the proposal
under review by the Ohio Board of Education to include ID in the science curriculum in that state.[5] Interestingly, a couple of minor changes
do appear in these paragraphs over the course of the six years since the
first piece appeared, which help us to address a second point in Meyer's
challenge.

In the 1996 op-ed pieces, Meyer claimed that "none of the standard high
school biology texts even mentions the Cambrian Explosion" and suggested
that science educators had omitted it deliberately. "Scientific literacy," he
opined "requires that students know all significant facts whether or not they
happen to support cherished theories" (Meyer 1996a, 1996b). The implication is that scientists are withholding information about the Cambrian
explosion in order to protect evolutionary theory. However, it is hardly the
case that scientists view the Cambrian radiation as an embarrassing, unsolv-
able problem for evolution, as ID theorists purport, and the suggestion of a
conspiracy of silence is absurd. One can find any number of discussions of
the Cambrian radiation in the scientific literature, and new studies regularly
increase our understanding of that interesting evolutionary episode. This
is no skeleton in the closet, kept hidden away from students, as even Meyer
is increasingly forced to admit. In his 1998 op-ed piece, he changed "none
of the standard high school biology texts" to "only one"; in the 2002 piece,
he was forced to modify it to "few." Science, we see, is quite open about its
theories.

ID theorists, by contrast, are very close-mouthed about their own views.
If evolution really cannot hope to explain the Cambrian explosion, and
ID theorists can do better, one would expect them to show how. However,
no "alternative theory" is forthcoming. ID leaders who are Young Earth
creationists – such as Paul Nelson, Percival Davis, and others – do not even accept the scientific dating of the Cambrian. However, even the Old Earthers,
such as Behe and presumably Meyer, have offered no positive account.

Is their view that the "at least fifty separate major groups of organisms"[6]
(note Meyer's pointed claim of separateness) were separately created at that
time? What about those phyla that arose before or afterward? And why the
invariable focus at the arbitrary level of the phylum; isn't it rather the origin

of species, which is what Darwin explained, that is more pertinent? What about the vast numbers of species that arose in the subsequent half-billion years, or in the prior three billion? According to ID theory, even the smallest increase in genetic information must be the result of the "insertion of design." Although their view would thus seem to require countless such insertions, they decline to say where, when, or how this happens. The biologist Kenneth Miller asked Dembski and Behe this question point blank during a debate at the American Museum of Natural History, and neither was willing to take a stand on even one specific point in time at which this supposedly occurred (Milner et al. 2002).

The pattern of vagueness and evasion regarding the specific theoretical commitments or possible tests of ID is pervasive. In response to my direct questions during the same debate, Behe refused to answer whether a proposed experiment would suffice to identify whether a system met his notion of "irreducible complexity" (he said he smelled a trap), and Dembski would not even take a stand on the age of the Earth (Milner et al. 2002). One could cite numerous similar examples. I have not seen the chapter that Meyer is writing on the Cambrian explosion for the present volume, but I encourage readers to check whether he departs from the pattern and offers any specific positive account. If ID is to have even a shot at being a real scientific alternative, one should expect to see some precise, testable (and eventually tested) hypotheses that answer the obvious questions: what was designed and what wasn't; and when, where, how, and by whom was design information supposedly inserted?[7]

Although his Discovery Institute biography describes Meyer as the author of "numerous technical articles," the group does not list or include any of these in its database of his writings, as one would expect if they involved ID research, but instead calls attention to his op-ed pieces. However, his influence runs far deeper than this would suggest. As one of the philosophers who dominate the ID movement, Meyer's work on the epistemological presuppositions of the evolution/creation debate has helped to define the core features of the movement from its very inception.

In "Scientific Tenets of Faith," Meyer argues that science is based upon "foundational assumptions of naturalism" that are as much a matter of faith as those of "creation theory" (Meyer 1986). His argument prefigures by several years the argument that would make Philip Johnson famous, that scientific naturalism is akin to religious dogma and that the assumptions of creation theory should supplant it.

Meyer makes the same error of imprecision that Johnson later would make on this point, failing to distinguish metaphysical from methodological naturalism. The former holds that the world is a closed system of physical causes and that nothing else exists. This rebuts another of Meyer's charges in his op-ed pieces, because evolutionary biology makes no claim about "strictly mindless, material forces" in such a metaphysical sense. Science holds to naturalism only in the more modest methodological sense – that is, in not allowing itself to appeal to miracles or other supernatural interventions that would violate natural causal regularities – and remains neutral with regard to metaphysical possibilities. Moreover, these methodological constraints of order and uniformity are not held dogmatically, but are based upon sound reasons that ground evidential inference (Pennock 1996).

Unmindful or perhaps unaware of this crucial distinction, Meyer writes of the "necessity of making intelligent foundational assumptions" that can "lend explanation and meaning to the necessary functions of Inquiry" (Meyer 1986), but thinks that these are just a matter of faith. As noted, he thinks the assumptions of creation theory are at least as good as those of science. Significantly, in making this point, Meyer draws a direct connection to the battle in Arkansas during the early 1980s regarding legislation mandating balanced treatment of evolution and a purportedly scientific theory of creation. Meyer claims that the naturalistic assumptions underlying science put it on a par with creation theory.[8]

[T]hese foundational assumptions are not unlike the much scorned "tenets of faith" whose detected presence in creation theory first disqualified it as legitimate science in an Arkansas federal court three years ago. This observation neither suggests nor repudiates a defense of creation theory as legitimate science. It does, however, assert that from the definition offered by the American Civil Liberties Union . . . science itself does not qualify as legitimate science. (Meyer 1986)

By neglecting the distinction noted above, Meyer fails to see that scientific naturalism is not taken on faith; rather, it is a working hypothesis that is justified, in part, by science's continued success. It is conceivable that in the long run it will fail, but so far the method shows no signs of weakening and every sign of increasing strength.

The claim of generic equivalence (which we see is false) with regard to the need for *some* presuppositional basis is only the initial part of the ID program. Rejecting naturalism and any evolutionary account as a basis for the possibility of human knowledge, Meyer and other ID theorists turn to the alternative biblical presuppositions that they believe must be put in place in order to ground claims of truth:

Given the current and historical difficulty human philosophic systems have faced in accounting for truth as autonomous from revelation, scientists and philosophers might be most receptive to systems of thought that find their roots in Biblical theology. (Meyer 1986)

That is to say, Meyer doubts that there could be any warranted basis for truth claims apart from revelation and Christian assumptions. Like ID advocate Alvin Plantinga (whose entire epistemology is based upon a Christian presuppositionalism), Meyer holds that human knowledge can be justified only on the assumption that God designed the human mind and that it transcends the material world.[9] He writes:

The Judeo-Christian scriptures have much to say about the ultimate source of human reason, the existence of a real and uniformly ordered universe, and the ability present in a creative and ordered human intellect to know that universe. Both the Old and New Testaments define these relationships such that the presuppositional base necessary to modern science is not only explicable but also meaningful. (Meyer 1986)

Appealing to a "real and uniformly ordered universe" is just what method-ological naturalism says scientists must do, but ID theorists are wrong to think that one must ground this constraint in scripture. Indeed, taking their biblical route actually subverts that necessary base of presumed or-der and uniformity, because it assumes, to the contrary, that it is broken by the Designer's creative interventions.

We shall return to a consideration of ID theory's proposal that a "theistic science" (as Johnson calls it) is a better presuppositional basis for warranted knowledge, but first let us briefly examine the claim that such a scriptural assumption is necessary not only to make science explicable, but also to make it "meaningful."

Why does all this matter? In *Tower of Babel* (Pennock 1999, Chapter 7), I explained how Johnson and others in his movement see not only a point of science but also the meaning of life itself as being at stake. Among other things, they believe that if evolution is true, then there is no ground for moral values. This is not a peripheral issue involving their motivation, but an essential part of their philosophical argument. That God created us for a purpose is, for them, the necessary foundation for true human moral-ity and proper social order. At the conclusion of the article just consid-ered, immediately following his statement about the scriptural presuppo-sitional grounding of their view of science, Meyer adumbrates the moral issue:

Moreover all of us would do well to reflect on the scriptural axiom that "in Him all things hold together," and further reflect on the serious consequences to a society and culture that divorce spiritual thought not only from moral considerations but scientific ones as well. (Meyer 1986)

We find a further elaboration of this Christian assumption of the ID view in an article Meyer wrote in collaboration with Charles Thaxton, another important early leader of the ID movement.

In "Human Rights: Blessed by God or Begrudged by Government?," Thaxton and Meyer focus not on abortion, divorce, homosexuality, or the other purported evils that Johnson discusses, but on the notion of human dignity as the basis for human rights. They see human dignity as arising necessarily from the idea that human being are the glory of God's Creation. Here is how they make the argument:

Historically, Western society has derived its belief in the dignity of man from its Judeo-Christian belief that man is the glory of God, made in his image. According to this view, human rights depend upon the Creator who made man with dignity, not upon the state. (Thaxton and Meyer 1987)

This perspective and language would show up several years later in the "Wedge Document" – the manifesto from the Center for the Renewal of Science and Culture, by then under Meyer's directorship, that laid out the ideological foundations and strategic plans of the ID movement. Thaxton and Meyer's article continues by contrasting the traditional Judeo-Christian view with what they say is the contemporary scientific view "that promulgates a less exalted view of man," in which he is merely a material being "cast up by chance in an ... impersonal universe" (Thaxton and Meyer 1987). Their thesis that the modern scientific worldview is a barren materialism that stands in opposition to the Judeo-Christian view also appears as the key point of the ID Wedge manifesto, which pledges "nothing less than the overthrow of materialism and its cultural legacies" and the renewal of "a broadly theistic understanding of nature."[10]

Thaxton and Meyer say that according to the modern view, "only man's material complexity distinguishes him from the other biological structures that inhabit the universe" (Thaxton and Meyer 1987), and they claim that this is inadequate to ground human rights. They have no truck with the possibility that moral rights could apply to nonhuman animals. Indeed, they don't want to consider man an animal at all; they believe it is critical that there be something that is "distinctively human," for otherwise it would "relegat[e] man to the level of animals" (Thaxton and Meyer 1987). Their goal of keeping human beings categorically distinct from animals goes hand in glove with their theological grounding of dignity, and from this it is for them but a small step to the rejection of biological evolution.

Thaxton and Meyer briefly consider the argument of those who promote "merely reiterating the Judeo-Christian doctrine of creation" as a "useful fiction," but reject it on the ground that no merely fictional doctrine will suffice to "rescue man from his current moral dilemma" (Thaxton and Meyer 1987). So, what will save man? Not belief alone. Nothing less than the *truth* of Divine creation. They put it this way:

Judaism and Christianity do not teach that the doctrine of man's creation in the Divine image establishes his dignity. They teach that the fact of man's creation has established human dignity. (Thaxton and Meyer 1987)

It is this teaching upon which their entire argument turns. To emphasize the point, they immediately restate it as their central, major thesis:

Only if man is (in fact) a product of special Divine purposes can his claim to distinctive or intrinsic dignity be sustained. (Thaxton and Meyer 1987)

This religious assumption, in one variation or another, stands at the very center of the ID worldview. It is behind Johnson's notion of "theistic realism." It is behind Dembski's insistence that the human mind transcends any possible material instantiation. It is the reason that ID can brook no compromise with evolution, since they see evolution as incompatible with what they take to be the basic fact of man's special creation. The "ill-conceived accommodation," as Dembski puts it (Dembski 1995, 3), that theistic evolutionists make is, according to ID theorists, nothing less than an intolerable surrender of their foundational assumption.

Thaxton and Meyer close their article with a purported contrast between the way human rights are honored in the United States and in the Soviet Union; they are inalienable here and dispensable there, they claim. This difference, they argue, is a direct result of a difference between a government based upon Christian theology and one grounded in scientific materialism. They write, "Soviet indifference to human rights is reasoned correctly from an erroneous perception of man called Marxism – a materialist perception [sic] that Karl Marx himself held to be scientific" (Thaxton and Meyer 1987). On the other hand, they believe that America is built on the idea that "dignity is built into man by his Creator" (Thaxton and Meyer 1987). They worry, however, that the acceptance of evolution and naturalism will undermine these values here and place us in the same position as the Soviets.

The orthodoxy of Judaism and Christianity contends that man has dignity because he has been created in the image of God. If the orthodox view is false, as is now widely assumed in the academic and legal professions, then one wonders how long it will be until we in the West reason correctly from a strictly scientific perception [sic] of human nature. (Thaxton and Meyer 1987)

There are more problems with Thaxton and Meyer's argument than we have time even to broach here. Even if one were to accept their cartoon analysis of the difference between the United States and the (now former) Soviet Union, there seems to be no good reason to think that a scientific view of human nature (or even metaphysical materialism, which is not the same thing) is incompatible with human rights. Nor does history bear out the implied claim that Judeo-Christian theism necessarily leads to a respect for human rights. More significantly, from a moral point of view, it seems quite wrong to accept their premise that moral rights are limited to human beings in contrast to all other beings. However, rather than pursue these points, I want to mention two other serious problems that are more directly related to our present concerns.

The first is the faulty assumption that being specially created in the image of God, or for some divine purpose, is sufficient to ground moral value. Ironically, their mistaken view is related to what is known as the naturalistic fallacy, though in their particular case it might be better termed the *supernaturalistic fallacy.* Even if one was created for X, it does not follow that one ought to do X. If one is divinely created in the image of an angry and vengeful God, it does not follow that one ought to be angry and vengeful, or that one has moral worth by virtue of being created in such an image. Similarly, one would not have moral worth by virtue of being divinely created in a loving and merciful image, but rather by being loving and merciful. Furthermore, one would be praiseworthy for having such moral virtues, irrespective or whether one is an evolved or a supernaturally created being. Another way to put the problem is that Thaxton and Meyer commit a version of the genetic fallacy. Moral dignity is a function of what virtues one has and how one comports oneself, not of how one came to be. Similarly, moral rights (and concomitant responsibilities) do not depend upon one's origins, but upon one's capacities and relationships. In opposition to this, Thaxton and Meyer's position is akin to the archaic view that the right to govern can only be granted by God – rather than, say, being justified by the will of the governed.

The second problem involves the way the Wedge Document indicts evolution in relation to the moral issue. Suppose we grant for the sake of argument that dignity and rights can be justified only if they are granted by God. Why do ID theorists think that that idea is threatened by evolution? It is because they see "Darwinism" as being on a par with Marxism. They use the term to mean "fully naturalistic evolution," by which they mean a metaphysical position that denies the existence of God. However, as discussed earlier, Darwinian evolution is a scientific view, not a metaphysical one. It is not atheistic, but rather agnostic about the existence of God. Evolutionary biology is naturalistic (or materialistic) in exactly the same way that physics is – or chemistry, or medicine, or plumbing.

If one steps back and asks what the philosophical import of Darwinian evolution is for classical arguments for the existence of God, the only thing one can say is that it shows that there is no need to appeal to divine design to explain biological adaptations. Putting this another way, in canvassing the modal options, it does not tell us that God is impossible, but only that God is not necessary; it leaves God as a possibility in which one may believe on faith. This conclusion about divine design is unacceptable to ID theorists. As we shall see, their entire argument aims to establish the necessity of transcendent design.

Meyer's most systematic treatment of the design inference was published recently in the *Journal of Interdisciplinary Studies*. "The Return of the God Hypothesis" (Meyer 1999) was the lead article in its issue of the journal and received the Oleg Zinam Award for best essay in *JIS* for 1999, so it is recognized as an important articulation of the ID position.[11]

Meyer begins by recounting a story about Napoleon Bonaparte's exchange with Pierre-Simon Laplace regarding the latter's *Treatise on Celestial Mechanics*. In reply to Bonaparte's question as to why God did not figure in his account, Laplace reputedly answered that he had had no need of that hypothesis. Meyer emphasizes that Laplace's "fully naturalistic account of celestial origins" departed from a "long-established theistic orientation," such as had been exemplified in Isaac Newton's account, which explained the solar system by appeal to "divine design" (Meyer 1999, 1).

Once again, the phrase "fully naturalistic account" signals the Wedge's metaphysical target. Meyer explains how science since Laplace seemed to support a "materialistic or naturalistic" worldview rather than a theistic one, by showing how matter "could in effect arrange itself without a pre-existent designer or Creator" (Meyer 1999, 2). By the close of the nineteenth century, he states, "both the evidential and philosophical basis of theistic arguments from nature had seemingly evaporated. Neither science nor philosophy had need of the God hypothesis" (Meyer 1999, 2). The Wedge movement hopes to bring death to materialism by reasserting the necessity of the God hypothesis.[12]

Meyer argues that it was a mistake for natural theologians to retreat in the face of science to the idea that design was located in the laws of nature, rather than in such "complex contrivances that could be understood by direct analogy to human creativity" (4), because it led to the relegation of divine design to the status of merely subjective belief. He explains:

One could still believe that a mind super-intended over the workings of nature, but one might just as well assert that nature and its laws existed on their own. Thus, by the end of the nineteenth century, natural theologians could no longer point to any specific artifact of nature that required intelligence as a necessary explanation. As a result, intelligent design became undetectable except through the eyes of faith. (4)

Much of this summary is quite correct, though contemporary theologians would probably argue that it is not a mistake, but rather far more proper from a religious point of view, to think that divine design must be accepted on faith instead of upon so-called "evidences" (a term from creation science that Meyer uses regularly). Natural theology, from this perspective, misunderstands the essence of religion in trying to emulate the natural sciences. The very definition of faith and its religious significance lies in believing without evidence, or even in spite of evidence to the contrary.

Another, indirect advantage of declining to conceptualize God as a scientific hypothesis is that it avoids pitting religion and science against one another. Meyer acknowledges this, noting that the standard twentieth-century theological position has been to deny a conflict between science and religion, most often by taking them as having complementary, nonoverlapping teachings. In keeping with the ID program, however, Meyer rejects any such accommodation. He aims to revive the earlier view that science and theistic belief are "mutually reinforcing" (3). Nor does he stop with a generic theism. The goal, as he puts it, is to show that "the testimony of nature (or science) actually supports important tenets of theism or the Judeo-Christian religion" (2).

So how does ID theory propose to do this? By reasserting the design argument, but this time with reference to a new set of contrivances that they claim to be inexplicable in principle by any naturalistic metaphysics, but that are purportedly easily explained by biblical theism.

Meyer first pays homage to the classic design argument from William Paley's *Natural Theology*, noting how he catalogued systems "that suggested the work of a superintending intelligence" by virtue of their "astonishing complexity and superb adaptation of means to ends," which purportedly "could not originate strictly through the blind forces of nature" (3–4). But rather than repeat Paley's examples, most of which sound ridiculous today, Meyer cites more recent puzzles from cosmology, physics, chemistry, and biology. He touts the "staggering" implications of the Big Bang, which

provide[s] a scientific description of what Christian theologians have long described in doctrinal terms as *creatio ex nihilo* – creation out of nothing (again, nothing physical). (8)

Meyer argues that ID theory supports "a Judeo-Christian understanding of Creation" (26) over all other metaphysical views. He argues that the Big Bang singularity is sufficient to establish Christian theism over naturalism, because naturalism purportedly cannot account for the origin of the four-dimensional universe itself. For an entity to explain that, Meyer says, it must transcend those four dimensions. He concludes: "In so far as God, as conceived by Judeo-Christian theists, possesses precisely such transcendent causal powers, theism provides a better explanation than naturalism for the singularity affirmed by Big Bang cosmology" (25). He makes the same argument against pantheism.

However, there is nothing precise about an appeal to God here; one could as easily appeal to any supernatural power, divine or otherwise, in the same vague manner – so to call this "evidence" is wishful thinking at best. Moreover, if Meyer really thinks that the Big Bang singularity provides *confirming* evidence of the Christian God, he would also have to agree that God would be *disconfirmed* should the Big Bang model by supplanted by, for example, the cyclic universe model, as is currently proposed by Paul Steinhardt and Neil Turok, in which time and space exist forever. In fact, if the two sorts of naturalism are kept straight, the possibility of God or of any alternative supernatural stand-in remains untouched by either model.

Leaving this aside, Meyer does say that the Big Bang does not by itself provide evidence for "the other attributes of God," such as intelligence and rationality, but he believes that ID theory can provide epistemic support for these, and more:

[T]he Big Bang theory provides for aspects of theistic belief, namely, theism's affirmation of a finite universe and a specifically *transcendent* Creator. Other types of scientific evidence may provide support for other attributes of a theistic God, or even other aspects of Biblical teaching. (26)

In order to support more specifically biblical teachings, Meyer turns to his arguments for special creation. Meyer claims that the purported "fine-tuning" for life of the fundamental physical parameters of the universe "strongly suggests design by a pre-existent intelligence" (9). Moreover, he suggests that the intelligent designer must be a personal agent, because "a completely impersonal intelligence is almost a contradiction in terms" (26). Once again, he concludes that this supports the Christian notion of God over any naturalistic or pantheistic view.

Meyer argues that theism also wins over a deist view,[13] because only theism, as an interventionist view, "can explain the origin of biological information as the result of God's creative activity (within a natural order that He otherwise sustains) at some point after His initial Creation" (Meyer 1999, 27). Meyer claims that ID theory can empirically demonstrate this necessity, citing another of his articles in which he argues against a form of theistic evolution (Meyer 1999a). (One would like to know how such claims square with statements that ID theorists make in other forums to the effect that as far as they can say, information may all be "front-loaded"; but we shall leave such inconsistencies aside.) It is in biology that ID theorists make what they believe is their strongest argument, citing the complexities of subcellular machines as the best evidence of intelligent design.

None of these specific arguments is original with Meyer, however, and whenever he writes or speaks of them he relies upon his fellow Wedge members Michael Behe and William Dembski. In "The Return of the God Hypothesis," he begins with Behe's notion of "irreducible complexity" and cites the bacterial flagellum, which has become their centerpiece example. He rehearses Behe's original argument that the flagellum could not in principle have arisen gradually by a Darwinian mechanism, because it relies for its functioning upon "the coordinated interaction of some forty complex protein parts," the absence of any one of which "would result in the complete loss of motor function" (Meyer 1999, 14–15).

I have previously shown (Pennock 1999) why Behe's notion of irreducible complexity fails as an in-principle argument against the Darwinian mechanism.[14] In an article in 2001, Behe conceded that a counterexample I gave did undermine the notion of irreducible complexity as he had defined it (Behe 2001). He pledged that a revised definition would repair the problem, but did not provide one at the time, nor has he in the years since. Moreover, I had showed conceptually how a gradual stepwise process using a simple natural scaffolding could produce an IC system (Pennock 2000). Since then, colleagues and I have experimentally demonstrated the evolution of an IC system (Lenski et al. 2003). Miller, Orr, Doolittle, Kitcher, Shanks and Joplin, and many others have published other criticisms of Behe's concept as well as of the specific examples he gave. At the time Meyer wrote his article, Behe's notion was still relatively fresh (not counting, of course, Paley, Ray, and others who made the same argument centuries earlier), but by now his variation of the idea has been thoroughly discredited.

The second argument from biology that Meyer cites involves William Dembski's notion of "specified complexity" and his "design inference." The claim is that proteins are not just complex but also "specified," and that such "complex specified information" (CSI) cannot arise naturally but only by an intelligent cause. Meyer puts the argument this way:

Since we know intelligent agents can (and do) produce functionally specified sequences of symbols or arrangements of matter (information content), intelligent agency qualifies as a sufficient causal explanation for the origin of this effect. And since ... naturalistic scenarios have proven universally inadequate for explaining the origin of information content, mind or creative intelligence now stands as the best and only entity with causal power to produce this feature of living systems. (Meyer 1999, 19)

He repeats the claim that intelligent agency is "empirically *necessary*" and "the only" known cause of information content three more times on the same page, but repetition does not improve the argument.

The appeal to our "uniform experience of intelligent design" as the cause of information simply begs the question. The goal of the ID argument is to undermine "naturalistic scenarios," by which ID theorists mean the adequacy of material causes. However, our uniform experience is of design by natural agents – almost invariably, human beings. Human beings, as far as all experience has shown, are made of ordinary natural materials, which is good evidence that natural processes *can* produce CSI. Thus, for ID theorists to cite human design as the basis for their inference to the necessity of supernatural design is to assume what they are trying to prove.[15]

In fact, although ID theorists do regularly claim simply to be making an inductive inference on uniformitarian grounds from our experience of the intelligent actions of other people, in other places where they spell it out in more detail, the argument turns out to be quite different. Dembski's technical argument is set up as an argument by elimination; if one can filter out "necessity" and "chance" as possible explanations of some phenomenon, then "design" wins by default as the sole remaining option. But intentional design, in the ordinary sense of the term that is relevant here, is orthogonal to the other two concepts, so the argument fails. Dembski has done no more than formalize the God of the gaps argument (Pennock 2004). However, there is no need here to go further into this or the many other problems with Dembski's argument: Dembski states that Behe's notion of irreducible complexity is a special case of complex specified information; so given that Behe's argument fails, Dembski's does also as a corollary.

Neither Dembski's specified complexity nor Behe's irreducible complexity – the Wedge movement's best shots – can support ID's astounding claim to have scientifically demonstrated the necessity of the God hypothesis.

However, the problems with ID are even more fundamental than the specific flaws in these arguments, as may be seen in Meyer's attempt to provide a generic justification for their approach.

Meyer argues that the natural sciences give rationally compelling support for the existence of God when understood in terms of ID theory's reformulation of the classic design argument. These "evidences" are "not a formal deductive proof of God's existence" (Meyer 1999, 13), he says, nor do they "depend upon analogical reasoning" (Meyer 1999, 19). Rather, he claims, they function as part of a scientific proof – an *abductive inference* or an *inference to the best explanation*.[16] Meyer summarizes the form of the argument as follows:

> DATA: The surprising fact A is observed.
> LOGIC: *But if B were true, then A would be a matter of course.*
> CONCLUSION: Hence, there is reason to suspect that B is true. (Meyer 1999, 21)

Let us initially grant for the sake of argument that the ID argument succeeds as an abductive inference of this sort. Does this allow ID theorists to draw the conclusion that they repeatedly trumpet, namely, that transcendent intelligent design is *the only* way to produce biological complexity? Clearly not, even on Meyer's account. The argument would give some reason to think the God hypothesis is true, but *not* that it is *necessary*.

In point of fact, however, Meyer has not succeeded even in fulfilling the form of the abductive inference he claims to be following. Look again at the logic he sets out. "B" would be the design hypothesis, and "A" would be the bacterial flagellum (or fine-tuning, or one of the other "evidences" that ID theorists cite). In what sense is the latter "a matter of course" given the former? Designers in general cannot create whatever they please. In our uniform experience, they all have limits of knowledge and power. Thus, we cannot say that creating a flagellum would even be possible for a designer, let alone "a matter of course." But suppose they avoid this problem by explicitly stating the God hypothesis, under the Judeo-Christian assumption of omniscience and omnipotence. That does not help, for we have no way to know what it would or would not please God to do. Are we to assume that God likes flagella? On what grounds? Is Meyer somehow privy to God's intentions? (It might be that ID advocates believe that scripture does provide revealed knowledge of this sort, but I am here taking them at their word that they do not appeal explicitly to the Bible in their arguments. Even if they were to make such an appeal, however, the same problem would arise in a different fashion because of issues of how to interpret the purportedly divine word – after all, aren't God's ways supposed to be inherently mysterious?) The key point is that the God hypothesis (or the euphemistic mere creation or design hypothesis) does not provide any explanatory expectations whatsoever regarding any of the purported "evidences." Thus, their abductive design inference cannot even get off the ground, let alone rise to the lofty metaphysical heights claimed by the Wedge.

Notice how different the God hypothesis is from the ordinary cases in which we infer that someone designed something – as in anthropology, for example. In such cases, we have a wealth of knowledge about human beings, their causal powers, their previous creations, and their possible intentions. When we evaluate the explanatory virtues of a hypothesis of intentional design in ordinary cases, we therefore have relevant information that can drive the inference. Drawing a conclusion that intelligence is behind a design is more difficult when we turn to other animals, but even in those cases – especially with the other primates – we have much background knowledge to which we can appeal. (It is telling that ID theorists never give examples of intelligent design by other animals – probably because, as we saw earlier, they believe that humans are unique in being created with a transcendent mind.)

What about the possibility of extraterrestrial intelligence? Appeal to the SETI project as a way to justify indirectly the scientific acceptability of the design inference is a staple of the Wedge, though it is by no means original to them.[17] Extraterrestrial intelligence would probably be hard to detect, but even in this case we have a basis for a possible inference, given the presumption that extraterrestrials would be natural beings like us with understandable intentions. However, unless we are willing to naturalize God, we have no grounds for any inference once we open the door to divine design.

The same sort of problem arises for the hypothesis of divine design if one tries to assess its explanatory virtues. For instance, is theism indeed a *simpler* hypothesis than Darwinian evolution, as Meyer says? He gives no argument to support this claim, and it seems obviously false on its face. Unlike the Darwinian mechanism, which explains how biological complexity can arise from simple processes, the design hypothesis simply pushes the problem of complexity back a step and exacerbates it. It is certainly not simpler to explain biological complexity by reference to a mysterious agent who is infinitely greater in complexity. The hypothesis fares no better with regard to other explanatory virtues (Pennock 1999, 2003).

In fact, their design inference is an inference to the best explanation in only the most attenuated sense. Meyer's discussion here is one of the few places in which one finds even a mention of specific explanatory virtues. In almost every other case, as noted earlier, when one examines the logic of their design inference, one finds nothing more than an argument from ignorance.

To conclude, let us tie ID's anti-evolutionism back to the moral problem, which Meyer and Thaxton posed in terms of justifying human rights:

[I]f the traditional understanding of man is correct, if it is not only doctrinal but factual, then governments can derive human rights from a dignity that actually exists. But if the traditional view is false and the modern scientific view prevails, then there is no dignity and human rights are a delusion, not only in Moscow but here in the West as well. (Thaxton and Meyer 1987)

As we have seen, this is a false dilemma in many ways; but for those who believe that the options are so stark, it is no wonder that there can be no accommodation to evolution. DNA must be designed; the "traditional" religious view must be factual. As we saw, Wedge members boldly claim that their science shows that it is, and that ID theory confirms not only theism over every other metaphysical view, but also, specifically, the traditional Christian notion of God and Creation.[18] However, as we also saw, the arguments they give indeed are "repetitive, imprecise, immodest, and very unsatisfactory," just as Dembski's interlocutor came to conclude. Methodological naturalism is neutral with regard to the God hypothesis and, in any case, human rights were never in jeopardy. Their reasons for rejecting evolution and the modern scientific view are unsound, as is their pre-modern alternative.

Finally, their reference to "Moscow" is also significant; it is common for the Wedge to link Communism and "Darwinism." However, it behooves us to recall that in the former Soviet Union, Darwinian evolution was rejected on ideological grounds. Because the Communist Party denounced the Darwinian view in favor of Lysenkoism, a variant of Lamarckism that was more in line with Party ideology, biological research was set back for a generation. ID-ology could have the same effect in this country, if it succeeds in its lobbying efforts. The Christian presuppositionalism that grounds the ID Wedge movement is a protected religious belief, but it cannot replace the foundation of modern science and it does not belong in the science classroom.

Notes

1. Dembski continues his regular use of the metaphors of war in his writings. In proposing his internet society <iscid.org> as a means of networking for Intelligent Design theorists, he explains: "Concentration of forces is a key principle of military tactics. Without it, troops, though willing and eager, wallow in indecision and cannot act effectively" (Dembski 2002).
2. After I began writing this article, the Discovery Institute dropped the loaded term "Renewal" from the name, so the CRSE is now the CSE – the Center for Science and Culture. Dembski's Biola address, however – also delivered while this article was being written – continues to cite the vision of a "cultural renewal" that supposedly will come with the defeat of materialism and naturalism as being ID supporters' primary motivation (Dembski 2002).

3. Among other efforts, he has been the spokesperson for the movement at a hearing on Curriculum Controversies in Biology before the U.S. Commission of Civil Rights in August 1998; at a briefing for some members of Congress and their staffs on Scientific Evidence of Intelligent Design and Its Implications for Public Policy and Education in May 2000; and at a hearing before the Ohio State Board of Education in March 2002, trying to get that body to include ID in the state's science curriculum.

4. One can find similar repetition in the writing of the other ID leaders, including Dembski and Behe, who have often recycled paragraphs without citing their original appearance.

5. Strangely, this op-ed piece lists the president of the Discovery Institute, Bruce Chapman, as the primary author.

6. Meyer uses this figure in the two 1996 op-ed pieces and in the 1998 piece, but changes it to forty in the 2002 piece.

7. It is not only we critics who point out that ID fails to qualify as science in this regard. In his Biola speech, Dembski mentioned one sympathetic geneticist who was intrigued with ID but who felt pessimistic about its prospects, writing: "If I knew how to scientifically approach the question you pose, I would quit all that I am doing right now, and devote the rest of my career in pursuit of its answer. The fact that I have no idea how to begin gathering scientific data that would engage the scientific community is the very reason that I don't share your optimism that this approach will work" (Dembski 2002). Dembski told his audience that he himself remained optimistic that ID had research potential, but tellingly, he admitted that he had no specific research proposals to offer, just some possible "research themes."

8. Though the terms are slightly different, Meyer's argument here is the same as that made by creation scientists in Arkansas, who also contended that the creation hypothesis is scientifically on a par with evolution.

9. Meyer sees a direct connection between naturalism and evolution that he believes undermines the possibility of knowledge. According to him, naturalism "view[s] the human mind as a composite of evolutionary adjustments responding to chemical and biological stimuli" (Meyer 1986), and he claims that on this account "the validity of human reason and natural science is destroyed" (Meyer 1986). For Meyer, as for Dembski, who later made a similar argument (Dembski 1990), the human mind must be metaphysically different in kind from the material world. However, neither Meyer nor Dembski has shown that there is anything to the notion of "knowing truth" that would require a proverbial ghost in the machine.

10. The complete Wedge document is available at <www.stephenjaygould.org/ctrl/ archive/wedge_document.html>. For an analysis of the document see Forrest 2001 and Forrest and Gross 2003.

11. By coincidence, this *JIS* issue also contained an article by two other ID proponents, Karl W. Giberson and Donald A. Yerxa. In "Providence and the Christian Scholar," *Journal of Interdisciplinary Studies* 11(1999): 123–40, they repeat the standard ID criticisms of methodological naturalism (MN), explaining that accepting MN at the level of one's discipline can "result in an incoherence" with the theism of one's faith, and arguing that the tension can be resolved by

rejecting the idea that the physical universe is closed to the possibility of divine action and by readmitting God's Providential action as a real explanatory category.

12. When lobbying for ID in the public schools, Wedge members sometimes deny that ID makes any claims about the identity of the designer. It is ironic that their political strategy leads them to deny God in the public square more often than Peter did. Meyer, as least, is more forthright.

13. That is, that "the information necessary to build life was present in the initial configuration of matter at the Big Bang" (Meyer 1999, 26).

14. Also, in Pennock 2001 I replied to some subsequent points that he made.

15. As I discuss elsewhere (Pennock 2003), Dembski actually seems to believe that his design inference proves that the human mind necessarily transcends natural processes. But he cannot offer this both as a conclusion and as a premise in his design inference.

16. Dembski makes the same claim in his own writings on the design inference.

17. The SETI analogy and the elements of Dembski's argument, for instance, go back at least as far as Young Earth creationist Norman Geisler (Pennock 1999, 251).

18. Strictly speaking, it would be better to say that this is *their* preferred notion of Christianity, since many Christian theologians would reject the position they advocate. In any case, my argument in this chapter is not against the existence of God but against ID's claim that they had confirmed God as a scientific hypothesis.

References

Behe, Michael J. 2001. Reply to my critics: A response to reviews of *Darwin's Black Box: The Biochemical Challenge to Evolution. Biology & Philosophy* 16: 685–709.

Chapman, Bruce, and Stephen C. Meyer. 2002. Darwin would love this debate. *Seattle Times,* June 10.

Dembski, William A. 1990. Converting matter into mind: Alchemy and the philosopher's stone in cognitive science. *Perspectives on Science and Christian Faith* 42(4): 202–26.

    1995. What every theologian should know about creation, evolution, and design. *Center for Interdisciplinary Studies Transactions* 3(2): 1–8.

    2002. Becoming a disciplined science: Prospects, pitfalls, and a reality check for ID. Paper read at the Research and Progress in Intelligent Design conference at Biola University, La Mirada, California. <www.arn.org/docs/dembski/wd_disciplinedscience.htm>

Forrest, Barbara. 2001. The wedge at work: How intelligent design creationism is wedging its way into the cultural and academic mainstream. In *Intelligent Design Creationism and Its Critics: Philosophical, Theological and Scientific Perspectives*, ed. R. T. Pennock. Cambridge, MA: MIT Press.

Forrest, Barbara, and Paul Gross. 2003. *Creationism's Trojan Horse: The Wedge of Intelligent Design.* New York: Oxford University Press.

Hartwig, Mark D., and Stephen C. Meyer. [1989] 1993. A note to teachers. In *Of Pandas and People*, ed. P. Davis and D. H. Kenyon. Dallas: Haughton.

Lenski, Richard, Charles Ofria, Robert T. Pennock, and Christoph Adami. 2003. The evolutionary origin of complex features. *Nature* 423 (May 8): 139–44.

Meyer, Stephen C. 1986. Scientific tenets of faith. *Perspectives on Science and Christian Faith* 38(1): 40–2.

1996a. "Don't ask, don't tell" in biology instruction. *The Washington Times*, July 4, 1996.

1996b. Limits of natural selection a reason to teach all theories. *The Tacoma New Tribune*, May 12, 1996.

1998. Let schools provide full disclosure. *The Spokesman-Review*, March 29, 1998.

1999. The return of the God hypothesis. *Journal of Interdisciplinary Studies* 11(1/2): 1–38.

1999a. Teleological evolution: The difference it doesn't make. In *Darwinism Defeated?*, ed. R. Clements. Vancouver, BC: Regents.

2001. Darwin in the dock: A history of Johnson's wedge. *Touchstone: A Journal of Mere Christianity* 14(3): 57.

Miller, Kenneth R. 1999. *Finding Darwin's God: A Scientist's Search for Common Ground between God and Evolution.* New York: Cliff Street Books.

Milner, Richard, Eugenie Scott, William A. Dembski, Robert T. Pennock, Kenneth R. Miller, and Michael Behe. 2002. American Museum of Natural History debate transcript. Available at <www.ncseweb.org/article.asp?category=15>.

Pennock, Robert T. 1996. Naturalism, evidence and creationism: The case of Phillip Johnson. *Biology and Philosophy* 11(4): 543–59.

1999. *Tower of Babel: The Evidence against the New Creationism.* Cambridge, MA: MIT Press.

2000. Lions and tigers and APES, Oh my! Creationism vs. evolution in Kansas. *AAAS Dialogue on Science, Ethics and Religion* <http://www.aaas.org/spp/dser/evolution/perspectives/pennock.htm>.

2001. Whose God? What science?: Reply to Behe. *Reports of the National Center for Science Education* 21(3–4): 16–19.

2004. God of the gaps: The argument from ignorance and the limits of methodological naturalism. In *Scientists Confront Creationism*, revised edition, ed. A. J. Petto and L. R. Godfrey. New York: Norton.

Thaxton, Charles B., and Stephen C. Meyer. 1987. Human rights: Blessed by God or begrudged by government? *Los Angeles Times* December 27.

PART II

COMPLEX SELF-ORGANIZATION

8

Prolegomenon to a General Biology

Stuart Kauffman

Lecturing in Dublin, one of the twentieth century's most famous physicists set the stage of contemporary biology during the war-heavy year of 1944. Given Erwin Schrödinger's towering reputation as the discoverer of the Schrödinger equation, the fundamental formulation of quantum mechanics, his public lectures and subsequent book were bound to draw high attention. But no one, not even Schrödinger himself, was likely to have foreseen the consequences. Schrödinger's *What Is Life?* is credited with inspiring a generation of physicists and biologists to seek the fundamental character of living systems. Schrödinger brought quantum mechanics, chemistry, and the still poorly formulated concept of "information" into biology. He is the progenitor of our understanding of DNA and the genetic code. Yet as brilliant as was Schrödinger's insight, I believe he missed the center. *Investigations* seeks that center and finds, in fact, a mystery.[1]

In my previous two books, I laid out some of the growing reasons to think that evolution was even richer than Darwin supposed. Modern evolutionary theory, based on Darwin's concept of descent with heritable variations that are sifted by natural selection to retain the adaptive changes, has come to view selection as the sole source of order in biological organisms. But the snowflake's delicate sixfold symmetry tells us that order can arise without the benefit of natural selection. *Origins of Order* and *At Home in the Universe* give good grounds to think that much of the order in organisms, from the origin of life itself to the stunning order in the development of a newborn child from a fertilized egg, does not reflect selection alone. Instead, much of the order in organisms, I believe, is self-organized and spontaneous. Self-organization mingles with natural selection in barely understood ways to yield the magnificence of our teeming biosphere. We must, therefore, expand evolutionary theory.

Yet we need something far more important than a broadened evolutionary theory. Despite any valid insights in my own two books, and despite the fine work of many others, including the brilliance manifest in the past three decades of molecular biology, the core of life itself remains shrouded from view. We know chunks of molecular machinery, metabolic pathways, means of membrane biosynthesis – we know many of the parts and many of the processes. But what makes a cell alive is still not clear to us. The center is still mysterious.

And so I began my notebook "Investigations" in December of 1994, a full half century after Schrödinger's *What Is Life?*, as an intellectual enterprise unlike any I had undertaken before. Rather bravely and thinking with some presumptuousness of Wittgenstein's famous *Philosophical Investigations*, which had shattered the philosophical tradition of logical atomism in which he had richly participated, I betook myself to my office at home in Santa Fe and grandly intoned through my fingers onto the computer's disc, "Investigations," on December 4, 1994. I sensed my long search would uncover issues that were then only dimly visible to me. I hoped the unfolding, ongoing notebook would allow me to find the themes and link them into something that was vast and new but at the time inarticulate.

Two years later, in September of 1996, I published a modestly well-organized version of *Investigations* as a Santa Fe Institute preprint, launched it onto the web, and put it aside for the time being. I found I had indeed been led into arenas that I had in no way expected, led by a swirl of ever new questions. I put the notebooks aside, but a year later I returned to the swirl, taking up again a struggle to see something that, I think, is right in front of us – always the hardest thing to see. *Investigations* is the fruit of these efforts. I would ask the reader to be patient with unfamiliar terms and concepts.

My first efforts had begun with twin questions. First, in addition to the known laws of thermodynamics, could there possibly be a fourth law of thermodynamics for open thermodynamic systems, some law that governs biospheres anywhere in the cosmos or the cosmos itself? Second, living entities – bacteria, plants and animals – manipulate the world on their own behalf: the bacterium swimming upstream in a glucose gradient that is easily said to be going to get "dinner"; the paramecium, cilia beating like a Roman warship's oars, hot after the bacterium; we humans earning our livings. Call the bacterium, paramecium, and us humans "autonomous agents," able to act on our own behalf in an environment.

My second and core question became, What must a physical system be to be an autonomous agent? Make no mistake, we autonomous agents mutually construct our biosphere, even as we coevolve in it. Why and how this is so is a central subject of all that follows.

From the outset, there were, and remain, reasons for deep skepticism about the enterprise of *Investigations*. First, there are very strong arguments to say that there can be no general law for open thermodynamic systems. The core argument is simple to state. Any computer program is an algorithm that, given data, produces some sequence of output, finite or infinite. Computer programs can always be written in the form of a binary symbol string of

1 and 0 symbols. All possible binary symbol strings are possible computer programs. Hence, there is a countable, or denumerable, infinity of computer programs. A theorem states that for most computer programs, there is no compact description of the printout of the program. Rather, we must just unleash the program and watch it print what it prints. In short, there is no shorter description of the output of the program than that which can be obtained by running the program itself. If by the concept of a "law" we mean a compact description, ahead of time, of what the computer program will print then for any such program, there can be no law that allows us to predict what the program will actually do ahead of the actual running of the program.

The next step is simple. Any such program can be realized on a universal Turing machine such as the familiar computer. But that computer is an open nonequilibrium thermodynamic system, its openness visibly realized by the plug and power line that connects the computer to the electric power grid. Therefore, and I think this conclusion is cogent, there can be no general law for all possible nonequilibrium thermodynamic systems.

So why was I conjuring the possibility of a general law for open thermodynamic systems? Clearly, no such general law can hold for all open thermodynamic systems.

But hold a moment. It is we humans who conceived and built the intricate assembly of chips and logic gates that constitute a computer, typically we humans who program it, and we humans who contrived the entire power grid that supplies the electric power to run the computer itself. This assemblage of late-twentieth-century technology did not assemble itself. We built it.

On the other hand, no one designed and built the biosphere. The biosphere got itself constructed by the emergence and persistent coevolution of autonomous agents. If there cannot be general laws for all open thermodynamic systems, might there be general laws for thermodynamically open but self-constructing systems such as biospheres? I believe that the answer is yes. Indeed, among those candidate laws is a candidate fourth law of thermodynamics for such self-constructing systems.

To roughly state the candidate law, I suspect that biospheres maximize the average secular construction of the diversity of autonomous agents and the ways those agents can make a living to propagate further. In other words, on average, biospheres persistently increase the diversity of what can happen next. In effect, as we shall see later, biospheres may maximize the average sustained growth of their own "dimensionality."

Thus, the enterprise of *Investigations* soon began to center on the character of the autonomous agents whose coevolution constructs a biosphere. I was gradually led to a labyrinth of issues concerning the core features of autonomous agents able to manipulate the world on their own behalf. It may be that those core features capture a proper definition of life and that definition differs from the one Schrödinger found.

To state my hypothesis abruptly and without preamble, I think an autonomous agent is a self-reproducing system able to perform at least one thermodynamic work cycle. It will require most of *Investigations* to unfold the implications of this tentative definition.

Following an effort to understand what an autonomous agent might be – which, as just noted, involves the concept of work cycles – I was led to the concepts of work itself, constraints, and work as the constrained release of energy. In turn, this led to the fact that work itself is often used to construct constraints on the release of energy that then constitutes further work. So we confront a virtuous cycle: Work constructs constraints, yet constraints on the release of energy are required for work to be done. Here is the heart of a new concept of "organization" that is not covered by our concepts of matter alone, energy alone, entropy alone, or information alone. In turn, this led me to wonder about the relation between the emergence of constraints in the universe and in a biosphere, and the diversification of patterns of the constrained release of energy that alone constitute work and the use of that work to build still further constraints on the release of energy. How do biospheres construct themselves or how does the universe construct itself?

The considerations above led to the role of Maxwell's demon, one of the major places in physics where matter, energy, work, and information come together. The central point of the demon is that by making measurements on a system, the information gained can be used to extract work. I made a new distinction between measurements the demon might make that reveal features of nonequilibrium systems that cannot be used to extract work, and measurements he might make of the nonequilibrium system that cannot be used to extract work. How does the demon know what features to measure? And, in turn, how does work actually come to be extracted by devices that measure and detect displacements from equilibrium from which work can, in principle, be obtained? An example of such a device is a windmill pivoting to face the wind, then extracting work by the wind turning its vanes. Other examples are the rhodopsin molecule of a bacterium responding to a photon of light or a chloroplast using the constrained release of the energy of light to construct high-energy sugar molecules. How do such devices come into existence in the unfolding universe and in our biosphere? How does the vast web of constraint construction and constrained energy release used to construct yet more constraints happen into existence in the biosphere? In the universe itself? The answers appear not to be present in contemporary physics, chemistry, or biology. But a coevolving biosphere accomplishes just this coconstruction of propagating organization.

Thus, in due course, I struggled with the concept of organization itself, concluding that our concepts of entropy and its negative, Shannon's information theory (which was developed initially to quantify telephonic traffic and had been greatly extended since then) entirely miss the central issues. What is happening in a biosphere is that autonomous agents are coconstructing and propagating organizations of work, of constraint construction, and of task completion that continue to propagate and proliferate diversifying organization.

This statement is just plain true. Look out your window, burrow down a foot or so, and try to establish what all the microscopic life is busy doing and building and has done for billions of years, let alone the macroscopic ecosystem of plants, herbivores, and carnivores that is slipping, sliding, hiding, hunting, bursting with flowers and leaves outside your window. So, I think, we lack a concept of propagating organization.

Then too there is the mystery of the emergence of novel functionalities in evolution where none existed before: hearing, sight, flight, language. Whence this novelty? I was led to doubt that we could prestate the novelty. I came to doubt that we could finitely prestate all possible adaptations that might arise in a biosphere. In turn, I was led to doubt that we can prestate the "configuration space" of a biosphere.

But how strange a conclusion. In statistical mechanics, with its famous liter box of gas as an isolated thermodynamic system, we can prestate the configuration space of all possible positions and momenta of the gas particles in the box. Then Ludwig Boltzmann and Willard Gibbs taught us how to calculate macroscopic properties such as pressure and temperature as equilibrium averages over the configuration space. State the laws and the initial and boundary conditions, then calculate; Newton taught us how to do science this way. What if we cannot prestate the configuration space of a biosphere and calculate with Newton's "method of fluxions," the calculus, from initial and boundary conditions and laws? Whether we can calculate or not does not slow down the persistent evolution of novelty in the biosphere. But a biosphere is just another physical system. So what in the world is going on? Literally, what in the world is going on?

We have much to investigate. At the end, I think we will know more than at the outset. But *Investigations* is at best a mere beginning.

It is well to return to Schrödinger's brilliant insights and his attempt at a central definition of life as a well-grounded starting place. Schrödinger's *What Is Life?* provided a surprising answer to his enquiry about the central character of life by posing a core question: What is the source of the astonishing order in organisms? The standard – and Schrödinger argued, incorrect – answer, lay in statistical physics. If an ink drop is placed in still water in a petri dish, it will diffuse to a uniform equilibrium distribution. That uniform distribution is an average over an enormous number of atoms or molecules and is not due to the behavior of individual molecules. Any local fluctuations in ink concentration soon dissipate back to equilibrium.

Could statistical averaging be the source of order in organisms? Schrödinger based his argument on the emerging field of experimental genetics and the recent data on X-ray induction of heritable genetic mutations. Calculating the "target size" of such mutations, Schrödinger realized that a gene could comprise at most a few hundred or thousand atoms.

The sizes of statistical fluctuations familiar from statistical physics scale as the square root of the number of particles, N. Consider tossing a fair coin 10,000 times. The result will be about 50 percent heads, 50 percent tails, with a fluctuation of about 100, which is the square root of 10,000. Thus, a typical fluctuation from 50:50 heads and tails is 100/10,000 or 1 percent. Let the number of coin flips be 100 million, then the fluctuations are its square root, or 10,000. Dividing, 10,000/100,000,000 yields a typical deviation of .01 percent from 50:50.

Schrödinger reached the correct conclusion: If genes are constituted by as few as several hundred atoms, the familiar statistical fluctuations predicted by statistical mechanics would be so large that heritability would be essentially impossible. Spontaneous mutations would happen at a frequency vastly larger than observed. The source of order must lie elsewhere.

Quantum mechanics, argued Schrödinger, comes to the rescue of life. Quantum mechanics ensures that solids have rigidly ordered molecular structures. A crystal is the simplest case. But crystals are structurally dull. The atoms are arranged in a regular lattice in three dimensions. If you know the positions of all the atoms in a minimal-unit crystal, you know where all the other atoms are in the entire crystal. This overstates the case, for there can be complex defects, but the point is clear. Crystals have very regular structures, so the different parts of the crystal, in some sense, all "say" the same thing. As shown below, Schrödinger translated the idea of "saying" into the idea of "encoding." With that leap, a regular crystal cannot encode much "information." All the information is contained in the unit cell.

If solids have the order required but periodic solids such as crystals are too regular, then Schrödinger puts his bet on aperiodic solids. The stuff of the gene, he bets, is some form of aperiodic crystal. The form of the aperiodicity will contain some kind of microscopic code that somehow controls the development of the organism. The quantum character of the aperiodic solid will mean that small discrete changes, or mutations, will occur. Natural selection, operating on these small discrete changes, will select out favorable mutations, as Darwin hoped.

Fifty years later, I find Schrödinger's argument fascinating and brilliant. At once he envisioned what became, by 1953, the elucidation of the structure of DNA's aperiodic double helix by James Watson and Francis Crick, with the famously understated comment in their original paper that its structure suggests its mode of replication and its mode of encoding genetic information.

Fifty years later we know very much more. We know the human genome harbors some 80,000 to 100,000 "structural genes," each encoding the RNA that, after being transcribed from the DNA, is translated according to the genetic code to a linear sequence of amino acids, thereby constituting a protein. From Schrödinger to the establishment of the code required only about twenty years.

Beyond the brilliance of the core of molecular genetics, we understand much concerning developmental biology. Humans have about 260 different cell types: liver, nerve, muscle. Each is a different pattern of expression of the 80,000 or 100,000 genes. Since the work of François Jacob and Jacques Monod thirty-five years ago, biologists have understood that the protein transcribed from one gene might turn other genes on or off. Some vast network of regulatory interactions among genes and their products provides the mechanism that marshals the genome into the dance of development.

We have come close to Schrödinger's dream. But have we come close to answering his question, What is life? The answer almost surely is no. I am unable to say, all at once, why I believe this, but I can begin to hint at an explanation. *Investigations* is a search for an answer. I am not entirely convinced of what lies within this book; the material is too new and far too surprising to warrant conviction. Yet the pathways I have stumbled along, glimpsing what may be a terra nova, do seem to me to be worth serious presentation and serious consideration.

Quite to my astonishment, the story that will unfold here suggests a novel answer to the question, What is life? I had not expected even the outlines of an answer, and I am astonished because I have been led in such unexpected directions. One direction suggests that an answer to this question may demand a fundamental alteration in how we have done science since Newton. Life is doing something far richer than we may have dreamed, literally something incalculable. What is the place of law if, as hinted above, the variables and configuration space cannot be prespecified for a biosphere, or perhaps a universe? Yet, I think there are laws. And if these musings be true, we must rethink science itself.

Perhaps I can point again at the outset to the central question of an autonomous agent. Consider a bacterium swimming upstream in a glucose gradient, its flagellar motor rotating. If we naively ask, "What is it doing?" we unhesitatingly answer something like, "It's going to get dinner." That is, without attributing consciousness or conscious purpose, we view the bacterium as acting on its own behalf in an environment. The bacterium is swimming upstream in order to obtain the glucose it needs. Presumably we have in mind something like the Darwinian criteria to unpack the phrase, "on its own behalf." Bacteria that do obtain glucose or its equivalent may survive with higher probability than those incapable of the flagellar motor trick, hence, be selected by natural selection.

An autonomous agent is a physical system, such as a bacterium, that can act on its own behalf in an environment. All free-living cells and organisms are clearly autonomous agents. The quite familiar, utterly astonishing feature of autonomous agents – *E. coli*, paramecia, yeast cells, algae, sponges, flat worms, annelids, all of us – is that we do, every day, manipulate the universe around us. We swim, scramble, twist, build, hide, snuffle, pounce.

Yet the bacterium, the yeast cell, and we all are just physical systems. Physicists, biologists, and philosophers no longer look for a mysterious élan vital, some ethereal vital force that animates matter. Which leads immediately to the central, and confusing, question: What must a physical system be such that it can act on its own behalf in an environment? What must a physical system be such that it constitutes an autonomous agent? I will leap ahead to state now my tentative answer: A molecular autonomous agent is a self-reproducing molecular system able to carry out one or more thermodynamic work cycles.

All free-living cells are, by this definition, autonomous agents. To take a simple example, our bacterium with its flagellar motor rotating and swimming upstream for dinner is, in point of plain fact, a self-reproducing molecular system that is carrying out one or more thermodynamic work cycles. So is the paramecium chasing the bacterium, hoping for its own dinner. So is the dinoflagellate hunting the paramecium sneaking up on the bacterium. So are the flower and flatworm. So are you and I.

It will take a while to fully explore this definition. Unpacking its implications reveals much that I did not remotely anticipate. An early insight is that an autonomous agent must be displaced from thermodynamic equilibrium. Work cycles cannot occur at equilibrium. Thus, the concept of an agent is, inherently, a non-equilibrium concept. So too at the outset it is clear that this new concept of an autonomous agent is not contained in Schrödinger's answer. Schrödinger's brilliant leap to aperiodic solids encoding the organism that unleashed mid-twentieth-century biology appears to be but a glimmer of a far larger story.

FOOTPRINTS OF DESTINY: THE BIRTH OF ASTROBIOLOGY

The telltale beginnings of that larger story are beginning to be formulated. The U.S. National Aeronautics and Space Agency has had a long program in "exobiology," the search for life elsewhere in the universe. Among its well-known interests are SETI, a search for extraterrestrial life, and the Mars probes. Over the past three decades, a sustained effort has included a wealth of experiments aiming at discovering the abiotic origins of the organic molecules that are the building blocks of known living systems.

In the summer of 1997, NASA was busy attempting to formulate what it came to call "astrobiology," an attempt to understand the origin, evolution, and characteristics of life anywhere in the universe. Astrobiology does not yet exist – it is a field in the birthing process. Whatever the area comes to be called as it matures, it seems likely to be a field of spectacular success and deep importance in the coming century. A hint of the potential impact of astrobiology came in August 1997 with the tentative but excited reports of a Martian meteorite found in Antarctica that, NASA scientists announced, might have evidence of early Martian microbial life. The White House organized the single-day "Space Conference," to which I was pleased to be invited. Perhaps thirty-five scientists and scholars gathered in the Old Executive Office Building for a meeting led by Vice President Gore. The vice president began the meeting with a rather unexpected question to the group: If it should prove true that the Martian rock actually harbored fossilized microbial life, what would be the least interesting result?

The room was silent, for a moment. Then Stephen Jay Gould gave the answer many of us must have been considering: "Martian life turns out to be essentially identical to Earth life, same DNA, RNA, proteins, code." Were it so, then we would all envision life flitting from planet to planet on our solar system. It turns out that a minimum transit time for a fleck of Martian soil kicked into space to make it to earth is about fifteen thousand years. Spores can survive that long under desiccating conditions.

"And what," continued the vice president, "would be the most interesting result?" Ah, said many of us, in different voices around the room: Martian life is radically different from Earth life.

If radically different, then . . .

If radically different, then life must not be improbable.

If radically different, then life may be abundant among the myriad stars and solar systems, on far planets hinted at by our current astronomy.

If radically different and abundant, then we are not alone.

If radically different and abundant, then we inhabit a universe rife with the creativity to create life.

If radically different, then – thought I of my just published second book – we are at home in the universe.

If radically different, then we are on the threshold of a new biology, a "general biology" freed from the confines of our known example of Earth life.

If radically different, then a new science seeking the origins, evolution, characteristics, and laws that may govern biospheres anywhere.

A general biology awaits us. Call it astrobiology if you wish. We confront the vast new task of understanding what properties and laws, if any, may characterize biospheres anywhere in the universe. I find the prospect stunning. I will argue that the concept of an autonomous agent will be central to the enterprise of a general biology.

A personally delightful moment arose during that meeting. The vice president, it appeared, had read *At Home in the Universe*, or parts of it. In *At Home*, and also in this book, I explore a theory I believe has deep merit, one that asserts that, in complex chemical reaction systems, self-reproducing molecular systems form with high probability.

The vice president looked across the table at me and asked, "Dr. Kauffman, don't you have a theory that in complex chemical reaction systems life arises more or less spontaneously?"

"Yes."

"Well, isn't that just sensible?"

I was, of course, rather thrilled, but somewhat embarrassed. "The theory has been tested computationally, but there are no molecular experiments to support it," I answered.

"But isn't it just sensible?" the vice president persisted.

I couldn't help my response, "Mr. Vice President, I have waited a long time for such confirmation. With your permission, sir, I will use it to bludgeon my enemies."

I'm glad to say there was warm laughter around the table. Would that scientific proof were so easily obtained. Much remains to be done to test my theory.

Many of us, including Mr. Gore, while maintaining skepticism about the Mars rock itself, spoke at that meeting about the spiritual impact of the discovery of life elsewhere in the universe. The general consensus was that such a discovery, linked to the sense of membership in a creative universe, would alter how we see ourselves and our place under all, all the suns. I find it a gentle, thrilling, quiet, and transforming vision.

MOLECULAR DIVERSITY

We are surprisingly well poised to begin an investigation of a general biology, for such a study will surely involve the understanding of the collective behaviors of very complex chemical reaction networks. After all, all known life on earth is based on the complex webs of chemical reactions – DNA, RNA, proteins, metabolism, linked cycles of construction and destruction – that form the life cycles of cells. In the past decade we have crossed a threshold that will rival the computer revolution. We have learned to construct enormously diverse "libraries" of different DNA, RNA, proteins, and other organic molecules. Armed with such high-diversity libraries, we are in a position to begin to study the properties of complex chemical reaction networks.

To begin to understand the molecular diversity revolution, consider a crude estimate of the total organic molecular diversity of the biosphere. There are perhaps a hundred million species. Humans have about a hundred thousand structural genes, encoding that many different proteins. If all the genes within a species were identical, and all the genes in different species were at least slightly different, the biosphere would harbor about ten trillion different proteins. Within a few orders of magnitude, ten trillion will serve as an estimate of the organic molecular diversity of the natural biosphere. But the current technology of molecular diversity that generates libraries of more or less random DNA, RNA, or proteins now routinely produces a diversity of a hundred trillion molecular species in a single test tube.

In our hubris, we rival the biosphere.

The field of molecular diversity was born to help solve the problem of drug discovery. The core concept is simple. Consider a human hormone such as estrogen. Estrogen acts by binding to a specific receptor protein; think of the estrogen as a "key" and the receptor as a "lock." Now generate sixty-four million different small proteins, called peptides, say, six amino acids in length. (Since there are twenty types of amino acids, the number of possible hexamers is 20^6, hence, sixty-four million.) The sixty-four million hexamer peptides are candidate second keys, any one of which might be able to fit into the same estrogen receptor lock into which estrogen fits. If so, any such second key may be similar to the first key, estrogen, and hence is a candidate drug to mimic or modulate estrogen.

To find such an estrogen mimic, take many identical copies of the estrogen receptor, affix them to the bottom of a petri plate, and expose them simultaneously to all sixty-four million hexamers. Wash off all the peptides that do not stick to the estrogen receptor, then recover those hexamers that do stick to the estrogen receptor. Any such peptide is a second key that binds the estrogen receptor locks and, hence, is a candidate estrogen mimic.

The procedure works, and works brilliantly. By 1990, George Smith at the University of Missouri used a specific kind of virus, a filamentous phage that infects bacteria. The phage is a strand of RNA that encodes proteins. Among these proteins is the coat protein that packages the head of the phage as part of an infective phage particle. George cloned random DNA sequences encoding random hexamer peptides into one end of the phage coat protein gene. Each phage then carried a different, random DNA sequence in its coat protein gene, hence made a coat protein with a random six amino acid sequence at one end. The initial resulting "phage display" libraries had about twenty million of the sixty-four million different possible hexamer peptides.

Rather than using the estrogen receptor and seeking a peptide estrogen mimic that binds the estrogen receptor, George Smith used a monoclonal antibody molecule as the analogue of the receptor and sought a hexamer peptide that could bind the monoclonal antibody. Monoclonal antibody technology allows the generation of a large number of identical antibody molecules, hence George could use these as identical mock receptors. George found that, among the twenty million different phage, about one in a million would stick to his specific monoclonal antibody molecules. In fact, George found nineteen different hexamers binding to his monoclonal antibody. Moreover, the nineteen different hexamers differed from one another, on average, in three of the six amino acid positions. All had high affinity for his monoclonal antibody target.

These results have been of very deep importance. Phage display is now a central part of drug discovery in many pharmaceutical and biotechnology companies. The discovery of "drug leads" is being transformed from a difficult to a routine task. Not only is work being pursued using peptides but also using RNA and DNA sequences. Molecular diversity has now spread to the generation of high-diversity libraries of small organic molecules, an approach called "combinatorial chemistry." The promise is of high medical importance. As we understand better the genetic diversity of the human population, we can hope to create well-crafted molecules with increased efficacy as drugs, vaccines, enzymes, and novel molecular structures. When the capacity to craft such molecules is married, as it will be in the coming decades, to increased understanding of the genetic and cellular signaling pathways by which ontogeny is controlled, we will enter an era of "postgenomic" medicine. By learning to control gene regulation and cell signaling, we will begin to control cell proliferation, cell differentiation, and tissue regeneration to treat pathologies such as cancer, autoimmune diseases, and degenerative diseases.

But George Smith's experiments are also of immediate interest, and in surprising ways that will bear on our later discussion of autonomous agents.

George's experiments have begun to verify the concept of a "shape space" put forth by George Oster and Alan Perelson of the University of California, Berkeley, and Los Alamos National Laboratory more than a decade earlier. In turn, shape space suggests "catalytic task space." We will need both to understand autonomous agents.

Oster and Perelson had been concerned about accounting for the fact that humans can make about a hundred million different antibody molecules. Why, they wondered. They conceived of an abstract shape space with perhaps seven or eight "dimensions." Three of these dimensions would correspond to the three spatial dimensions, length, height, and width of a molecular binding site. Other dimensions might correspond to physical properties of the binding sites of molecules, such as charge, dipole moment, and hydrophobicity.

A point in shape space would represent a molecular shape. An antibody binds its shape complement, key and lock. But the precision with which an antibody can recognize its shape complement is finite. Some jiggle room is allowed. So an antibody molecule "covers" a kind of "ball" of complementary shapes in shape space. And then comes the sweet argument. If an antibody covers a ball, an actual volume, in shape space, then a finite number of balls will suffice to cover all of shape space. A reasonable analogy is that a finite number of Ping-Pong balls will fill up a bedroom.

But how big of a Ping-Pong ball in shape space is covered by one antibody? Oster and Perelson reasoned that in order for an immune system to protect an organism against disease, its antibody repertoire should cover a reasonable fraction of shape space. Newts, with about ten thousand different antibody molecules, have the minimal known antibody diversity. Perelson and Oster guessed that the newt repertoire must cover a substantial fraction, say about $1/e$ – where e is the natural base for logarithms – or 37 percent of shape space. Dividing 37 percent by 10,000 gives the fractional volume of shape space covered by one antibody molecule. It follows that 100,000,000 such balls, thrown at random into shape space and allowed to overlap one another, will saturate shape space. So, 100 million antibody molecules is all we need to recognize virtually any shape of the size scale of molecular binding sites.

And therefore the concept of shape space carries surprising implications. Not surprisingly, similar molecules can have similar shapes. More surprisingly, very different molecules can have the same shape. Examples include endorphin and morphine. Endorphin is a peptide hormone. When endorphin binds the endorphin brain receptor, a euphoric state is induced. Morphine, a completely different kind of organic molecule, binds the endorphin receptor as well, with well-known consequences. Still more surprising, a finite number of different molecules, about a hundred million, can constitute a universal shape library. Thus, while there are vastly many different proteins, the number of effectively different shapes may only be on the order of a hundred million.

If one molecule binding to a second molecule can be thought of as carrying out a "binding task," then about a hundred million different molecules may constitute a universal toolbox for all molecular binding tasks. So if we can now create libraries with 100 trillion different proteins, a millionfold in excess of the universal library, we are in a position to begin to study molecular binding in earnest.

But there may also be a universal enzymatic toolbox. Enzymes catalyze, or speed up, chemical reactions. Consider a substrate molecule undergoing a reaction to a product molecule. Physical chemists think of the substrate and product molecules as lying in two potential "energy wells," like a ball at the bottom of one of two adjacent bowls. A chemical reaction requires "lifting" the substrate energetically to the top of the barrier between the bowls. Physically, the substrate's bonds are maximally strained and deformed at the top of this potential barrier. The deformed molecule is called the "transition state." According to transition state theory, an enzyme works by binding to and stabilizing the transition state molecule, thereby lowering the potential barrier of the reaction. Since the probability that a molecule acquires enough energy to hop to the top of the potential barrier is exponentially less as the barrier height increases, the stabilization of the transition state by the enzyme can speed up the reaction by many orders of magnitude.

Think of a catalytic task space, in which a point represents a catalytic task, where a catalytic task is the binding of a transition state of a reaction. Just as similar molecules can have similar shapes, so too can similar reactions have similar transition states, hence, such reactions constitute similar catalytic tasks. Just as different molecules can have the same shapes, so too can different reactions have similar transition states, hence constitute the "same" catalytic task. Just as an antibody can bind to and cover a ball of similar shapes, an enzyme can bind to and cover a ball of similar catalytic tasks. Just as a finite number of balls can cover shape space, a finite number of balls can cover catalytic task space.

In short, a universal enzymatic toolbox is possible. Clues that such a toolbox is experimentally feasible come from many recent developments, including the discovery that antibody molecules, evolved to bind molecular features called epitopes, can actually act as catalysts.

Catalytic antibodies are obtained exactly as one might expect, given the concept of a catalytic task space. One would like an antibody molecule that binds the transition state of a reaction. But transition states are ephemeral. Since they last only fractions of a second, one cannot immunize with a transition state itself. Instead, one immunizes with a stable analogue of the transition shape; that is, one immunizes with a second molecule that represents the "same" catalytic task as does the transition state itself. Antibody molecules binding to this transition state analogue are tested. Typically, about one in ten antibody molecules can function as at least a weak catalyst for the corresponding reaction.

These results even allow a crude estimate of the probability that a randomly chosen antibody molecule will catalyze a randomly chosen reaction. About one antibody in a hundred thousand can bind a randomly chosen epitope. About one in ten antibodies that bind the transition state analogue act as catalysts. By this crude calculation, about one in a million antibody molecules can catalyze a given reaction.

This rough calculation is probably too high by several orders of magnitude, even for antibody molecules. Recent experiments begin to address the probability that a randomly chosen peptide or DNA or RNA sequence will catalyze a randomly chosen reaction. The answer for DNA or RNA appears to be about one in a billion to one in a trillion. If we now make libraries of a hundred trillion random DNA, RNA, and protein molecules, we may already have in hand universal enzymatic toolboxes. Virtually any reaction, on the proper molecular scale of reasonable substrates and products, probably has one or more catalysts in such a universal toolbox.

In short, among the radical implications of molecular diversity is that we already possess hundreds of millions of different molecular functions – binding, catalytic, structural, and otherwise.

In our hubris, we rival the biosphere.

In our humility, we can begin to formulate a general biology and begin to investigate the collective behaviors of hugely diverse molecular libraries. Among these collective behaviors must be life itself.

LIFE AS AN EMERGENT COLLECTIVE BEHAVIOR OF COMPLEX CHEMICAL NETWORKS

In the summer of 1996, Philip Anderson, a Nobel laureate in physics, and I accompanied Dr. Henry MacDonald, incoming director of NASA, Ames, to NASA headquarters. Our purpose was to discuss a new linked experimental and theoretical approach to the origin-of-life problem with NASA Administrator Dan Golden and his colleague, Dr. Wesley Huntress. I was excited and delighted.

As long ago as 1971, I had published my own first foray into the origin-of-life problem as a young assistant professor in the Department of Theoretical Biology at the University of Chicago. I had wondered if life must be based on template replicating nucleic acids such as DNA or RNA double helices and found myself doubting that standard assumption. Life, at its core, depends upon autocatalysis, that is, reproduction. Most catalysis in cells is carried out by protein enzymes. Might there be general laws supporting the possibility that systems of catalytic polymers such as proteins might be self-reproducing? Proteins are, as noted, linear sequences of twenty kinds of standard amino acids. Consider, then, a first copy of a protein that has the capacity to catalyze a reaction by which two fragments of a potential second copy of that same protein might be ligated to make the second copy of the whole protein. Such a protein, *A*, say, thirty-two amino acids long, might act on two fragments, say, fifteen amino acids and seventeen amino acids in length, and ligate the two to make a second copy of the thirty-two amino acid sequence.

But if one could imagine a molecule, *A*, catalyzing its own formation from its own fragments, could one not imagine two proteins, *A* and *B*, having the property that *A* catalyzes the formation of *B* by ligating *B*'s fragments into a second copy of *B*, while *B* catalyzes the formation of *A* by catalyzing the ligation of *A*'s fragments into a second copy of *A*? Such a little reaction system would be *collectively autocatalytic*. Neither *A* alone, nor *B* alone, would catalyze its own formation. Rather the *AB* system would jointly catalyze its reproduction from *A* and *B* fragments. But if *A* and *B* might achieve collective autocatalysis, might one envision a system with tens or hundreds of proteins, or peptides, that were collectively autocatalytic?

Might collective autocatalysis of proteins or similar polymers be the basic source of self-reproduction in molecular systems? Or must life be based on template replication, as envisioned by Watson and Crick, or as envisioned even earlier by Schrödinger in his aperiodic solid with its microcode? In view of the potential for a general biology, what, in fact, are the alternative bases for self-reproducing molecular systems here and anywhere in the cosmos? Which of these alternatives is more probable, here and anywhere?

By 1971 I had asked and found a preliminary answer to the following question: In a complex mixture of different proteins, where the proteins might be able to serve as candidates to ligate one another into still larger amino acid sequences, what are the chances that such a system will contain one or more collectively autocatalytic sets of molecules? The best current guess is that, as the molecular diversity of a reaction system increases, a critical threshold is reached at which collectively autocatalytic, self-reproducing chemical reaction networks emerge spontaneously.

If this view is correct, and the kinetic conditions for rapid reactions can be sustained, perhaps by enclosure of such a reproducing system in a bounding membrane vesicle, also synthesized by the system, the emergence of self-reproducing molecular systems may be highly probable. No small conclusion this: Life abundant, emergent, expected. Life spattered across megaparsecs, galaxies, galactic clusters. We as members of a creative, mysteriously unfolding universe. Moreover, the hypothesis is richly testable and is now under the early stages of testing.

One way or another, we will discover a second life – crouched under a Mars rock, frozen in time; limpid in some pool on Titan, in some test tube in Nebraska in the next few decades. We will discover a second life, one way or another.

What monumental transformations await us, proudly postmodern, mingled with peoples on this very globe still wedded to archetypes thousands of years old.

THE STRANGE THING ABOUT THE THEORY OF EVOLUTION

We do not understand evolution. We live it with moss, fruit, fin, and quill fellows. We see it since Darwin. We have insights of forms and their formation, won from efforts since Aristotle codified the embryological investigations that over twenty-five centuries ago began with the study of deformed fetuses in sacrificial animals.

But we do not understand evolution.

"The strange thing about the theory of evolution," said one of the Huxleys (although I cannot find which one), "is that everyone thinks he understands it." How very well stated in that British fashion Americans can admire but not emulate. ("Two peoples separated by a common language," as Churchill dryly put it.)

The strange thing about the theory of evolution is that everyone thinks he understands it. How very true. It seems, of course, so simple. Finches hop around the Galapagos, occasionally migrating from island to island. Small and large beaks serve for different seeds. Beaks fitting seeds feed the young. Well-wrought beaks are selected. Mutations are the feedstock of heritable variation in a population. Populations evolve by mutation, mating, recombination, and selection to give the well-marked varieties that are, for Darwin, new species. Phylogenies bushy in the biosphere. "We're here, we're here," cry all for their typical four-million-year-stay along the four-billion-year pageant.

"We're here!"

But how?

How, in many senses. First, Darwin's theory of evolution is a theory of descent with modification. It does not yet explain the genesis of forms, but the trimmings of the forms, once they are generated. "Rather like achieving an apple tree by trimming off all the branches," said a late-nineteenth-century skeptic.

How, in the fundamental sense: Whence life in the first place? Darwin starts with life already here. Whence life is the stuff of all later questions about whence the forms to sift.

How, in still a different sense. Darwin assumed gradualism. Most variation would be minor. Selection would sift these insensible alterations, a bit more lift, a little less drag, until the wing flew faultless in the high-hoped sky, a falcon's knot-winged, claw-latching dive to dine.

But whence the gradualism itself? It is not God given, but true, that organisms are hardly affected by most mutations. Most mutations do have little effect, some have major effects. In *Drosophila*, many mutants make small modifications in bristle number, color, shape. A few change wings to legs, eyes to antennae, heads to genitalia. Suppose that all mutations were of dramatic effect. Suppose, to take the limiting philosophical case, that all mutations were what geneticists call "lethals." Since, indeed, some mutations are lethals, one can, a priori, imagine creatures in which all mutations were lethal prior to having offspring. Might be fine creatures, too, in the absence of any mutations, these evolutionary descendants of, well, of what? And progenitors of whom? No pathway to or from these luckless ones.

Thus, evolution must somehow be crafting the very capacity of creatures to evolve. Evolution nurtures herself! But not yet in Darwin's theory, nor yet in ours.

Take another case – sex. Yes, it captures our attention, and the attention of most members of most species. Most species are sexual. But why bother? Asexuals, budding quietly wherever they bud, require only a single parent. We plumaged ones require two, a twofold loss in fitness.

Why sex? The typical answer, to which I adhere, is that sexual mating gives the opportunity for genetic recombination. In genetic recombination, the double chromosome complement sets, maternal and paternal homologues, pair up, break and recombine to yield offspring chromosomes the left half of which derives from one parental chromosome, the right half of which derives from the other parental chromosome.

Recombination is said to be a useful "search procedure" in an evolving population. Consider, a geneticist would say, two genes, each with two versions, or alleles: A and a for the first gene, B and b for the second gene. Suppose A confers a selective advantage compared to a, and B confers an advantage with respect to b. In the absence of sex, mating, and recombination, a rabbit with A and b would have to wait for a mutation to convert b to B. That might take a long time. But, with mating and recombination, a rabbit with A on the left end of a maternal chromosome, and B on the right end of the homologous paternal chromosome might experience recombination. A and B would now be on a single chromosome, hence be passed on to the offspring. Recombination, therefore, can be a lot faster than waiting for mutation to assemble the good, AB chromosome.

But it is not so obvious that recombination is a good idea after all. At the molecular level, the recombination procedure is rather like taking an airplane and a motorcycle, breaking both in half, and using spare bolts to attach the back half of the airplane to the front half of the motorcycle. The resulting contraption seems useless for any purpose.

In short, the very usefulness of recombination depends upon the gradualness that Darwin assumed. In later chapters I will discuss the concept of a "fitness landscape." The basic idea is simple. Consider a set of all possible frogs, each with a different genotype. Locate each frog in a high-dimensional "genotype space," each next to all genotypes that differ from it by a single mutation. Imagine that you can measure the fitness of each frog. Graph the fitness of each frog as a height above that position in genotype space. The resulting heights form a fitness landscape over the genotype space, much as the Alps form a mountainous landscape over part of Europe.

In the fitness landscape, image, mutation, recombination, and selection can conspire to pull evolving populations upward toward the peaks of high fitness. But not always. It is relatively easy to show that recombination is only a useful search procedure on smooth fitness landscapes. The smoothness of a fitness landscape can be defined mathematically by a correlation function giving the similarity of fitnesses, or heights, at two points on the landscape separated by a mutational distance. In the Alps, most nearby points are of similar heights, except for cliffs, but points fifty kilometers apart can be of very different heights. Fifty kilometers is beyond the correlation length of the Alps.

There is good evidence that recombination is only a useful search strategy on smooth, highly correlated landscapes, where the high peaks all cluster near one another. Recombination, half airplane – half motorcycle, is a means to look "between" two positions in a high-dimensional space. Then if both points are in the region of high peaks, looking between those two points is likely to uncover further new points of high fitness, or points on the slopes of even higher peaks. Thereafter, further mutation, recombination, and selection can bring the adapting population to successively higher peaks in the high-peaked region of the genotype space. If landscapes are very rugged and the high peaks do not cluster into smallish regions, recombination turns out to be a useless search strategy.

But most organisms are sexual. If organisms are sexual because recombination is a good search strategy, but recombination is only useful as a search strategy on certain classes of fitness landscapes, where did those fitness land-scapes come from? No one knows.

The strange thing about evolution is that everyone thinks he under-stands it.

Somehow, evolution has brought forth the kind of smooth landscapes upon which recombination itself is a successful search strategy.

More generally, two young scientists, then at the Santa Fe Institute, proved a rather unsettling theorem. Bill Macready and David Wolpert called it the "no-free-lunch theorem." They asked an innocent question. Are there some search procedures that are "good" search procedures, no matter what the problem is? To formalize this, Bill and David considered a mathematical convenience – a set of all possible fitness landscapes. To be simple and concrete, consider a large three-dimensional room. Divide the room into very small cubic volumes, perhaps a millimeter on a side. Let the number of these small volumes in the room be large, say a trillion. Now consider all possible ways of assigning integers between one and a trillion, to these small volumes. Any such assignment can be thought of as a fitness landscape, with the integer representing the fitness of that position in the room.

Next, formalize a search procedure as a process that somehow samples M distinct volumes among the trillion in the room. A search procedure spec-ifies how to take the M samples. An example is a random search, choosing the M boxes at random. A second procedure starts at a box and samples its neighbors, climbing uphill via neighboring boxes toward higher integers. Still another procedure picks a box, samples neighbors and picks those with lower integers, then continues.

The no-free-lunch theorem says that, averaged over all possible fitness landscapes, no search procedure outperforms any other search procedure. What? Averaged over all possible fitness landscapes, you would do as well trying to find a large integer by searching randomly from an initial box for your M samples as you would climbing sensibly uphill from your initial box.

The theorem is correct. In the absence of any knowledge, or constraint, on the fitness landscape, on average, any search procedure is as good as any other.

But life uses mutation, recombination, and selection. These search pro-cedures seem to be working quite well. Your typical bat or butterfly has managed to get itself evolved and seems a rather impressive entity. The no-free-lunch theorem brings into high relief the puzzle. If mutation, recombi-nation, and selection only work well on certain kinds of fitness landscapes, yet most organisms are sexual, and hence use recombination, and all or-ganisms use mutation as a search mechanism, where did these well-wrought fitness landscapes come from, such that evolution manages to produce the fancy stuff around us?

Here, I think, is how. Think of an organism's niche as a way of making a living. Call a way of making a living a "natural game." Then, of course, natural games evolve with the organisms making those livings during the past four billion years. What, then, are the "winning games"? Naturally, the winning games are the games the winning organisms play. One can almost see Darwin nod. But what games are those? What games are the games the winners play?

Ways of making a living, natural games, that are well searched out and well mastered by the evolutionary search strategies of organisms, namely, mutation and recombination, will be precisely the niches, or ways of making a living, that a diversifying and speciating population of organisms will manage to master. The ways of making a living presenting fitness landscapes that can be well searched by the procedures that organisms have in hand will be the very ways of making a living that readily come into existence. If there were a way of making a living that could not be well explored and exploited by organisms as they speciate, that way of making living would not become populated. Good jobs, like successful jobholders, prosper.

So organisms, niches, and search procedures jointly and self-consistently co-construct one another! We make the world in which we make a living such that we can, and have, more or less mastered that evolving world as we make it. The same is true, I will argue, for an econosphere. A web of economic activities, firms, tasks, jobs, workers, skills, and learning, self-consistently came into existence in the last forty thousand years of human evolution.

The strange thing about the theory of evolution is that everyone thinks he understands it. But we do not. A biosphere, or an econosphere, self-consistently co-constructs itself according to principles we do not yet fathom.

LAWS FOR A BIOSPHERE

But there must be principles. Think of the Magna Carta, that cultural enterprise founded on a green meadow in England when John I was confronted by his nobles. British common law has evolved by precedent and determinations to a tangled web of more-or-less wisdom. When a judge makes a new determination, sets a new precedent, ripples of new interpretation pass to near and occasionally far reaches of the law. Were it the case that every new precedent altered the interpretation of all old judgments, the common law could not have coevolved into its rich tapestry. Conversely, if new precedents never sent out ripples, the common law could hardly evolve at all.

There must be principles of coevolutionary assembly for biospheres, economic systems, legal systems. Coevolutionary assembly must involve coevolving organizations flexible enough to change but firm enough to resist change. Edmund Burke was basically right. Might there be something deep here? Some hint of a law of coevolutionary assembly?

Perhaps. I begin with the simple example offered by Per Bak and his colleagues some years ago – Bak's "sand pile" and "self-organized criticality." The experiment requires a table and some sand. Drop the sand slowly on the table. The sand gradually piles up, fills the tabletop, piles to the rest angle of sand, then sand avalanches begin to fall to the floor.

Keep adding sand slowly to the sand pile and plot the size distribution of sand avalanches. You will obtain many small avalanches and progressively fewer large avalanches. In fact, you will achieve a characteristic size distribution called a "power law." Power law distributions are easily seen if one plots the logarithm of the number of avalanches at a given size on the *y*-axis, and the logarithm of the size of the avalanche on the *x*-axis. In the sand pile case, a straight line sloping downward to the right is obtained. The slope is the power law relation between the size and number of avalanches.

Bak and his friends called their sand pile "self-organized critical." Here, "critical" means that avalanches occur on all length scales, "self-organized" means that the system tunes itself to this critical state.

Many of us have now explored the application of Bak's ideas in models of coevolution. With caveats that other explanations may account for the data, the general result is that something may occur that is like a theory of coevolutionary assembly that yields a self-organized critical biosphere with a power law distribution of small and large avalanches of extinction and speciation events. The best data now suggest that precisely such a power law distribution of extinction and speciation events has occurred over the past 650 million years of the Phanerozoic. In addition, the same body of theory predicts that most species go extinct soon after their formation, while some live a long time. The predicted species lifetime distribution is a power law. So too are the data.

Similar phenomena may occur in an econosphere. Small and large avalanches of extinction and speciation events occur in our technologies. A colleague, Brian Arthur, is fond of pointing out that when the car came in, the horse, buggy, buggy whip, saddlery, smithy, and Pony Express went out of business. The car paved the way for an oil and gas industry, paved roads, motels, fast-food restaurants, and suburbia. The Austrian economist Joseph Schumpeter wrote about this kind of turbulence in capitalist economies. These Schumpeterian gales of creative destruction appear to occur in small and large avalanches. Perhaps the avalanches arise in power laws. And, like species, most firms die young; some make it to old age – Storre, in Sweden, is over nine hundred years old. The distribution of firm lifetimes is again a power law.

Here are hints – common law, ecosystems, economic systems – that general principles govern the coevolutionary coconstruction of lives and livings, organisms and natural games, firms and economic opportunities. Perhaps such a law governs any biosphere anywhere in the cosmos.

I suggest other candidate laws for any biosphere in the course of *Investigations*. As autonomous agents coconstruct a biosphere, each must manage to categorize and act upon its world in its own behalf. What principles might govern that categorization and action, one might begin to wonder. I suspect that autonomous agents coevolve such that each makes the maximum diversity of reliable discriminations upon which it can act reliably as it swims, scrambles, pokes, twists, and pounces. This simple view leads to a working hypothesis: Communities of agents will coevolve to an "edge of chaos" between overrigid and overfluid behavior. The working hypothesis is richly testable today using, for example, microbial communities.

Moreover, autonomous agents forever push their way into novelty – molecular, morphological, behavioral, organizational. I formalize this push into novelty as the mathematical concept of an "adjacent possible," persistently explored in a universe that can never, in the vastly many lifetimes of the universe, have made all possible protein sequences even once, bacterial species even once, or legal systems even once. Our universe is vastly nonrepeating; or, as the physicists say, the universe is vastly nonergodic. Perhaps there are laws that govern this nonergodic flow. I suggest that a biosphere gates its way into the adjacent possible at just that rate at which its inhabitants can just manage to make a living, just poised so that selection sifts out useless variations slightly faster than those variations arise. We ourselves, in our biosphere, econosphere, and technosphere, gate our rate of discovery. There may be hints here too of a general law for any biosphere, a hoped-for new law for self-constructing systems of autonomous agents. Biospheres, on average, may enter their adjacent possible as rapidly as they can sustain; so too may econospheres. Then the hoped-for fourth law of thermodynamics for such self-constructing systems will be that they tend to maximize their dimensionality, the number of types of events that can happen next.

And astonishingly, we need stories. If, as I suggest, we cannot prestate the configuration space, variables, laws, initial and boundary conditions of a biosphere, if we cannot foretell a biosphere, we can, nevertheless, tell the stories as it unfolds. Biospheres demand their Shakespeares as well as their Newtons. We will have to rethink what science is itself. And C.P. Snow's "two cultures," the humanities and science may find an unexpected, inevitable union.

Note

1. This chapter is a slightly modified version of the first chapter of Stuart Kauffman's *Investigations* (New York: Oxford University Press, 2000) and is reprinted here by kind permission of the author and Oxford University Press. The only modifications are to remove some references to other parts of the original book.

9

Darwinism, Design, and Complex Systems Dynamics

Bruce H. Weber and David J. Depew

1. THE ARGUMENT FROM INCREDULITY

When Saint Augustine was a young man he could not imagine how evil could possibly have come into the world. Later, he came to realize that this line of questioning was dangerous and could bring religion into disrepute (*Confessions* VI, Chapter 5; VII, Chapters 3–5). His was a wise realization. 'How possibly' types of questions may do effective rhetorical work in the early stages of a line of inquiry or argumentation, but the types of answers they engender are open to refutation by the answers to other types of questions, such as 'why actually' or 'why necessarily' questions. 'Why actually' questions – or at least an important class of them that are asked by scientists – have a remarkable way of shutting down 'how possibly' questions.

When looking at the functional complexity of the living world, 'how possibly' questions emerge rather intuitively. As Michael Ruse has pointed out, the traditional argument from design actually has two steps (Ruse 2002). The first move is from observed functional complexity to the notion that such phenomena appear to have been designed, the argument *to* design. During this process, alternative, natural explanations are rejected using the criterion of incredulity: how *possibly* could the vertebrate eye appear as a result of random events, even under natural processes and laws? This was the line of argument taken by William Paley some two centuries ago (Paley 1802). In effect, he dared anyone to come up with a fully natural explanation of biological functional complexity. The remainder of the project of natural theology is to argue *from* design to the existence of a designer, and even to deduce attributes of the designer God from the properties of His designed Creation.

Darwin, who as an undergraduate had read and admired Paley, took Paley's dare and made his research program that of finding a natural explanation of *apparent* design by way of biological adaptation. Darwin's explanatory mechanism was natural selection acting upon the random but heritable variations of reproducing organisms in particular environments. This selective process produced, over generational time, adaptive traits in lineages, as well as diversification – descent with modification. Much of Darwin's "long argument" was to show that natural selection *could* account for the empirical claim of a common descent for all living beings.[1] Many of Darwin's sympathizers, John Stuart Mill among them, thought the argument no stronger than that (Mill 1874, 328). But Darwin was not just responding to the 'how possibly' question posed by Paley; he was also providing a conceptual framework within which 'why actually' questions could be coherently formulated and answered. The Darwinian research tradition developed through a succession of research programs over the course of the twentieth century. Under the aegis of rubrics such as genetical Darwinism or the Modern Evolutionary Synthesis, questions about adaptation, geographical distribution, speciation, and related matters were asked and answered with ever greater empirical and theoretical richness. The Synthesis may not have been a *complete* theory of evolution, but it has resolved enough questions to create a presumption in its favor. Thus, even though contention over the very idea of evolution remained in the general culture, the project of natural theology receded from view.

In recent times, however, there have been attempts to revive natural theology based on new notions derived from physics (the anthropic principle) and from biochemistry (irreducible complexity). It is the latter approach, advanced by Michael Behe and a number of other "intelligent design theorists," that most directly attempts to recover Paley's argument and that in consequence will concern us (Behe 1996, 2001).[2] Behe summarizes the functional complexity at the molecular level of a variety of biological phenomena, such as the clotting of blood, the locomotion of bacteria, the immune system, the biochemistry that underlies vision, and even the origin of life. He concedes early on that he accepts natural selection and common descent; he even concedes that there are plausible accounts of how the vertebrate eye could have evolved (see, for example, Futuyma 1998; Gilbert 2000). He thus bears witness to the changed presumptions in favor of selectionist explanations. Having done so, however, Behe shifts the argument to the greater complexity and functional intricacy of the *molecules* that underwrite macroscopic biological adaptations. How *possibly*, he asks, could this particular biochemical feature or trait, which requires X number of components, each with a precise structure and function, have arisen by natural selection, when a loss or defect in any one of the components must cause a loss of the function of the trait? Any system, that is, with $X-1$ components will have no function, no fitness – nothing upon which natural selection can act. Indeed, Behe claims that "many biochemical systems cannot be built up by natural selection working on mutation; no direct, gradual route exists to these irreducibly complex systems" (Behe 1996, p. 202).

Behe's argument from incredulity takes the following schematic form:

(1) Natural selection has to be gradual, linear, and sequential if it is to result in adapted traits.
(2) Observed molecular functional complexity of Y is inconsistent with (1).
(3) Natural selection cannot possibly account for Y.

Behe tacitly assumes that the molecular components of cells, if not cells themselves, *are* little machines, artifacts with moving parts. Since the answer to the question of whether natural selection could possibly have been the causal explanation of the appearance of a particular adaptive trait is negative, and since Behe denies that there is any other possible natural explanation, he reasons by disjunctive syllogism that the only other possible explanation is intelligent design. We can call this "the argument to the only alternative." The rhetorical advantage of this strategy is that, having eliminated the only other possibility, it is not necessary to provide any positive argument *for* design. But the strategy will persuade only if three conditions are fulfilled: (1) there are no empirically adequate answers to the 'how possibly' questions Behe poses; (2) cells and their molecular components are no more than machines; and (3) there is a simple dichotomy at the level of possible explanations. We will consider Behe's argument in the light of these three conditions. We will find reason to doubt whether any of them is strong enough to sustain Behe's argument.

2. THE ARGUMENT AGAINST INCREDULITY

When considering 'how possibly' questions, one needs to think about the kinds of *systems* in which changes occur over time (i.e., the nature and dynamics of these systems). Intelligent Design (ID) theorists, such as Behe, presuppose that the only possible systems are ones that are highly decomposable. This is implicit in the strong analogy they draw between biological or biochemical systems and man-made machines; the latter presume a linear assembly model of systems. Unfortunately, contemporary hyper-adaptationist versions of Darwinism – arguments that look for an adaptationist explanation for virtually every trait or evolutionary phenomenon – meet ID on the same ground by employing similar classes of models (Dawkins 1986; Dennett 1995).[3] In effect, the ID theorists and the hyper-adaptationists are taking in each other's laundry when they both use the terminology of design, even as they differ in the *source* of the design. This way of framing the debate between evolutionists and creationists foreshortens the space of what will count as possible explanations. Unfortunately, too, this is something both camps seem to prefer. Both assume that there are only two alternatives and argue that the other alternative is impossible. A knockout blow is the aim of both a Dennett and a Behe, a Dawkins and a Dembski.

But deploying 'how possibly' questions in order to foreclose any but one answer to 'why actually' questions is a very risky business. Because 'how possibly' claims seldom, if ever, reach the status of 'why necessarily' answers in favor of the disjunct they prefer, they can be upended by plausible 'why actually' hypotheses that put the burden of proof back onto the questioner. Suppose for a moment that 'why actually' scenarios or hypotheses are given as plausible answers to questions posed by an ID theorist about the origin of a particular biological mechanism. Then either (1) a retreat from the claim of the 'irreducible complexity' of that trait must be made, or (2) it must be acknowledged – especially if the number of such retreats accumulates – that the assumption of decomposable systems acted upon by gradual, linear, sequential selection is not valid. Both eventualities weaken the ID case. If evolutionary theory is merely a stalking horse for materialism, then ID arguments do have a certain rhetorical point. However, the problem is that ID theorists foreclose the 'how possibly' question too quickly, slyly converting the 'why actually' question into a dichotomy of either external design or selection.

What if natural systems are not strongly decomposable but in fact are generated by and composed of parallel processes? What if such natural systems can "limp along" when they open new survival space with the beginning of the emergence of novel properties, and what if these properties can be polished by selection to a shine of apparent design? Such scenarios, if biochemically plausible, could take the force out of the appeal to irreducible complexity and the argument from incredulity. That is why well-documented processes, such as gene duplication, are relevant. They allow for parallel and divergent evolution of enzyme functions that can open new energy sources, produce new metabolic pathways and complex functions, and cope with potentially lethal mutations (Shanks and Joplin 1999; Thornhill and Ussery 2000; Deacon 2003).

Behe argues that "the impotence of Darwinian theory in accounting for the molecular basis of life is evident... from the complete absence in the professional scientific literature of any detailed models by which complex biochemical system could have been produced" (Behe 1996, 187). Yet literature published before Behe wrote, and subsequent papers in the professional scientific literature, have provided just such 'how possibly' scenarios; and further research in some cases is shifting to the 'why actually' issues.

For example, a chemically plausible route for the evolutionary origin of the Krebs citric acid cycle has been proposed. It might serve as a more general model for the emergence of complex metabolic pathways (Melendex-Hevia, Waddell, and Cascante 1996). Behe could respond to such arguments with the admission that this particular system, the Krebs cycle, has turned out not to be irreducibly complex, since an explanation of its emergence can be plausibly advanced. Behe has taken just such a position about the emergent phenomena of patterns of chemical reactions in the Belousov-Zhabotinsky (BZ) reaction, which was originally developed to model the functional complexity of the Krebs cycle (Belousov 1958; Behe 2000; Shanks 2001). It turns out that the BZ reaction can be understood in terms of theories of nonlinear chemical kinetics and nonequilibrium thermodynamics, even though the emergent phenomena were not anticipated by the decomposability assumption of conventional chemical descriptions (Tyson 1976, 1994). The BZ reaction can be used as a model to explore the dynamics of other complex systems, including biological ones (Shanks and Joplin 1999; Shanks 2001). Behe does concede that the origin of *some* biochemical systems may also be explainable by natural processes, and hence by processes not fitting his definition of irreducible complexity (Behe 2000). But wait a minute. We were assured by Behe that *no* complex biochemical system, including metabolic pathways such as the Krebs cycle, had been given a Darwinian explanation in response to his 'how possibly' questions (Behe 1996). There are many classes of Darwinian explanations, and Darwinian explanations are only a subclass of naturalistic explanations. All we need at present is a plausible naturalistic explanation, whether Darwinian or not. As similar arguments are extended to other metabolic pathways, however, the scope of the phenomena that Behe claims to explain by ID shrinks. Well, Behe will say, even if metabolism is capable of naturalistic explanation, still there are other very complex phenomena that surely resist explanation. But at a certain point, it is this very assertion that is in dispute.

Consider the following case, in which explicitly Darwinian naturalism figures. Behe has claimed that "*no one on earth has the vaguest idea how the coagulation cascade came to be*" (Behe 1996, 97, emphasis in original). But there are, it turns out, plausible scenarios by which crude, primitive blood clotting *could* have occurred in organisms that employed proteins from other functions to achieve a marginal stoppage of bleeding (Doolittle and Feng 1987; Xu and Doolittle 1990; Doolittle 1993; Miller 1999). Through gene duplication and shuffling of the subregions of genes coding for domains within the proteins produced, such protein functions could expand, diverge, and become targets of selection that could lead to improved clotting. Proposals such as this can be tested by making predictions about both the specific relationships of sequences of the various genes within the pathway and to sequences of genes for proteins in more primitive organisms that might have been the origin of the original genetic information. This kind of work is proceeding and is beginning to provide the basis for shifting to the question of 'why actually' such a complex cascade has actually arisen in evolution by natural selection.

Again, Behe asserts that "the *only* way a cell could make a flagellum is if the structure were already coded for in its DNA" (Behe 1996, 192, emphasis added). The bacterial flagella that Behe considers have over forty protein components, and it would seem that he has a strong case here. Yet only thirty-three proteins are needed for fully functional flagella in some bacteria not considered by Behe. Even so, such a number constitutes an explanatory challenge. Clues to how such a system might have evolved, employing processes of protein "recruitment" and gene duplication followed by divergence, are provided by the roles of ion gradients in the organization of bacterial metabolism and their possible role in self-organizational and emergent phenomena, a subject to which we will return (Harold 1991; Harold 2001).

Another example Behe uses is the truly complex vertebrate immune system of over ten thousand genes. He claims that "the scientific literature has no answers to the origin of the immune system" (Behe 1996, 138). But shortly after Behe published his book, the discovery of the RAG transposases and transposons in contemporary vertebrate immune systems suggested possible routes by which "reverse transcription" (similar to the way in which the HIV virus converts its RNA message into DNA and incorporates it into T-cells) could have worked, in addition to gene duplication and domain shuffling through RNA splicing mechanisms, to generate the immune system (Agrawal, Eastman, and Schatz 1998; Agrawal 2000; Kazazian 2000).

There is, to be sure, a long way to go in developing our understanding of how immunity came into existence. But it should be clear that the reasonable way to proceed is through pursuing the combined research programs of molecular genetics, genomics, proteomics, and informatics rather than simply by declaring the immune system to be too complex for any other than ID explanations (if indeed they can be called explanations). In time, Behe may be forced to concede that the immune system, as well as blood clotting and metabolic pathways, are subject to naturalistic explanations and, although complex, are not in the end irreducibly complex. We are just now entering a new age in understanding the relationship between genes and biochemical pathways and systems. Why does Behe think that, just as this new cascade of research gets under way, he can put a stop to the whole business?

Finally, let us return to Paley's example of the eye. Behe concedes, as mentioned earlier, that the eye as an organ might have a Darwinian explanation. But he argues that the biochemistry of vision is too complex to have any other than a design explanation. What can be said about this? Photosensitive enzyme systems exist in bacteria, functioning in energy metabolism, and indeed may be quite ancient (Harold 1986; Nicholls and Ferguson 1992). Proteins that function in the lens turn out to be similar to the sequences of amino acids in heat shock proteins, and to various metabolic enzymes that function in the liver (Wistow 1993), suggesting once again a process of *parallel processing* via gene duplication and divergence as a new function becomes selectively advantageous. More dramatically, recent developments in our understanding of the molecular basis of biological developmental processes reveals that certain genes responsible for the development of eyes, such as the *pax 6* gene, arose only once in the history of life and, although not coding for eyes per se, code for the processes that lead to the construction of eyes in various lineages (Gehring 1998; Gilbert 2000, 705; Carroll, Grenier, and Weatherbee 2001). Comparisons of the genes regulating development in different species hold the promise of giving us clues to the origin and evolution of complex, multicellular organisms and their organs (Carroll, Grenier, and Weatherbee 2001).

A recurring theme in the specific research areas just mentioned is that evolutionary processes occur through parallel processing rather than by sequentially adding one perfected component to another. Thus evolutionary biologists, *other than those committed to an extreme form of adaptationism*, have explanatory resources that ID theorists deny (see, for example, Thornhill and Ussery 2000). (That is why ID creationists and hyper-adaptationists have a common interest in narrowing what counts as evolutionary theory and as Darwinism.) Even if organisms are conceived as decomposable artifacts – a position against which we will argue later – they are not analogous to the assembly protocols for watches, in which the components are specifically designed and perfected for unique functions and then put together *in a specified sequence*. Interestingly, engineers now find that in designing complex machines (such as a processor composed of over forty million transistors), it is more efficacious to have the components combined in a functional pattern than to assemble them from perfect components. Indeed, the best strategy is to get as quickly as possible to a crudely functional whole and then to remove and replace less reliable subsystems. Challet and Johnson estimate that one-half of the components could be significantly imperfect and yet the system as a whole would function reliably (Challet and Johnson 2002). With complexity comes redundancy and parallelism that can give functionality; and with functionality comes pressure for improved components over time. Thus, even allowing ID theory its artifact metaphor, it poses its 'how possibly' arguments the wrong way around. The function of the whole system *does not* depend upon perfectly designed and articulated components.[4]

The questions that need to be addressed, then, are what types of systems are living organisms – and ecosystems as well? What are the dynamical principles of such systems? Do design and selection exhaust the possibilities of explanatory space in such systems, or are there other relevant natural processes, such as self-organization? To these issues we now turn our attention.

3. SELF-ORGANIZATION AND EMERGENCE IN COMPLEX NATURAL SYSTEMS

Both ID and hyper-adaptationist Darwinian (HD) theorists share assumptions about the nature of organisms and systems. These shared assumptions imply that self-organization is not a serious contending explanatory resource for apparent design in natural systems. ID and HD theorists both view cells and organisms as machines with independent components that can be either designed separately or selected for their contribution to optimal fitness. By contrast, the sort of complex system dynamical perspective for which we have argued views cells and organisms as lineages of developing entities in which everything changes at the same time, in parallel fashion, during life cycles and over generational time. In this view, organisms are open thermodynamic systems with gradients stabilized by the actions of macromolecules that have a degree of autopoeisis; their parts do not aggregate but rather develop by differentiation (Depew and Weber 1995; Weber and Depew 1996, 2001). They do not passively, and only incidentally, conform to the requirements of the Second Law of Thermodynamics. They exploit that law in becoming the kinds of systems they are.

Behe's key claim is that there is no way that natural explanations can possibly account for the origin of life. Hence, for Behe, life itself must be irreducibly complex, and so the product of intelligent design. We agree with him about the complexity. But we would prefer to frame the question as one about the emergence of life through a complex *process* rather than as the product of a specific *event*, or series of events, as Behe (and his hyper-adaptationist counterparts) assume.

Our own guiding ideas about the origin of life are integral to our approach to how complex systems dynamics might enrich Darwinian theory. Darwin himself famously sought to sidestep the issue of the origin of life by placing it outside of his explanatory framework. But in the late twentieth century, biologists have increasingly sought to articulate proposals describing how living things came to be within a broader Darwinian framework. Some of them, placing emphasis on replication and the role of genes in directing all other biological processes, have put the focus on naked, replicating RNA as the starting point of life, with the rest of cellular structures and processes viewed as "survival machines" for the genes (Dawkins 1976; Dawkins 1989). A quite different view arises, however, when the problem of the emergence of life is addressed in the context of complexity theory (for recent, reader-friendly introductions to complexity theory, see Casti 1994; Taylor 2001).

The complex systems dynamics to which we have drawn attention includes the energetic driving force of far-from-equilibrium thermodynamics (that is, far from the equilibrium state in which there is no change in the energy of the system as a whole). This gives rise to matter-energy gradients and to internal structures, breaking symmetries and producing internal order and organization at the expense of increases in entropy (disorder and/or degraded energy) in the environment (Schrödinger 1944; Nicolis and Prigogine 1977; Prigogine 1980; Wicken 1987; Prigogine 1997). A consequence of such a state of affairs is the appearance of nonlinear (nonadditive, potentially amplifying) interactions between components and processes in such systems through which organized structures emerge. The appearance of higher-order, macroscopic (whole, global, or collective) properties and structures from lower-order microscopic (part or individual) components is called "self-organization" though perhaps "system organization" would be a more apt phrase.

Stuart Kauffman is one of the theorists who have modeled the dynamics of such systems by exploring, through computer simulations, the dynamics of systems in which nonlinearity obtains and self-organization can occur (Kauffman 1993, 1995, 2000). ID theorists, including Behe, have attacked Kauffman as if he were the *only* advocate of self-organizing complexity, and as if his simulations were cut off from any physical or biological reality (Behe 1996, 156, 189–92). (Similar objections have been raised against the magnum opus of Stephen Wolfram [Wolfram 2002]). Kauffman's approach is dismissed by Behe as irrelevant, since, according to Behe, self-organization is only a mathematical phenomenon that occurs in computer simulations under specific initial and boundary conditions – as well as rules of component interaction – that are set by the programmer.

To this criticism there are two responses. First, any computer simulation requires constraints that are introduced by the programmer; the question is whether the constraints are realistic. Kauffman does argue that the constraints that he introduces in various modeling applications are plausible and reflect the essence, if not the minute detail, of energetic and molecular properties and propensities of the natural system – or at least that they reflect the consequences of such systems being far from equilibrium and having nonlinearity (Kauffman 1993, 1995, 2000). It is unreasonable to argue that Kauffman has to avoid any constraints to totally random interactions in his models, since analogous constraints do exist in nature.

Second, there is the response that self-organization is not just a mathematical, theoretical concept, but rather is a real phenomenon. In summarizing a series of recent studies in cell biology, Tom Misteli recently argued that "[t]hese studies have permitted the direct and quantitative testing of the role of self-organization in the formation of cytoskeletal networks, thus elevating the concept of self-organization from a largely theoretical consideration to a cell biological *reality*" (Misteli 2001, 182, emphasis added). Self-organizing metabolic and signaling networks that display systemwide dynamics are viewed by Richard Strohman as playing a coequal role with genes in controlling human disease phenotypes (Strohman 2002). An even more wide-ranging review of the phenomena of self-organization and its appropriate simulations has recently appeared, according to which self-organization, rather than limiting the action of natural selection, enhances its efficacy through economical use of information and by providing the formation of whole patterns upon which selection can act (Camazine et al. 2001). We have argued in a similar vein that, although selection and self-organization can be conceptualized as interacting in a variety of logically possible ways, these two principles can and should be viewed as complementary *phenomena* (Weber and Depew 1996).

Energy, primarily from the sun, would be expected to drive the chemistry of the primitive Earth to greater molecular complexity under the constraints of atomic and molecular properties and propensities under far-from-equilibrium conditions (Wicken 1987; Williams and Frausto da Silva 1996, 1999). At some point, there would have arisen sufficient chemical complexity and catalysis of reactions that regions on the abiotic Earth would have produced chemical self-organization, in which some chemical components would have acted as catalysts for reaction sequences that produced more of themselves – the phenomenon of autocatalysis. Under such conditions, at least some amino acids, which are the building blocks of proteins, as well as the purine and pyrimidine bases, which are components of nucleic acids, would have been produced. Even polymers of these can be stabilized by various mechanisms. Even though such polymers would not contain "specified information," they could be weakly catalytic and would contribute to the overall autocatalytic system (Kauffman 1993; Weber 1998).

Kauffman has applied his flexible NK model (where N = the number of components and K = the number of interactions between components) to model both the range of possible sequences of proteins and their autocatalytic interactions. Kauffman modeled a "catalytic task space" that represents all the conceivable chemical reactions (on the order of a few million) and mapped that onto the protein sequence space. In an array or ensemble of random sequences of amino acids in proteins, a large number of chemical reactions would appear that would be weakly catalyzed by subpopulations of protein sequences. Then a chemical-type selection (favoring enhanced catalytic activity) would act on protein sequences in such a way that it would enhance the catalytic efficiency of the catalysts and contribute to the overall thermodynamic efficiency of the system. Even if there were no physical or chemical boundary for such autocatalytic systems – but very much more likely, if there were closure bounded by a chemical barrier from the rest of the chemical environment – such systems could become sufficiently interactive to achieve "catalytic closure." They would generate, that is to say, a system of autocatalytic chemical transformations that would become self-contained – able to generate all its constituents by pulling in available chemical substrates and to exhibit weakly heritable information, even in the absence of macromolecular memory molecules such as RNA.

It is likely that such catalytic closure did not occur in dilute sea water (contra the naked RNA approach), but instead within a barrier composed of molecules with water-avoiding and water-seeking ends (amphiphiles). These would have served the same function as that provided by modern cellular membranes (Morowitz, Deamer, and Smith 1991; Morowitz 1992). Plausible sources for such molecules and their spontaneous self-organization into membranous vesicles have been demonstrated (Deamer and Pashley 1989; Deamer and Harang 1990). Such vesicles open possible pathways for organic molecules likely to be present to insert themselves in the "proto-membrane" and capture light energy driving proton gradients, which in turn could drive metabolic reactions through phosphate chemistry, including polymerization reactions (Mitchell 1961; Williams 1961; Weber 1998; Harold 2001). Catalytic closure would most likely have occurred over an ensemble of contiguous vesicles rather than just one. If RNA polymers did not already exist, they could reasonably be expected to form under the chemical environment of these "proto-cells" (Joyce and Orgel 1993; Weber 1998). Even if RNA polymers arose prior to such proto-cells, the effect of self-catalyzing RNA has probably been exaggerated in some literature (see, for example, Dawkins 1989; Dawkins 1999). Initially, proteins and RNA strands might interact not only because of propensities for interaction between certain bases and amino acids, but also because such interactions might be favored by mutual stabilization against hydrolysis (Carter and Kraut 1974; Wicken 1987; Davydov 1998).

What is envisioned here is a crude but complex system with all the functional components of a living cell in place: proto-enzymes made of proteins; nucleic acids interacting with proteins but not yet serving as templates; a membranous barrier, but one not yet made of phospholipids; rudimentary energy-capture mechanisms; and a proto-metabolism. Although such a system possibly – even probably – lies at the origin of life and the appearance of its "design," it would still have been markedly imperfect from a design perspective (Weber 1998). We would expect, however, that parallel events would have intensified articulation of the protein – nucleic acid interactions that led ultimately to nucleic acids that serve primarily as the templates for proteins. In this way, information about catalysts would have grown increasingly stable and have been internalized in the system. An increased exploration of the catalytic task space can be expected to have occurred as more specific sequences acquired functions. This, in turn, would have expanded the repertoire of metabolic reactions, leading to more and better constituents and to an expansion of metabolism and the emergence of key pathways. Thus specified information would emerge *along with* the emergence of life and could continue to expand during biological evolution (Weber 1998; Collier and Hooker 1999; Schneider 2000; Adami, Ofria, and Collier 2000).

In this account, there would not be a single cellular ancestor of all living things, as in the RNA-first scenario, but instead a complex ensemble of ensembles of true cellular forms that would have complex relationships. These would look rather like the recent deductions that Woese has made by comparing protein and nucleic acid sequences from contemporary bacteria and archea and inferring an ancestral *process of emergence* rather than a specific cellular ancestor common to all currently living beings (Woese 1998, 2002).

Up to the point at which a true memory mechanism for replicating nucleic acids as templates of proteins arose, the emergence of proto-cellular forms would have been effected by an interplay of chemical self-organization and chemical selection principles. Once true cells with replication of nucleic acids *as genomes* were present, however, true biological or natural selection would have emerged for the first time. We have argued that such a view of the emergence of life entails an approach in which natural selection itself is an emergent phenomenon and not "just a theory." One advantage of this view of natural selection as working with self-organization is that natural selection turns out to be a far more realistic phenomenon than is allowed by ID theorists.

At the same time, our view also has something negative to say about natural selection as envisioned by hyper-adaptationists such as Dawkins and Dennett. Natural selection is tied to the kinds of dissipative systems that support it and instantiate it; it is not a substrate-neutral principle or an algorithm that can range over an indefinite, open series of entities that reproduce, vary, and differentially survive (Dawkins 1979; Dennett 1995). Further, we have argued that what held for the emergence of life probably obtains also for its subsequent evolution, so that there would continue to be an interplay of selection and self-organization in which self-organizational processes are still occurring in a highly stabilized way within the cell and in the developmental process (Weber and Depew 2001).

Our point in saying these things is that the sciences of complexity have the potential to enrich the Darwinian research tradition and that phenomena of self-organization can and do play an important role in evolutionary phenomena, especially those involving emergence (Johnson 2001). In this case, the tradition of Darwinism is less directly vulnerable to ID-type arguments than are hyper-adaptationist versions of Darwinism that seem to play into either/or thinking. Let us make quite clear once again, however, that the scenario we have sketched and other scenarios like it are *not* presented as 'why actually' arguments. They are presented as 'how possibly' arguments that have a much higher likelihood of resulting in 'why actually' arguments than does ID. Behe's claim is that modern Darwinism is not capable of producing *any* 'how possibly' arguments about the biochemical underpinnings of evolutionary arguments, let alone of moving on to 'why actually' arguments. Addressing ourselves to this claim, we have exhibited 'how possibly' arguments that are germane and that, increasingly, are being considered, formulated, and tested by Darwinians holding a variety of theoretical commitments. ID proponents can be commended for articulating the problems of emergence and of major evolutionary transformations that some Darwinians have discounted or deferred addressing. Still, there is little reason to think that they are contributing to the solution of those problems.

4. NATURAL THEOLOGY REDUX

By the twentieth century, the project of natural theology had lost much of its momentum. It was seen not only as irrelevant to the pursuit of scientific research, but also as unnecessary, or even worse, for faith and theology (Barth and Brunner 1947; McGrath 2001). Ruse has argued that Christian theology is logically compatible with evolutionary biology – even of the most pan-adaptationist, "ultra-Darwinian" stripe – if we are willing to abandon natural theology, which has had a very contingent and episodic relationship to Christian theology, historically considered (Ruse 2001; Ruse 2002). John Haught argues that facing Darwinian evolution squarely can bring rewards to theology by forcing theology to take a fresh look at the problem of evil and encouraging it to develop a theology of nature (Haught 2000). Haught and others have suggested that to go forward creatively in theology is to build upon the concept of *kenosis* (God's self-emptying in creating; see the essays in Polkinghorne 2001). In such a theology, we would not expect God to leave footprints and fingerprints all over nature. Nor would we expect God to intervene in the "functional integrity of nature" even while sustaining nature and its laws (Van Till 1996). Indeed, autopoetic processes of the type explored by complex systems dynamical theory, along with natural selection, could contribute to a theology of nature consistent with a trinitarian framework (Gregersen 1998). Both traditional Darwinism and the type of enriched Darwinism that we espouse are compatible with such a new program. Yet they remain neutral toward it because they are also compatible with some forms of naturalism (but not with "building-block materialism").

The natural theology of ID theory is opposed to these innovations. It makes natural theology of a certain Paleyesque kind central to Christian theology. (It should be noted that only in England and the United States has the theological tradition ever relied heavily on natural theology of any sort; even the tradition of Aquinas was dormant, or actively suppressed, from the fourteenth through the sixteenth century; and when the Jesuits revived it, it was countered by Jansenists and Calvinists alike. The tradition is currently very recessive, even in Anglo-American theology.) The danger of ID, considered as a theological position rather than in the scientific light in which we have discussed it here, is that it potentially implies a limiting conception of God (while adding nothing to the pursuit of scientific exploration). These facts suggest that from a theological as well as a scientific perspective, the presumption should be in favor of methodological naturalism – the working hypothesis that a scientific explanation for a puzzling phenomenon will be found that does not invoke a source of functional design outside of nature. It is important to add, however, that it is not logically necessary that methodological naturalism must lead to metaphysical naturalism, or materialism, which must deny any type of theology. There is no theological reason to expect nature to be marked with *vestigia dei* that are to be discerned and interpreted by recognizing "irreducible complexity" or "explanatory filters" as a basis for inferring design (Behe 1996, 2002; Dembski 1998, 1999, 2002).

The theologian George Murphy has commented, "God does not compel the belief of skeptics by leaving puzzles in creation which science can't solve" (Murphy 2002, 9). Just as Saint Augustine learned that certain lines of questioning could be harmful to his religion, ID theorists may learn to their chagrin that a modern-day resurrection of natural theology does not advance the theistic perspective they seek to restore to academe (Johnson 2000). Instead they may expose theism even more nakedly to the withering winds of radical secularism. It has happened before – and will probably happen again.

Notes

1. Darwin's *Origin of Species* contains its fair share of 'how possibly' arguments. In the case of the eye, for example, Darwin seeks to reverse the presumption against a natural explanation. See Darwin 1859, pp. 186–94.
2. The anthropic line of argument was pioneered by William Whewell in his Bridgewater Treatise, *Astronomy and General Physics Considered with Reference to Natural Theology* (1833). Whewell developed this line of argument as an explicit *alternative* to Paleyesque natural theology, of which he was critical.
3. It is possible that the recent revival of Paleyesque arguments has been stimulated by the increasingly strong shift toward hyper-adaptation among Darwinians, especially in works about "Darwinism" that circulate in the public sphere as trade books. Other motives include the failure of more biblically based versions of creationism to find support in the courts for teaching biology without a Darwinian bias. See Depew 1998.
4. The emergence of machines that involve some self-assembly, combined with subsequent paring away of less-than-fully-efficient parts, might seem to fly in the face of the assumption that machines are, by nature, decomposable and assembled in a linear way. This is not so. The linear aspect comes in the paring-down stage. This is just the sort of thing natural selection can do, moreover, and so does not support ID as the only alterative.

References

Adami, C., C. Ofria, and T. C. Collier. 2000. Evolution of biological complexity. *Proceedings of the National Academy of Science (USA)* 97: 4463–68.

Agrawal, A. 2000. Transposition and evolution of antigen-specific immunity. *Science* 290: 1715–16.

Agrawal, A., Q. M. Eastman, and D. G. Schatz. 1998. Transposition mediated by RAG1 and RAG2 and its implications for the evolution of the immune system. *Nature* 394: 744–51.

Barth, K., and E. Brunner. 1947. *Natural Theology*. London: SCM Press.

Behe, M. J. 1996. *Darwin's Black Box: The Biochemical Challenge to Evolution*. New York: The Free Press.

2000. Self-organization and irreducibly complex systems: A reply to Shanks and Joplin. *Philosophy of Science* 67: 155–62.

2001. Reply to my critics: A response to reviews of *Darwin's Black Box: The Biochemical Challenge to Evolution*. *Biology and Philosophy* 16: 685–709.

Camazine, S., J-L. Deneubourg, N. R. Franks, J. Sneyd, Guy Theraulaz, and E. Bonabeau. 2001. *Self-Organization in Biological Systems*. Princeton, NJ: Princeton University Press.

Carroll, S. B., J. K. Grenier, and S. D. Weatherbee. 2001. *From DNA to Diversity: Molecular Genetics and the Evolution of Animal Design*. Malden, MA: Blackwell Science.

Carter, C.W., Jr., and J. Kraut. 1974. A proposed model for interaction of polypeptides with RNA. *Proceedings of the National Academy of Science USA* 71: 283–7.

Casti, J. L. 1994. *Complexification*. New York: Harper Collins.

Challet, D., and N. F. Johnson. 2002. Optimal combinations of imperfect objects. *Physical Review Letters* 89: 028701.

Collier, J. D., and C. A. Hooker. 1999. Complexly organized dynamical systems. *Open Systems and Information Dynamics* 6: 241–302.

Darwin, C. R. 1859. *On the Origin of Species by Means of Natural Selection, or the Preservation of Favored Races in the Struggle for Life*. London: John Murray.

Davydov, O. V. 1998. Amino acid contribution to the genetic code structure: End-atom chemical rules of doublet composition. *Journal of Theoretical Biology* 193: 679–90.

Deacon, T. 2003. The hierarchic logic of emergence: Untangling the interdependence of evolution and self-organization. In *Evolution and Learning: The Baldwin Effect Reconsidered*, ed. B. H. Weber and D. J. Depew. Cambridge, MA: MIT Press, pp. 273–308.

Deamer, D. W., and E. Harang. 1990. Light-dependent pH gradients are generated in liposomes containing ferrocyanide. *BioSystems* 24: 1–4.

Deamer, D. W., and R. M. Pashley. 1989. Amphiphilic components of the Murchison carbonaceous condrite: Surface properties and membrane formation. *Origin of Life and Evolution of the Bioshpere* 19: 21–38.

Dembski, W. A. 1998. *The Design Inference: Eliminating Chance through Small Probabilities*. Cambridge: Cambridge University Press.

1999. *Intelligent Design: The Bridge between Science and Theology*. Downers Grove, IL: InterVarsity Press.

2002. *No Free Lunch: Why Specified Complexity Cannot Be Purchased without Intelligence*. Lanham, MD: Rowman and Littlefield.

Dennett, D. 1995. *Darwin's Dangerous Idea: Evolution and the Meanings of Life*. New York: Simon and Schuster.

Depew, D. J. 1998. Intelligent design and irreducible complexity. A rejoinder. *Rhetoric and Public Affairs* 1: 571–8.

Depew, D. J., and B. H. Weber. 1995. *Darwinism Evolving: Systems Dynamics and the Genealogy of Natural Selection*. Cambridge, MA: MIT Press.

Doolittle, R. F. 1993. The evolution of vertebrate blood coagulation. *Thrombosis Haemostasis* 70: 24–8.

Doolittle, R. F., and D. F. Feng. 1987. Reconstructing the history of vertebrate blood coagulation from a consideration of the amino acid sequences of clotting proteins. *Cold Spring Harbor Symposium on Quantitative Biology* 52: 869–74.

Futuyma, D. J. 1998. *Evolutionary Biology*, 3rd ed. Sunderland, MA: Sinauer.

Gehring, W. J. 1998. *Master Control Genes in Development and Evolution: The Homeobox Story.* New Haven, CT: Yale University Press.

Gilbert, S. F. 2000. *Developmental Biology*, 6th ed. Sunderland, MA: Sinauer.

Gregersen, N. H. 1998. The idea of creation and the theory of autopoietic processes. *Zygon* 33: 333–67.

Harold, F. M. 1986. *The Vital Force: A Study of Bioenergetics.* New York: W. H. Freeman.

____ 1991. Biochemical topology: From vectorial metabolism to morphogenesis. *Bioscience Reports* 11: 347–82.

____ 2001. *The Way of the Cell: Molecules, Organisms and the Order of Life.* New York: Oxford University Press.

Haught, J. F. 2000. *God after Darwin: A Theology of Evolution.* Boulder, CO: Westview Press.

Johnson, P. E. 2000. *The Wedge of Truth: Splitting the Foundations of Naturalism.* Downers Grove, IL: InterVarsity Press.

Johnson, S. 2001. *Emergence: The Connected Lives of Ants, Brains, Cities, and Software.* New York: Scribner.

Joyce, G. F., and L. E. Orgel 1993. Prospects for understanding the origin of the RNA world. In *The RNA World*, ed. R. F. Gesteland and J. F. Atkins. Cold Spring Harbor, NY: Cold Spring Harbor Laboratory Press, pp. 1–24.

Kazazian, H. H., Jr. 2000. L1 retrotransposons shape the mammalian genome. *Science* 289: 1152–53.

McGrath, A. E. 2001. *A Scientific Theology: Nature.* Grand Rapids, MI: Eerdmans.

Meléndez-Hevia, E., T. G. Waddell, and M. Cascante. 1996. The puzzle of the Krebs citric acid cycle: Assembling the pieces of chemically feasible reactions, and opportunism in the design of metabolic pathways during evolution. *Journal of Molecular Biology* 43: 293–303.

Miller, K. R. 1999. *Finding Darwin's God: A Scientist's Search for Common Ground between God and Evolution.* New York: HarperCollins.

Mills, J. S. 1874. *A System of Logic*, 8th ed. London: Longmans, Green.

Misteli, T. 2001. The concept of self-organization in cellular architecture. *Journal of Cell Biology* 155: 181–5.

Morowitz, H. 1992. *Beginnings of Cellular Life: Metabolism Recapitulates Biogenesis.* New Haven, CT: Yale University Press.

Morowitz, H., D. W. Deamer, and T. Smith. 1991. Biogenesis as an evolutionary process. *Journal of Molecular Evolution* 33: 207–8.

Murphy, G. L. 2002. Intelligent design as a theological problem. *Covalence* 4(2): 1–9.

Nicholls, D. G., and S. J. Ferguson. 1992. *Bioenergetics 2.* San Diego: Academic Press.

Nicolis, G., and I. Prigogine. 1977. *Self-Organization in Nonequilibrium Systems: From Dissipative Structures to Order Through Fluctuations.* New York: Wiley.

Paley, W. 1802. *Natural Theology, or Evidences of the Existence and Attributes of the Deity Collected from the Appearances of Nature.* London: Fauldner.

Polkinghorne, J. 2001. *The Works of Love: Creation as Kenosis.* Grand Rapids, MI: Eerdmans.

Prigogine, I. 1980. *From Being to Becoming: Time and Complexity in the Physical Sciences.* San Francisco: Freeman.

1997. *The End of Certainty: Time, Chaos, and the New Laws of Nature.* New York: The Free Press.

Ruse, M. 2001. *Can a Darwinian Be a Christian? The Relationship between Science and Religion.* Cambridge: Cambridge University Press.

2002. *Darwin and Design.* Cambridge, MA: Harvard University Press.

Schneider, T. D. 2000. Evolution of biological information. *Nucleic Acids Research* 28: 2794–99.

Schrödinger, E. 1944. *What Is Life? The Physical Aspect of the Living Cell.* Cambridge: Cambridge University Press.

Shanks, N. 2001. Modeling biological systems: The Belousov-Zhabotinsky reaction. *Foundations of Chemistry* 3: 33–53.

Shanks, N., and K. H. Joplin. 1999. Redundant complexity: A critical analysis of intelligent design in biochemistry. *Philosophy of Science* 66: 268–98.

Strohman, R. 2002. Maneuvering in the complex path from genotype to phenotype. *Science* 296: 701–3.

Taylor, M. C. 2001. *The Moment of Complexity: Emerging Network Culture.* Chicago: University of Chicago Press.

Thornhill, R. H., and D. W. Ussery 2000. A classification of possible routes of Darwinian evolution. *Journal of Theoretical Biology* 203: 111–16.

Tyson, J. L. 1976. *The Belousov-Zhabotinski Reaction; Lecture Notes in Biomathematics 10.* Berlin: Springer.

1994. What everyone should know about the Belousov-Zhabotinsky reaction. In *Frontiers in Mathematical Biology*, ed. S. A. Levin. New York: Springer, pp. 569–87.

Van Till, H. J. 1996. Basil, Augustine, and the doctrine of creation's functional integrity. *Science and Christian Belief* 8: 21–38.

Weber, B. H. 1998. Emergence of life and biological selection from the perspective of complex systems dynamics, in *Evolutionary Systems: Biological and Epistemological Perspectives on Selection and Self-Organization*, ed. G. Van de Vijver, S. N. Salthe, and M. Delpos. Dordrecht: Kluwer, pp. 59–66.

Weber, B. H., and D. J. Depew. 1996. Natural selection and self-organization: Dynamical models as clues to a new evolutionary synthesis. *Biology and Philosophy* 11: 33–65.

2001. Developmental systems, Darwinian evolution, and the unity of science. *In Cycles of Contingency*, ed. S. Oyama, R. Gray, and P. Griffiths. Cambridge, MA: MIT Press, 239–53.

Wicken, J. S. 1987. *Evolution, Information and Thermodynamics: Extending the Darwinian Program.* New York: Oxford University Press.

Williams, R. J. P., and J. J. R. Fraústo da Silva. 1996. *The Natural Selection of the Chemical Elements: The Environment and Life's Chemistry.* Oxford: Oxford University Press.

1999. *Bringing Chemistry to Life: From Matter to Man.* Oxford: Oxford University Press.

Wistow, G. 1993. Lens crystallins: Gene recruitment and evolutionary dynamism. *Trends in Biochemical Science* 18: 301–6.

Woese, C. 1998. The universal ancestor. *Proceedings of the National Academy of Science (USA)* 95: 6854–9.

2002. On the evolution of cells. *Proceedings of the National Academy of Science (USA)* 99: 8742–7.

Wolfram, S. 2002. *A New Kind of Science.* Champaign, IL: Wolfram Media.

Xu, X., and R. F. Doolittle. 1990. Presence of a vertebrate fibrinogen-like sequence in an echinoderm. *Proceedings of the National Academy of Science (USA)* 87: 2097–101.

10

Emergent Complexity, Teleology, and the Arrow of Time

Paul Davies

1. THE DYING UNIVERSE

In 1854, in one of the bleakest pronouncements in the history of science, the German physicist Hermann von Helmholtz claimed that the universe must be dying. He based his prediction on the Second Law of Thermodynamics, according to which there is a natural tendency for order to give way to chaos. It is not hard to find examples in the world about us: people grow old, snowmen melt, houses fall down, cars rust, and stars burn out. Although islands of order may appear in restricted regions (e.g., the birth of a baby, crystals emerging from a solute), the disorder of the environment will always increase by an amount sufficient to compensate. This one-way slide into disorder is measured by a quantity called entropy. A state of maximum disorder corresponds to thermodynamic equilibrium, from which no change or escape is possible (except in the sense of rare statistical fluctuations). Helmholtz reasoned that the quantity of entropy in the universe as a whole remorselessly rises, presaging an end state in the far future characterized by universal equilibrium, following which nothing of interest will happen. This state was soon dubbed the "heat death of the universe."

Almost from the outset, the prediction of cosmic heat death after an extended period of slow decay and degeneration was subjected to theological interpretation. The most famous commentary was given by the philosopher Bertrand Russell in his book *Why I Am Not a Christian*, in the following terms:[1]

All the labors of the ages, all the devotion, all the inspiration, all the noonday brightness of human genius are destined to extinction in the vast death of the solar system, and the whole temple of man's achievement must inevitably be buried beneath the debris of a universe in ruins. All these things, if not quite beyond dispute, are yet so nearly certain that no philosophy which rejects them can hope to stand. Only within the scaffolding of these truths, only on the firm foundation of unyielding despair, can the soul's habitation henceforth be safely built.

The association of the Second Law of Thermodynamics with atheism and cosmic pointlessness has been an enduring theme. Consider, for example, this assessment by the British chemist Peter Atkins:[2]

We have looked through the window on to the world provided by the Second Law, and have seen the naked purposelessness of nature. The deep structure of change is decay; the spring of change in all its forms is the corruption of the quality of energy as it spreads chaotically, irreversibly and purposelessly in time. All change, and time's arrow, point in the direction of corruption. The experience of time is the gearing of the electrochemical processes in our brains to this purposeless drift into chaos as we sink into equilibrium and the grave.

As Atkins points out, the increase in entropy imprints upon the universe an arrow of time, which manifests itself in many physical processes, the most conspicuous of which is the flow of heat from hot to cold; we do not encounter cold bodies getting colder and spontaneously giving up their heat to warm environments. The irreversible flow of heat and light from stars into the cold depths of space provides a cosmic manifestation of this simple "hot to cold" principle. On the face of it, it appears that this process will continue until the stars burn out and the universe reaches a uniform temperature. Our own existence depends crucially on a state of thermodynamic disequilibrium occasioned by this irreversible heat flow, since much life on Earth is sustained by the temperature gradient produced by sunshine. Microbes that live under the ground or on the sea bed utilize thermal and chemical gradients from the Earth's crust. These too are destined to diminish over time, as thermal and chemical gradients equilibrate. Other sources of energy might provide a basis for life, but according to the Second Law, the supply of free energy continually diminishes until, eventually, it is all exhausted. Thus the death of the universe implies the death of all life, sentient and otherwise. It is probably this gloomy prognosis that led Steven Weinberg to pen the famous phrase, "The more the universe seems comprehensible, the more it also seems pointless."[3]

The fundamental basis for the Second Law is the inexorable logic of chance. To illustrate the principle involved, consider the simple example of a hot body in contact with a cold body. The heat energy of a material substance is due to the random agitation of its molecules. The molecules of the hot body move on average faster than those of the cold body. When the two bodies are in contact, the fast-moving molecules communicate some of their energy to the adjacent slow-moving molecules, speeding them up. After a while, the higher energy of agitation of the hot body spreads across into the cold body, heating it up. In the end, this flow of heat brings the two bodies to a uniform temperature, and the average energy of agitation is the same throughout. The flow of heat from hot to cold arises entirely because chaotic molecular motions cause the energy to diffuse democratically among all the participating particles. The initial state, with the energy distributed in a lopsided way between the two bodies, is relatively more ordered than the final state, in which the energy is spread uniformly throughout the system. One way to see this is to say that more information is needed to describe the initial state – namely, two numbers, the temperatures of the two bodies – whereas the final state can be described with only one number – the common final temperature. The loss of information occasioned by this transition may be quantified by the entropy of the system, which is roughly equal to the negative of the information content. Thus as information goes down, entropy, or disorder, goes up.

The transition of a collection of molecules from a low to a high entropy state is analogous to the shuffling of a deck of cards. Imagine that the cards are extracted from the package in suit and numerical order. After a period of random shuffling, the cards will very probably be jumbled up. The transition from the initial ordered state to the final disordered one is due to the chaotic nature of the shuffling process. So the Second Law is really just a statistical effect of a rather trivial kind. It essentially declares that a disordered state is much more probable than an ordered one – for the simple reason that there are numerically many more disordered states than ordered ones, so that when a system in an ordered state is randomly rearranged, it is very probably going to end up less ordered than it was before. Thus blind chance lies at the basis of the Second Law of Thermodynamics, just as it lies at the basis of Darwin's theory of evolution. Since chance – or contingency, as philosophers call it – is the opposite of law and order, and hence of purpose, it seems to offer powerful ammunition to atheists who wish to deny any overall cosmic purpose or design. If the universe is nothing but a physical system that began (for some mysterious reason) in a relatively ordered state, and is inexorably shuffling itself into a chaotic one by the irresistible logic of probability theory, then it is hard to discern any overall plan or point.

2. REACTION TO THE BLEAK MESSAGE OF THE SECOND LAW OF THERMODYNAMICS

Reaction to the theme of the dying universe began to set in the nineteenth century. Philosophers such as Henri Bergson[4] and theologians such as Teilhard de Chardin[5] sought ways to evade or even refute the Second Law of Thermodynamics. They cited evidence that the universe was in some sense getting better and better rather than worse and worse. In Teilhard de Chardin's rather mystical vision, the cosmic destiny lay not in an inglorious heat death but in an enigmatic "Omega Point" of perfection. The progressive school of philosophy saw the universe as unfolding to ever greater richness and potential. Soon after, the philosopher Alfred North Whitehead[6] (curiously, the coauthor with Bertrand Russell of *Principia Mathematica*) founded the school of process theology on the notion that God and the universe are evolving together in a progressive rather than a degenerative manner.

Much of this reaction to the Second Law had an element of wishful thinking. Many philosophers quite simply hoped and expected the law to be
wrong. If the universe is apparently running down – like a heat engine running out of steam, or a clock unwinding – then perhaps, they thought, nature
has some process up its sleeve that can serve to wind the universe up again.
Some sought this countervailing tendency in specific systems. For example,
it was commonly supposed at the turn of the twentieth century that life
somehow circumvents the strictures of thermodynamics and brings about
increasing order. This was initially sought through the concept of vitalism –
the existence of a life force that somehow bestowes order on the material
contents of living systems. Vitalism eventually developed into a more scientific version, what became known as organicism – the idea that complex
organic wholes might have organizing properties that somehow override
the trend into chaos predicted by thermodynamics.[7] Others imagined that
order could come out of chaos on a cosmic scale. This extended to periodic
resurrections of the cyclic universe theory, according to which the entire cosmos eventually returns to some sort of pristine initial state after a long period
of decay and degeneration. For example, during the 1960s it was suggested
by the cosmologist Thomas Gold[8] that one day the expanding universe may
start to recontract, and that during the contraction phase, the Second Law of
Thermodynamics would be reversed ("time will run backwards"), returning
the universe to a state of low entropy and high order. The speculation was
based on a subtle misconception about the role of the expanding universe
in the cosmic operation of the Second Law (see the following discussion).
It turns out that the expansion of the universe crucially serves to provide
the necessary thermodynamic disequilibrium that permits the entropy in
the universe to rise, but this does *not* mean that a reversal of the expansion will cause a reversal of the entropic arrow. Quite the reverse: a rapidly
contracting universe would drive the entropy level upward as effectively as
a rapidly expanding one. In spite of this blind alley, the hypothesis that the
directionality of physical processes might flip in a contracting universe was
also proposed briefly by Hawking,[9] who then abandoned the idea,[10] calling
it his "greatest mistake." Yet the theory refuses to lie down. Only this year, it
was revived yet again by L. S. Schulman.[11]

The notion of a cyclic universe is, of course, an appealing one, and
one that is deeply rooted in many ancient cultures; it persists today in
Hinduism, Buddhism, and Aboriginal creation myths. The anthropologist
Mircea Eliade[12] termed it "the myth of the eternal return." In spite of
detailed scrutiny, however, the Second Law of Thermodynamics remains
on solid scientific ground. So solid, in fact, that the astronomer Arthur
Eddington felt moved to write,[13] "if your theory is found to be against the
second law of thermodynamics I can give you no hope; there is nothing for
it but to collapse in deepest humiliation." Today, we know that there is nothing anti-thermodynamic about life. As for the cyclic universe theory, there is

no observational evidence to support it (indeed, there is some rather strong evidence to refute it).[14]

3. THE TRUE NATURE OF COSMIC EVOLUTION

In this chapter I wish to argue, not that the Second Law is in any way suspect, but that its significance for both theology and human destiny has been overstated. Some decades after Helmholtz's dying universe prediction, astronomers discovered that the universe is expanding. This changes the rules of the game somewhat. To give a simple example, there is good evidence that 300,000 years after the Big Bang that started the universe off, the cosmic matter was in a state close to thermodynamic equilibrium. This evidence comes from the detection of a background of thermal radiation that pervades the universe, thought to be the fading afterglow of the primeval heat. The spectrum of this radiation conforms exactly to that of equilibrium at a common temperature. Had the universe remained static at the state it had reached after 300,000 years, it would in some respects have resembled the state of heat death described by Helmholtz. However, the expansion of the universe pulled the material out of equilibrium, allowing heat to flow and driving complex physical processes. The universe cooled as it expanded, but the radiation cooled more slowly than the matter, opening up a temperature gap and allowing heat to flow from one to the other. (The temperature of radiation when expanded varies inversely in proportion to the scale factor, whereas the temperature of nonrelativistic matter varies as the inverse square of the scale factor.) In many other ways too, thermodynamic disequilibrium emerged from equilibrium, most notably in the formation of stars, which radiate their heat into the darkness of space. This directionality is the "wrong way" from the point of view of a naïve application of the Second Law (which predicts a transition from disequilibrium to equilibrium), and it shows that even as entropy rises, new sources of free energy are created.

I must stress that this "wrong way" tendency in no way conflicts with the letter of the Second Law. To see why this is so, an analogy may be helpful. Imagine a gas confined in a cylinder beneath a piston, as in a heat engine. The gas is in thermodynamic equilibrium at a uniform temperature. The entropy of the gas is at a maximum. Now suppose that the gas is compressed by driving the piston forward; it will heat up, as a consequence of Boyle's Law. If the piston is now withdrawn again, restoring the gas to its original volume, the temperature will fall once more. In a reversible cycle of contraction and expansion, the final state of the gas will be the same as the initial state. What happens is that the piston must perform some work in order to compress the gas against its pressure, and this work appears as heat energy in the gas, raising its temperature. In the second part of the cycle, when the piston is withdrawn, the pressure of the gas pushes the piston out and returns exactly the same amount of energy as the piston had injected. The temperature of the gas therefore falls to its starting value when the piston returns to its starting position.

However, in order for the cycle to be reversible, the piston must move very slowly relative to the average speed of the gas molecules. If the piston is moved suddenly, the gas will lag behind in its response, and this will cause a breakdown of reversibility. This is easy to understand. If the piston moves fast when it compresses the gas, there will be a tendency for the gas molecules to crowd up beneath the piston. As a result, the pressure of the gas beneath the piston will be slightly greater than the pressure within the body of the gas, and so the piston will have to do rather more work to compress the gas than would have been the case had it moved more slowly. This will result in more energy being transferred from the advancing piston to the gas than would otherwise have been the case. Conversely, when the piston is suddenly withdrawn, the molecules have trouble keeping pace and lag back somewhat, thus reducing the density and pressure of the gas adjacent to the piston. The upshot is that the work done by the gas on the piston during the outstroke is somewhat less than the work done by the piston on the gas during the instroke. The overall effect is a net transfer of energy from the piston to the gas, and the temperature, hence the entropy, of the gas rises with each cycle. Thus, although the gas was initially in a state of uniform temperature and maximum entropy, after the piston moves the entropy nevertheless rises. The point is, of course, that to say the entropy of the gas is a maximum is to say that it has the highest value *consistent with the external constraints* of the system. But if those constraints change – because of the rapid motion of the piston, for example – then the entropy can go higher. During the movement phase, then, the gas will change from a state of equilibrium to a state of disequilibrium. This comes about not because the entropy of the gas falls – it never does – but because the maximum entropy of the gas increases, and, moreover, it increases faster than the actual entropy. The gas then races to "catch up" with the new constraints.

We can understand what is going on here by appreciating the fact that the gas within a movable piston and cylinder is not an isolated system. To make the cycle run, there has to be an external energy source to drive the piston, and it is this source that supplies the energy that raises the temperature of the gas. If the total system – gas plus external energy source – is considered, then the system is clearly not in thermodynamic equilibrium to start with, and the rise in entropy of the gas is unproblematic. The entropy of the gas cannot go on rising forever. Eventually, the energy source will run out and the piston and cylinder device will stabilize in a final state of maximum entropy for the total system.

The confusion sets in when the piston-and-cylinder expansion and contraction is replaced by the cosmological case of an expanding and (maybe, one day) contracting universe. Here the role of the piston-and-cylinder arrangement is played by the gravitational field. The external energy supply is provided by the gravitational energy of the universe. This has some odd features, because gravitational energy is actually negative. Think, for example, of the solar system. One would have to do work to pluck a planet from its orbit around the sun. The more material concentrates, the lower the gravitational energy becomes. Imagine a star that contracts under gravity; it will heat up and radiate more strongly, thereby losing heat energy and making its gravitational energy more negative in order to pay for it. Thus the principle that a system will seek out its lowest energy state causes gravitating systems to grow more and more inhomogeneous with time. A smooth distribution of gas, for example, will grow clumpier with time under the influence of gravitational forces. Note that this is the opposite trend from the case of a gas, in which gravitation may be ignored. In that case, the Second Law of Thermodynamics predicts a transition toward uniformity. This is only one sense in which gravitation somehow goes "the wrong way."

It is tempting to think of the growth of clumpiness in gravitating systems as a special case of the Second Law of Thermodynamics – that is, to regard the initial smooth state as a low-entropy (or ordered) state, and the final clumpy state as a high-entropy (or disordered) one. It turns out that there are some serious theoretical obstacles to this simple characterization. One such obstacle is that there seems to be no lower bound on the energy of the gravitational field. Matter can just go on shrinking to a singular state of infinite density, liberating an infinite amount of energy on the way. This fundamental instability in the nature of the gravitational field forbids any straightforward treatment of the thermodynamics of self-gravitating systems. In practice, an imploding ball of matter would form a black hole, masking the ultimate fate of the collapsing matter from view. So from the outside, there is a bound on the growth of clumpiness. We can think of a black hole as the equilibrium end state of a self-gravitating system. This interpretation has been confirmed by Stephen Hawking, who proved that black holes are not strictly black, but glow with thermal radiation.[15] The Hawking radiation has exactly the form corresponding to thermodynamic equilibrium at a characteristic temperature.

If we sidestep the theoretical difficulties of defining a rigorous notion of entropy for the gravitational field and take some sort of clumpiness as a measure of disorder, then it is clear that a smooth distribution of matter represents a low-entropy state as far as the gravitational field is concerned, whereas a clumpy state, perhaps including black holes, is a high-entropy state. Returning to the theme of the cosmic arrow of time, and remembering the observed fact that the universe began in a remarkably smooth state, we may conclude that the matter was close to its maximum entropy state, but that the gravitational field was in a low-entropy state. The explanation for the arrow of time that describes the Second Law of Thermodynamics lies therefore in an explanation of how the universe attained the smooth state it had at the Big Bang. Penrose[16] has attempted to quantify the degree of surprise associated with this smooth initial state. In the case of, say, a normal gas, there is a basic relationship between the entropy of its state and the probability that the state would be selected from a random list of all possible states. The lower the entropy, the less probable would be the state. This link is exponential in nature, so that as soon as one departs from a state close to equilibrium (i.e., maximum entropy), the probability plummets. If one ignores the theoretical obstacles and just goes ahead and applies this same exponential statistical relationship to the gravitational field, it is possible to assess the "degree of improbability" that the universe should be found initially in such a smooth gravitational state. In order to do this, Penrose compared the actual entropy of the universe to the value it would have had if the Big Bang had coughed out giant black holes rather than smooth gas. Using Hawking's formula for the entropy of a black hole, Penrose was able to derive a discrepancy of 10^{30} between the actual entropy and the maximum possible entropy of the observable universe. Once this huge number is exponentiated, it implies a truly colossal improbability that the universe should start out in the observed relatively smooth state. In other words, the initial state of the universe is staggeringly improbable.

What should we make of this result? Should it be seen as evidence of design? Unfortunately, the situation is complicated by the inflationary universe scenario, which postulates that the universe jumped in size by a huge factor during the first split second. This would have the effect of smoothing out initial clumpiness. But this simply puts back the chain of explanation one step, because at some stage one must assume that the universe is in a less-than-maximum entropy state, and hence in an exceedingly improbable state. The alternative – that the universe began in its maximum entropy state – is clearly absurd, because it would then already have suffered heat death.

4. THE COSMOLOGICAL ORIGIN OF TIME'S ARROW

The most plausible physical explanation for the improbable initial state of the universe comes from quantum cosmology, as expounded by Hawking, Hartle, and Gell-Mann.[17] In this program, quantum mechanics is applied to the universe as a whole. The resulting "wave function of the universe" then describes its evolution. Quantum cosmology is beset with technical mathematical and interpretational problems, not the least of which is what to make of the infinite number of different branches of the wave function, which describes a superposition of possible universes. The favored resolution is the many-universes interpretation, according to which each branch of the wave function represents a really existing parallel reality, or alternative universe.

The many-universes theory neatly solves the problem of the origin of the arrow of time. The wave function as a whole can be completely time-symmetric, but individual branches of the wave function will represent universes with temporal directionality. This has been made explicit in the time-symmetric quantum cosmology of Hartle and Gell-Mann,[18] according to which the wave function of the universe is symmetric in time and describes a set of recontracting universes that start out with a Big Bang and end up with a big crunch. The wave function is the same at each temporal extremity (bang and crunch). However, this does not mean that time runs backward in the recontracting phase of each branch, à la Gold. To be sure, there are some branches of the wave function in which entropy falls in the recontracting phase, but these are exceedingly rare among the total ensemble of universes. The overwhelming majority of branches correspond to universes that either start out with low entropy and end up with high entropy, or vice versa. Because of the overall time symmetry, there will be equal proportions of universes with each direction of asymmetry. However, an observer in any one of these universes will by definition call the low-entropy end of the universe the Big Bang and the high-entropy end the big crunch. Without the temporal asymmetry implied, life and observers would be impossible, so there is an anthropic selection effect, with those branches of the universe that are thermodynamically bizarre (starting and ending in equilibrium) going unseen. Thus the ensemble of all possible universes shows no favored temporal directionality, although many individual branches do, and within those branches observers regard the "initial" cosmic state as exceedingly improbable. Although the Hartle–Gell-Mann model offers a convincing first step in explaining the origin of the arrow of time, it is not without its problems.[19]

To return to the description of our own universe (or our particular branch of the cosmological wave function), it is clear that the state of the universe in its early stages was one in which the matter and radiation were close to thermodynamic equilibrium, but the gravitational field was very far from equilibrium. The universe started, so to speak, with its gravitational clock wound up, but with the rest in an unwound state. As the universe expanded, there was a transfer of energy from the gravitational field to the matter, similar to that in the piston-and-cylinder arrangement. In effect, gravity "wound up" the rest of the universe. The matter and radiation started out close to maximum entropy consistent with the constraints, but then the constraints changed (the universe expanded). Because the rate of expansion was very rapid relative to the physical processes concerned, a lag opened up between the maximum possible entropy and the actual entropy, both of which were rising. In this way, the universe was pulled away from thermodynamic equilibrium by the expansion. Note that the same effect would occur if the universe contracted again, just as the instroke of the piston serves to raise the entropy of the confined gas. So there is no thermodynamic basis for supposing that the arrow of time will reverse should the universe start to contract.

The history of the universe, then, is one of entropy rising but chasing a moving target, because the expanding universe is raising the maximum possible entropy at the same time. The size of the entropy gap varies sharply as a function of time. Consider the situation one second after the big bang. (I ignore here the situation before the first second, which is complicated but crucial in determining some important factors, such as the asymmetry between matter and antimatter in the universe.) The universe consisted of a soup of subatomic particles – such as electrons, protons, and neutrons – and radiation. Apart from gravitons and neutrinos, which decoupled from the soup well before the first second owing to the weakness of their interactions, the rest of the cosmic stuff was more or less in equilibrium. However, all of this changed dramatically during the first 1,000 seconds or so. As the temperature fell, it became energetically favorable for protons and neutrons to stick together to form the nuclei of the element helium. All of the neutrons got gobbled up in this way, and about 25 percent of the matter was turned into helium. However, protons outnumbered neutrons, and most of the remaining 75 percent of the nuclear matter was in the form of isolated protons – the nuclei of hydrogen. Hydrogen is the fuel of the stars. It drives the processes that generate most of the entropy in the universe today, mainly by converting slowly into helium. So the lag behind equilibrium conditions is this: the universe would really "prefer" to be made of helium (it is more stable), but most of it is trapped in the form of hydrogen. I say "trapped" because, after a few minutes, the temperature of the universe fell below that required for nuclear reactions to proceed, and it had to wait until stars were formed before the conversion of hydrogen into helium could be resumed. Thus the expansion of the universe generated a huge entropy gap – a gap between the actual and the maximum possible entropy – during the first few minutes, when the equilibrium form of matter changed (due to the changing constraints occasioned by the cosmological expansion and the concomitant fall in temperature) from a soup of unattached particles to that of composite nuclei like helium. It was this initial few minutes that effectively "wound up" the universe, giving it the stock of free energy and establishing the crucial entropy gap needed to run all the physical processes, such as star burning, that we see today – processes that sustain interesting activity, such as life. The effect of starlight emission is to slightly close the entropy gap, but all the while the expanding universe serves to widen it. However, the rate of increase of the maximum possible entropy during our epoch is modest compared to what it was in the first few minutes after the Big Bang – partly because the rate of expansion is much less, but also because the crucial nuclear story was all over in a matter of minutes. (The gap-generating processes occasioned by the expansion of the universe today are all of a less significant nature.) I haven't done the calculation, but I suspect that today starlight emission generates more entropy than the rise in the maximum entropy caused by the expansion, so that the gap is probably starting to close, although it has a long way to go yet, and it could start to open up again if the dominant processes in the universe eventually proceed sufficiently slowly that they once more lag behind the pace of expansion. (This may happen if, as present observational evidence suggests, the expansion rate of the universe accelerates.)

5. THE ULTIMATE FATE OF THE UNIVERSE

Of course one wants to know which tendency will win out in the end. Will the expanding universe continue to provide free energy for life and other processes, or will it fail to keep pace with the dissipation of energy produced by entropy-generating processes such as starlight emission? This question has been subjected to much study following a trail-blazing paper by Freeman Dyson.[20] First, I should point out that the ultimate fate of the universe differs dramatically according to whether or not it will continue to expand forever. If there is enough matter in the universe, it will eventually reach a state of maximum expansion, after which it will start to contract at an accelerating rate until it meets its demise in a "big crunch" some billions of year later. In a simple model, the big crunch represents the end of space, time, and matter. The universe will not have reached equilibrium before the big crunch, so there will be no heat death, but rather death by sudden obliteration. Some theological speculations about the end state of a collapsing universe have been made by Tipler.[21]

In the case of an ever-expanding universe, the question of its final state is a subtle matter. The outcome depends both on the nature of the cosmological expansion and on assumptions made about the ultimate structure of matter and the fundamental forces that operate in the universe. If the expansion remains more or less uniform, then when the stars burn out and nuclear energy is exhausted, gravitational energy from black holes could provide vast resources to drive life and other activity for eons. Over immense periods of time, even black holes evaporate, via the Hawking effect. According to one scenario, all matter and all black holes eventually decay, leaving space filled with a dilute background of photons and neutrinos, diminishing in density all the time. Nevertheless, as Dyson has argued, even in a universe destined for something like the traditional heat death, it is possible for the integrated lifetime of sentient beings to be limitless. This is because, as energy sources grow scarce, such beings could hibernate for longer and longer periods of time in order to conserve fuel. It is a mathematical fact that, if they are able to carefully husband ever-dwindling supplies of fuel, the total "awake" duration for a community of sentient beings may still be infinite.

Another possibility is that the expansion of the universe will not be uniform. If large-scale distortions, such as shear, become established, then they can supply energy through gravitational effects. Mathematical models suggest that this supply of energy may be limitless. However, following the publication of the results from WMAP, a satellite that is carefully mapping the heat radiation left over from the Big Bang, a consensus has begun to build that the expansion of the universe is now dominated by so-called dark energy. In its simplest form, dark energy is a universal gravitational repulsive force that increases with time, serving to accelerate the rate of expansion until it settles on a final value in the far future. The effect of accelerating expansion is to slowly eliminate any large-scale nonuniformity, and to create a cosmological "event horizon" that limits the volume of space over which any observer has causal access. This also places an absolute upper bound on the total entropy that this volume of space may possess. After an immense duration of time, what is effectively "the universe" for any localized observer would approach this final state of theoretically possible maximum entropy. The "universe" thus suffers a fate that is really just a modern variant of the Helmholtz cosmic heat death of the nineteenth century. In this case, the end state of the universe would be one of empty space at a temperature of about 10^{-28}K. It is unlikely that Dyson's "immortality" scenario would work in such a universe.

Still another scenario is that, from the dying remnants of this universe, another "baby" universe may be created because of exotic quantum vacuum effects. Thus, even though our own universe may be doomed, others might be produced spontaneously, or perhaps artificially, to replace them as abodes for life. In this manner, the collection of universes, dubbed the "multiverse," may be eternal and ever-replenishing, even though any given universe is subject to heat death. I have summarized these many and varied scenarios in my book *The Last Three Minutes*.[22]

6. A LAW OF INCREASING COMPLEXITY?

Clearly, modern cosmology paints a far more complicated picture of the fate of the universe than the simple heat death scenario of the nineteenth century. But leaving this aside, the question arises as to the relevance of the Second Law of Thermodynamics to cosmic change in any case. While it is undeniably true that the entropy of the universe increases with time, it is not clear that entropy is the most significant indicator of cosmic change. If you ask a cosmologist for a brief history of the universe, you will get an answer along the following lines. In the beginning, the universe was in a very simple state – perhaps a uniform soup of elementary particles at a common temperature, or even just expanding empty space. The rich and complex state of the universe as observed today did not exist at the outset. Instead, it emerged in a long and complicated sequence of self-organizing and self-complexifying processes, involving symmetry breaking, gravitational clustering, and the differentiation of matter. All this complexity was purchased at an entropic price. As I have explained, the rapid expansion of the universe just after the Big Bang created a huge entropy gap, which has been funding the accumulating complexification ever since, and which will continue to do so for a long while yet. Thus the history of the universe is not so much one of entropic degeneration and decay as a story of the progressive enrichment of systems on all scales, from atoms to galaxies.

It is tempting to postulate a universal principle of increasing complexity to go alongside the degenerative principle of the Second Law of Thermodynamics. These two principles would not be in conflict, since complexity is not the opposite of entropy. The growth of complexity can occur alongside an increase in entropy. Indeed, the time irreversibility introduced into macroscopic processes by the Second Law of Thermodynamics actually provides the opportunity for a law of increasing complexity, since if complexity were subject to time-reversible rules, there would then be a problem about deriving a time-asymmetric law from them. (This is a generalized version of the dictum that death is the price that must be paid for life.) In a dissipative system, an asymmetric trend toward complexity is unproblematic, and examples of such a trend are well known from the study of far-from-equilibrium (dissipative) processes.[23]

Many scientists have flirted with the idea of a quasi-universal law of increasing complexity, which would provide another cosmic arrow of time. For example, Freeman Dyson[24] has postulated a principle of maximum diversity, according to which the universe in some sense works to maximize its richness, making it the most interesting system possible. In the more restrictive domain of biosystems, Stuart Kauffman has discussed a sort of fourth law of thermodynamics. In *Investigations*, he writes: "I suspect that biospheres maximize the average secular construction of the diversity of autonomous agents, and ways those agents can make a living to propagate further. In other words, biospheres persistently increase the diversity of what can happen next. In effect, biospheres may maximize the average sustained growth of their own dimensionality."[25] I have summarized various earlier attempts at formulating a principle of progressive cosmological organization in my book *The Cosmic Blueprint*.[26] These efforts bear a superficial resemblance to the medieval theological concept of the best of all possible worlds. However, they are intended to be physical principles, not commentaries on the human condition. It may well be the case that the laws of nature are constructed in such a way as to optimize one or more physical quantities, while having negligible implications for the specific activities of autonomous agents. Thus we can imagine that the laws of nature operate in such a way as to facilitate the emergence of autonomous agents and to bestow upon them the maximum degree of freedom consistent with biological and physical order, while leaving open the specifics of much of the behavior of those agents. So human beings are free to behave in undesirable ways, and the fact that human society may involve much misery – and is clearly not "the best of all possible worlds" – is entirely consistent with a principle of maximum richness, or diversity, or organizational potential.

Attractive though the advance of complexity (or organization, or richness, or some other measure of antidegeneration) may be, there are some serious problems with the idea, not least that of arriving at a satisfactory definition of precisely what quantity it is that is supposedly increasing with time. A further difficulty manifests itself in the realm of biology. The evolution of the Earth's biosphere seems to be a prime example of the growth of organized complexity. After all, the biosphere today is far more complex than it was 3.5 billion years ago. Moreover, extant individual organisms are more complex than the earliest terrestrial microbes. However, as Gould has stressed,[27] we must be very careful how we interpret these facts. If life started out simple, and has randomly explored the space of possibilities, it is no surprise that it has groped on average toward increased complexity. But this is not at all the same as displaying a systematic trend or drive toward greater complexity.

In order to illustrate the distinction, I have drawn two graphs (Figure 10.1) showing possible ways in which the complexity of life might change with time. In both cases, the mean complexity increases, but in (a) the increase is essentially just a diffusion effect, rather like a spreading wave packet in quantum mechanics. There is no progressive "striving" toward greater complexity. Clearly, the "wave packet" cannot diffuse to the left, because it is bounded by a state of minimal complexity (the least complexity needed for life to function). At any given time, there will be a most complex organism, and it is no surprise if later most-complex organisms are more complex than earlier most-complex organisms, merely due to the spreading of the tail of the distribution. The fact that today, *Homo sapiens* occupies the extreme end of the complexity distribution should not prejudice us into thinking that evolutionary history is somehow pre-oriented to the emergence of intelligence, or brain size, or whatever complex trait we deem to be significant. So the rise of complexity would be rather trivial in the case illustrated by (a); it does not illustrate any deep cosmic principle at work, but merely the operation of elementary statistics. By contrast, (b) shows what a progressive trend toward greater complexity might look like. The "wave packet" still diffuses, but the median also moves steadily to the right, in the direction of greater complexity. If this latter description were correct, it would suggest that a principle of increasing complexity is at work in biological evolution.

What does the fossil record show? Gould[27] argues forcefully that it supports (a) and confounds (b). His is the standard neo-Darwinian position. It is the essence of Darwinism that nature has no foresight – it cannot look ahead and anticipate the adaptive value of specific complex features. By contrast, the concept of "progressive" trends has an unnervingly teleological – even Lamarckian – flavor to it, making it anathema to most biologists.[28]

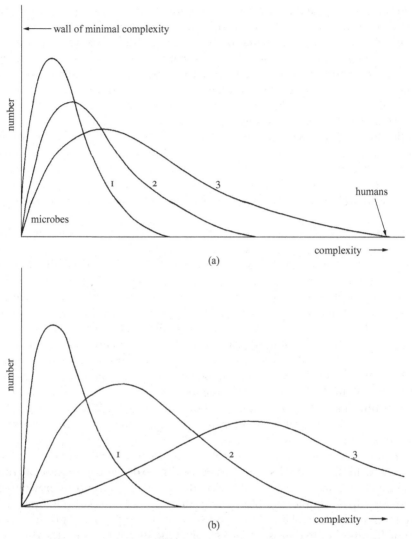

FIGURE 10.1. Ladder of progress? Biological complexity increases over time, but is there a systematic trend or just a random diffusion away from a "wall of simplicity"? The diffusion model, supported by Gould, is shown in (a). Curves 1, 2, and 3 represent successive epochs. Life remains dominated by simple microbes, but the tail of the distribution edges to the right. If there were a definite drive toward complexity, the curves would look more like those in (b).

On the experimental front, however, the fossil record is scrappy, and there have been challenges to Gould's position. For example, it has been pointed out[29] that the rate of encephalization (brain to body mass ratio) is on an accelerating trend. Moreover, the simplest microbes known seem to display far more complexity than the minimal possible, although this is open to the objection that we may not yet have identified the simplest autonomous organisms.

Many SETI researchers unwittingly support (b) in anticipating that intelligence and advanced technological communities will arise more or less automatically once life gets started. Not surprisingly perhaps, Gould, in common with many biologists, is a critic of SETI, believing that intelligence is exceedingly unlikely to develop on another planet, as there is no evolutionary trend, he maintains, that canalizes complexity in the direction of large brains.

The clearest example of a trend toward complexity comes not from biology but from astrophysics. A snapshot of the early universe was provided by the COBE satellite and more recently by WMAP, both of which mapped tiny variations in the heat radiation left over from the Big Bang. This radiation is extremely smooth across the sky, indicating that the early universe was highly uniform, as discussed earlier. The amplitude of the temperature ripples detected are only a few parts per million. The hot spots represent the first glimmerings of large-scale cosmological structure. Over time, the hotter, denser regions of the universe have grown at the expense of the rest. Gravity has caused matter to accumulate into galaxies, and the galaxies into clusters. Within galaxies, matter has concentrated into stars. It seems intuitively obvious that the elaborate structure of the Milky Way galaxy, with its spiral arms and myriad stars, is far more complex than the diffuse swirling nebula that produced it.

However, quantifying this growth in complexity is hard. Over the last few decades, there have been several attempts to try to develop a measure of gravitational complexity that will describe the irreversible trend in gravitating systems – from simple, smooth initial states to more complex, clumpy final states. As already explained, the end state resulting from this gravitational concentration of matter is the black hole. One idea is to try to describe the growth of clumpiness in terms of entropy, since it would then have an explanation in a suitably generalized Second Law of Thermodynamics. Penrose has made some specific proposals along these lines.[30] Unfortunately, no satisfactory definition of gravitational entropy has yet emerged.

What sort of principle would a law of increasing complexity be? First, it is obvious that there can be no absolute and universal law of this sort, because it is easy to think of counterexamples. Suppose that Earth were hit tomorrow by a large asteroid that wiped out all "higher" life forms. That event would reset the biological complexity of the planet to a seriously lower level. However, we might still be able to show that, following such a setback, and otherwise left alone, life rides an escalator of complexity growth. Another caveat concerns the ultimate fate of the universe. So long as there is an entropy gap, the universe can go on creating more and more complexity. However, in some scenarios of the far future, matter and black holes eventually decay and evaporate, leaving a very dilute soup of weakly interacting subatomic particles gradually cooling toward absolute zero and expanding to a state of zero density. The complexity of such a state seems much lower than the complexity today (although the presently observable universe will have grown to an enormous volume by then, so the total complexity may still be rather large). Therefore, in the end, complexity may decline. However, this does not preclude a more restrictive law of increasing complexity – one subject to certain prerequisites.

If some sort of complexity law were indeed valid, what would be its nature? Complexity theorists are fond of saying that they seek quasi-universal principles that would do for complexity what the laws of thermodynamics do for entropy. But would such a law emerge as a *consequence* of known physics, or do the laws of complexity *complement* the laws of physics? Are they emergent laws? Robert Wright[31] has argued that the history of evolution drives an increase in complexity through the logic of nonzero sumness – that is, individual units, whether cells, animals, or communities, may cooperate in ways that produce a net benefit for all. Evolution will exploit such win-win interactions and select for them. The question is then whether the nonzero sum principle is something that augments the laws of nature or follows from them. If it augments them, whence comes this higher-level law? If it is a consequence of the laws of nature, then will such a principle follow from a random set of natural laws, or is there something special about the actual laws of the real universe that facilitates the emergence of nonzero sum interactions?

Whatever the answers to these difficult questions, it is clear that any general principle of advancing complexity would reintroduce a teleological element into science, against the trend of the last three hundred years. This is not a new observation. A hint of teleology is discernible in the approach of many investigators who have suggested that complexity may grow with time.[32] For example, Kauffman[33] writes of "we, the expected." But I shall finish with my favorite quote of Freeman Dyson,[34] who has stated, poetically, that "in some sense the universe must have known we were coming."

CONCLUSION

Many physical processes are irreversible, and they thereby bestow upon the universe an arrow of time. This temporal directionality is usually traced to the Second Law of Thermodynamics, according to which order tends to give way to chaos, leading to an inexorable decay and degeneration in physical systems. The ultimate end state of this increasing disorder is the so-called cosmic heat death. However, this picture of a dying universe, which for 150 years has provoked nihilistic and atheistic commentary, may be overly simplistic. Modern cosmology reveals that the universe is expanding, and that it is subject to subtle gravitational effects that complicate the traditional "order into chaos" theme. Moreover, the discovery of self-organizing and self-complexifying processes in nature suggests that alongside the degenerative arrow of time there exists a creative arrow, pointing in the direction of increasing richness, diversity, and potential. There is no conflict between these two arrows. The ultimate fate of the universe remains open.

Notes

1. Bertrand Russell, *Why I Am Not a Christian* (New York: Allen and Unwin, 1957), p. 107.
2. Peter Atkins, "Time and Dispersal: The Second Law," in *The Nature of Time*, ed. Raymond Flood & Michael Lockwood (Oxford: Basil Blackwell, 1986), p. 98.
3. Steven Weinberg, *The First Three Minutes*, updated edition (New York: Harper and Row, 1988).
4. Henri Bergson, *Creative Evolution*, trans. A. Mitchell (London: Macmillan, 1964).
5. See, for example, C. Raven, *Teilhard de Chardin: Scientist and Seer* (New York: Harper and Row, 1962).
6. A. N. Whitehead, *Process and Reality: An Essay in Cosmology*, corrected edition, ed. D. R. Griffin and D. W. Sherburne (New York: The Free Press, 1978).
7. For a review of vitalism and related concepts, see Rupert Sheldrake, *The Presence of the Past* (London: Collins, 1988).
8. *The Nature of Time*, ed. T. Gold (Ithaca, NY: Cornell University Press, 1967).
9. S. W. Hawking, R. Laflamme, and G. W. Lyons, "The Origin of Time Asymmetry," *Physical Review* D47 (1993): 5342.
10. S. W. Hawking, "The No Boundary Condition and the Arrow of Time," in *Physical Origins of Time Asymmetry*, ed. J. J. Halliwell, J. Perez-Mercader, and W. H. Zurek (Cambridge: Cambridge University Press, 1994), p. 346.
11. L. S. Schulman, "Opposite Thermodynamic Arrows of Time" (forthcoming).
12. M. Eliade, *The Myth of the Eternal Return*, trans. W. R. Trask (New York: Pantheon, 1954).
13. A. S. Eddington, "The End of the World from the Standpoint of Mathematical Physics," *Nature* 127 (1931): 447.
14. P. C. W. Davies and J. Twamley, "Time Symmetric Cosmology and the Opacity of the Future Light Cone," *Classical and Quantum Gravity* 10 (1993): 931.
15. S. W. Hawking, "Particle Creation by Black Holes," *Communications in Mathematical Physics* 43 (1975): 199.
16. Roger Penrose, "Singularities and Time Asymmetry," in *General Relativity: An Einstein Centenary Survey*, ed. S. W. Hawking and W. Israel (Cambridge: Cambridge University Press, 1979), p. 581.
17. For a popular account, see Stephen Hawking, *A Brief History of Time* (New York: Bantam Books, 1988).

18. M. Gell-Mann and J. B. Hartle, "Time Symmetry and Asymmetry in Quantum Mechanics and Quantum Cosmology," in *Physical Origins of Time Asymmetry*, ed. J. J. Halliwell, J. Perez-Mercader, and W. H. Zurek (Cambridge: Cambridge University Press, 1994), p. 311.
19. See Davies and Twamley, "Time Symmetric Cosmology."
20. Freeman Dyson, "Time without End: Physics and Biology in an Open Universe," *Reviews of Modern Physics* 51 (1979): 447.
21. Frank Tipler, *The Physics of Immortality* (New York: Doubleday, 1994).
22. See, for example, Paul Davies, *The Last Three Minutes* (New York: Basic Books, 1994) and the references cited therein.
23. For a review, see Ilya Prigogine and Isabelle Stengers, *Order out of Chaos* (London: Heinemann, 1984).
24. Freeman Dyson, *Disturbing the Universe* (New York: Harper and Row, 1979), p. 250.
25. Stuart Kauffman, *Investigations* (Oxford: Oxford University Press, 2000), p. 153.
26. Paul Davies, *The Cosmic Blueprint* (London: Heinemann, 1987).
27. Stephen Jay Gould, *Life's Grandeur* (London: Jonathan Cape, 1996).
28. See, for example, Matthew H. Nitecki, *Evolutionary Progress* (Chicago: University of Chicago Press, 1988).
29. James Dale Russell, "Exponential Evolution: Implications for Intelligent Extraterrestrial Life," *Advances in Space Research* 3 (1983): 95.
30. See Penrose, "Singularities and Time Asymmetry."
31. Robert Wright, *Nonzero* (New York: Pantheon, 2000).
32. It is, in fact, possible to give expression to apparent "anticipatory" qualities in nature without reverting to causative teleology in the original sense. I have discussed this topic in a recent paper, "Teleology without Teleology: Purpose through Emergent Complexity," in *Evolutionary and Molecular Biology: Scientific Perspectives on Divine Action*, ed. Robert John Russell, William R. Stoeger, S. J., and Francisco J. Ayala (Vatican City: Vatican Observatory Publications; Berkeley: Center for Theology and the Natural Sciences, 1998), p. 151.
33. Stuart Kauffman, *At Home in the Universe* (Oxford: Oxford University Press, 1995).
34. Dyson, *Disturbing the Universe*, p. 250.

11

The Emergence of Biological Value

James Barham

1. INTRODUCTION

All the things we think of as paradigmatic cases of design – novels, paintings, symphonies, clothes, houses, automobiles, computers – are the work of human hands guided by human minds. Thus, design might be defined as matter arranged by a mind for a purpose that it values. But this raises the question, what are minds? Presumably, the activity of brains. The problem with this answer, however, is that brains themselves give every appearance of being designed. Most contemporary thinkers view brains as neurons arranged for the purpose of thinking in much the same way that, say, mousetraps are springs and levers arranged for the purpose of killing mice. But if that is so, then who arranged the neurons? Who or what values thinking, and whose purpose does it serve?

It is generally supposed that there are only two ways to answer these questions. One way has come to be known as *Intelligent Design*. On this view, our brains were designed by other minds existing elsewhere – say, in another galaxy or on another plane of being. But if these other minds are also instantiated in matter, then we have the same problem all over again. If not, then we have an even more difficult problem than the one we started with. To invoke immaterial minds to explain the design of material ones is surely a case of *obscurum per obscurius*.

The other way is what I shall call the *Mechanistic Consensus*. In summary, the Mechanistic Consensus holds that (1) the known laws of physics and chemistry, together with special disciplines such as molecular biology, fully explain how living things work, and (2) the theory of natural selection explains how these laws have come to cooperate with one another to produce the appearance of design in organisms. According to the Mechanistic Consensus, design is not objectively real but merely an optical illusion, like the rising and setting of the sun. On this view, living matter is nothing special. It is just chemistry shaped by natural selection.

In this paper, I will argue that the Mechanistic Consensus is wrong. It is wrong because, conventional wisdom to the contrary, (1) present day physics and chemistry do *not* provide the conceptual resources for a complete understanding of how living things work, and (2) natural selection does *not* provide an adequate means of naturalizing the normative teleology in living things. However, in spite of this failure of the Mechanistic Consensus, I will argue that we are still not forced to the Intelligent Design position, because there exists a third way of explaining the appearance of design in living things.

One of the hallmarks of a machine is that the relationship between its function and its material constitution is arbitrary. Intelligent Design and the Mechanistic Consensus agree that organisms are machines in this sense, consisting of matter that is inert insofar as its function is concerned. Both schools of thought view biological functions as something imposed on inert matter from the outside, by the hand of God or by natural selection, as the case may be. But what if the analogy between organisms and machines were fundamentally flawed? Suppose that the teleological and normative character of living things really derived from an essential connection between biological function and the spontaneous activity of living matter. In that case, such a connection might give rise to systems that prefer or value some of their own possible states over other, energetically equivalent ones, and that strive to attain these preferred states under the constraint of external conditions in accordance with means-ends logic. Then, instead of being an illusion, as the Mechanistic Consensus claims, the purpose and value seemingly inherent in the functional actions of living things might be objectively real. If all of this were so – and I will argue that it is – then living matter would be special, after all. Although we have little idea as yet in what this specialness consists, in the last section of this chapter I will briefly consider the implications of some promising lines of contemporary research in nonlinear dynamics and condensed matter physics for understanding the emergence of biological value.

2. THE MECHANIST'S DILEMMA

Living things give every appearance of purposiveness. It is entirely natural to describe biological processes as functions that operate according to means-ends logic. Functional ends or goals constitute norms with respect to which the means chosen may be judged good or bad, right or wrong, successful or unsuccessful. Furthermore, organisms must be capable of choosing means appropriate to their ends – that is, of being right – at least some of the time. For example, in order to live, a cell must move in the right direction when it encounters a nutrient gradient. The very existence of life presupposes the possibility of correct functioning. On the other hand, organisms are also necessarily capable of error. What appears to be a nutrient gradient may in fact turn out to be a lure or a poison. Functions are inherently capable of malfunctioning. Right or wrong, organisms behave according to functional logic. *This* is done so that *that* may happen. *A* is preferred; *B* is necessary for *A*; therefore, *B* is chosen. All function conforms to this pattern. Which is to say that, in pursuing their ends, organisms are not propelled by causes; rather, they act for reasons.

Behavior answering to these ordinary-language descriptions clearly exists. It is easily confirmed through elementary empirical observation. True, the teleological and normative language used in the previous paragraph to describe functional behavior might be dismissed as pretheoretical and without scientific value. But this claim presupposes the existence of an alternative theoretical language into which these descriptions can be translated without loss. This language must itself be rigorously purged of all traces of teleology and normativity. Does such a language in fact exist?

Open any cell biology textbook to any page, and what will you find? Talk of regulation, control, signals, receptors, messengers, codes, transcription, translation, editing, proofreading, and many other, similar terms. It is true that this technical vocabulary is an indispensable aid in describing many previously undreamed-of empirical phenomena. Molecular biology has greatly extended the scope and precision of our knowledge, and the terminology it has developed is an integral part of that accomplishment. But the fact remains that these concepts are no less normative than those of everyday speech. Adherents of the Mechanistic Consensus are untroubled by this defect, because they insist that it is only a matter of convenience. A metaphor like "second messenger," they say, is employed only to avoid intolerably verbose descriptions of the mechanistic interactions that underlie the appearances. Such a *façon de parler* is a promissory note redeemable in the hard currency of physics and chemistry. But as with any IOU, the notes issued by molecular biology are only as good as the guarantors backing them up. If the other sciences cannot pay them either, then the promises are worthless. For this reason, it behooves us to take a closer look at the conceptual solvency of the Mechanistic Consensus.

First, we are told that living things are made of ordinary matter and nothing but ordinary matter. And it is true that biological molecules are composed mostly of a handful of elements (CHNOPS), along with traces of some others, all long familiar to chemists. Certainly, there are no unknown elements in living things that are not present in the periodic table. Second, we are assured that the interactions between these elements *in vivo* are basically the same as those *in vitro* described by present day physics and chemistry. This is a more doubtful claim, to which I shall return later, but for now, let us grant this, too. Even so, there remains a fundamental difficulty.

The difficulty is that, while all the individual reactions in the cell may be described in ordinary physical terms as tending toward an energy minimum, the same cannot be said of the way in which the reactions are organized. When a signal molecule (say, a hormone) interacts with its receptor (a protein), what happens may be more or less understood in terms of biochemistry. But biochemistry has no conceptual resources with which to explain the meaning and the purpose of this reaction – the very things that constitute the reaction *as* a signal, and not just a meaningless jostling of matter. What makes the living cell profoundly different from ordinary inorganic matter is the way in which each reaction is coordinated with all the others for the good of the whole. There is no doubt that this coordination itself transcends the explanatory resources of biochemistry, because it operates according to functional logic, not just according to physical law (Pattee 1982; Rosen 1991; Jonker et al. 2002).

From a purely physical point of view – at least so far as our present state of knowledge is concerned – there is no reason why a reaction that is good for the organism, rather than one that is bad for it, should occur. The very categories of good and bad have no place in physics or chemistry as currently understood, and yet they are at the very heart of life. Every reaction in the cell is more than just a reaction, it is a functional action. Such an action constitutes a choice among states that are energetically equivalent so far as the ordinary laws of physics are concerned. Such preferred states are achieved, not by minimizing energy, but by doing work – that is, by directing internally stored energy here or there according to needs that are normative for the cell. Just as the laws of physics permit me to direct my automobile left or right at an intersection, so too they permit a cell to travel up or down a chemical gradient. There is no use seeking the explanation for such decisions in the physical forces impinging on me or on the cell. It is not physics (at least, not any presently understood physics) that explains purposive action; rather, it is the situational logic of functional action that governs the decisions of cells as integrated wholes (Albrecht-Buehler, 1990; Alt, 1994; Lauffenburger and Horwitz, 1996).

During the past fifty years or so, we have developed a highly sophisticated theoretical framework to explain how such coordinated, goal-directed action works – namely, the theory of feedback and cybernetic control. This theoretical understanding has made possible the construction of complex, self-regulating mechanical systems that operate according to a functional logic similar to that in living things and that fulfill a wide variety of human purposes. There is no doubt that this body of theory provides a great deal of insight into the internal operation of biological systems as well. But there remains a glaring problem. In the case of the machine, we decide what counts as its goal states, and *we* arrange its parts accordingly. Who or what does these things in the cell?

It is often assumed that invoking the concept of information will somehow solve this problem. It is true that all living things utilize information in some sense (Loewenstein 1999). However, this observation merely labels the problem; it contributes little or nothing to its solution. The reason is that, by definition, information is essentially semantic. Without meaning, there is no information; there are just spatial or temporal patterns. For a pattern to constitute information, we must posit a cognitive agent for which the pattern is meaningful. What, then, is semantic information? One plausible answer is: *a correlation between events and functional actions without tight thermodynamic coupling.*

The proviso is important, because correlations that are the direct result of the laws of physics do not constitute information. Information is only possible where choice exists. For choice to exist, the causes of the correlated events must be orthogonal to each other (Nagel 1998). Causes are said to be orthogonal if they are independent of each other insofar as the laws of physics are concerned – that is, if the existence of one does not necessitate the existence of the other. This is indeed the case throughout the living cell (Monod 1972; Pattee 2001; Polanyi 1969). In short, if the correlation between events in the cell were the direct result of the minimization of energy due to tight thermodynamic coupling, then it would make no sense to speak of their occurring on the basis of information. Since that is not the case, it does make sense to speak in this way. Without tight thermodynamic coupling, an event may act as a trigger of a functional action. In that case, the meaning of such an event may be interpreted as a sign of the presence of conditions favorable to the action. In effect, information is an event that tells a biological function: act now, and you will succeed (Barham 1996). Note, however, that the question of how such a correlation between events and goal-directed actions is possible is essentially the same problem that we have been discussing all along – that of explaining the design or normative teleology inherent in life. Shannonian information theory is of no help at all in solving this problem. It simply assumes intelligent agents at either end of the communication channel; it makes no pretense of explaining how physical patterns can acquire meaning in the first place. For this reason, in its present theoretical articulation, the concept of information is an integral part of the problem. It contributes little or nothing to its solution.

If the functional logic of the cell is irreducible to physical law as we currently understand it, then there would appear to be only two ways to explain it naturalistically. Either the teleological design of living things is, at bottom, a matter of chance; or else there is some unknown qualitative difference inherent in the material constitution of organisms that gives them an intrinsic functional integrity. The first option is appealing to the mechanistic biologist, but it is very hard for the physicist to swallow because of the fantastic improbability of living things from a statistical-mechanical point of view, as has often been pointed out (Eden 1967; Elsasser 1998; Lecomte du Noüy 1948; Schoffeniels 1976; Yockey 1992). The second option has attracted a number of physicists who have thought seriously about life (Denbigh 1975;

Elsasser 1998; Schrödinger 1992), but it is unpalatable to most biologists because to them it smacks of prescientific "vitalism."

This, then, is the Mechanist's Dilemma. Is life a statistical miracle? Or is the Mechanistic Consensus defective in some fundamental way? I will examine the first horn of this dilemma in the next section and the other one in section 4.

3. NORMATIVITY AND NATURAL SELECTION

According to the Mechanistic Consensus, the things that happen in organisms do not really happen for a purpose; it only looks that way. In reality, things just happen. Period. What happens in the organism is no different from what happens in the test tube. Enzymes cleave or bond their substrates according to the well-known laws of physics and chemistry. A catalyst is a catalyst is a catalyst. How, then, do mechanists explain the appearance of purposiveness in living things?

They say that some of the things that happen by chance in an organism have the consequence that they enhance the organism's fitness. This means that the probability of the organism's surviving to reproduce within a given set of environmental conditions is increased by the physical or chemical event in question. When this happens, the propensity for that event to occur will be transmitted to the next generation. Then, this event will tend to recur and to have the same consequence in the offspring, so long as the same environmental conditions exist, and likewise in the offspring's offspring. In this way, the representation of the original event in the overall population will gradually increase. At the limit, an event that first occurred in a single organism will spread to all members of a species. In that case, it will appear as though these organisms had been designed for their environment with respect to the event in question. But in reality, all that has happened is that the process of natural selection has locked into place an event that originally occurred by chance insofar as its fit with the environment is concerned.

It is widely assumed that this explanatory scheme gets rid of all the troublesome teleology in biology, but this is a mistake. Natural selection provides only the appearance of reduction, not the reality, as may be seen from a number of considerations. To begin with, we may note that the notions of survival and reproduction undergird the entire Darwinian schema and are not themselves explained by it. But these concepts already remove us from the terra firma of physical interactions and land us right back in the teleological soup. It is sometimes claimed that the stability of a chemical compound constitutes "survival" or that crystal growth is a primitive form of "reproduction," but these metaphors merely obscure the point at issue. Chemical compounds and crystals just seek their energy minimum given a set of contraints, whereas the intelligent responsiveness of an organism to its environment and the complex coordination of events involved in cell division transcend energy minimization. The latter, distinctively biological phenomena already contain the normative feature of striving to achieve particular preferred states by directing energy in some ways rather than in other, energetically equivalent ways. But this is the very thing we are trying to explain. Survival and reproduction demarcate the boundary between the living and the nonliving, and so are far from the unproblematic mechanistic concepts that a successful reduction would require.

Another problem is the way in which selection theory employs the notion of chance. In order for Darwinian reduction to go through, we must assume that an organism's parts are essentially independent variables, each of which is free to change at random with respect to the other parts and with respect to the whole organism's needs. But if organisms really were made of inert, functionally uncorrelated parts, then evolution would be impossible owing to combinatorial explosion. There has simply not been enough time since the Big Bang for even a single protein molecule to be created in this way with any reasonable probability, much less an entire cell – much less the whole inconceivably complex, functionally integrated organic world we see around us. If organisms were literally machines, they would indeed be miraculous – on this point, the Intelligent Design critique of Darwinism is perfectly sound. If organisms were really made of inert parts bearing no intrinsic relation to function, then we would indeed have to assume that they were designed by a humanlike intelligence, because that is the only conceivable way for functionally integrated wholes made of such parts to come into existence.

However, this does not mean that we are forced to accept the Intelligent Design conclusion. Instead, we may reject the premise. This means treating the "design inference" (Dembski 1998) as a reductio ad absurdum of the proposition that organisms are machines. By dropping this assumption, we may view organisms as active and fully integrated systems in which a change in one part leads to appropriate changes cascading throughout the system in accordance with functional logic. In this case, the possibility of evolutionary transformation begins to make sense from a physical point of view, *but now Darwinism has forfeited all of its reductive power.* We have simply assumed the functional organization of the cell, which is the very thing that we claimed to be able to explain by means of the theory of natural selection.

Darwinists often complain that such criticisms are based on a misunderstanding. It is not chance, they say, that bears the explanatory weight in their theory, it is the selection principle. Natural selection is said to act as a ratchet, locking into place the functional gains that are made, so that each new trait can be viewed as a small incremental step with an acceptable probability. But what Darwinists forget is that the way a ratchet increases probabilities and imposes directionality is *through its own structure.* In this context, the structure of the ratchet is simply the functional organization of life. Darwinists are entitled to claim that the explanatory burden of their theory lies upon the selection ratchet, thus avoiding the combinatorial explosion problem, only provided that they also acknowledge that the structure of this ratchet consists precisely in the intrinsic functional correlations among the parts of the organism. But if they do this, then they must also admit that they have merely assumed the very functional organization that they claimed to be able to explain, thus sneaking teleology in by the back door.

Finally, it is often claimed (e.g., Depew and Weber 1998) that the normativity of biological functions can be fully naturalized in terms of Wright's (1998) analysis, in which a function is a part of a system that exists because of the role that it plays within the system. In the case of biological functions, normative functions are traits that have been selected. That is, if a given trait does f, and f happens to cause the trait to be favored in the selection process, then f becomes the "proper function" (Millikan 1998) of that trait. But this analysis reduces the problem of naturalizing normativity to a matter of agreeing on a terminological convention; it has nothing to do with scientific explanation in the usual sense.

Of course, science often looks to history to explain how the present state of a system came into being, but the present causal powers of a system must nevertheless be explicable in terms of the system's present state. After all, "history" is just a convenient shorthand way of referring to the whole sequence of dynamical states of a given system, the past transformations of which have led to the system's present state. But this sequence in itself does not explain the present properties and causal powers of the system; rather, these are explained by the present physical state of the system, which is the only thing that is actual. Living systems are physical systems, and there is no reason to believe them to be exempt from this fundamental metaphysical principle. Therefore, we must conclude that it is something in the present state of a biological function, not its selection history per se, that accounts for its normativity. We must not confuse the present effects of history with history itself.

In summary, the massive coherence and coordination of the parts of biological systems, all intricately correlated to support those systems in existence as organized wholes, must arise either by chance or by some ordering principle conforming to functional logic. Elementary considerations of statistical mechanics and probability theory suffice to exclude the chance hypothesis.[1] Therefore, there must exist an ordering principle. This principle is logically prior to selection, since novel biological forms must already exist before they can be selected. Indeed, all viable novel forms are always already entrained into a fully integrated functional system before selection occurs. Therefore, variations in living form are the cause of differential reproduction, not the effect. This means that the theory of natural selection tacitly presupposes the functional integrity and adaptability of organisms. Which is another way of saying that Darwinism begs the question of teleology.

4. MATERIAL EMERGENCE AND THE GROUND OF
NORMATIVITY IN NATURE

So far, I have argued that neither the known laws of physics and chemistry nor the theory of natural selection succeeds in explaining the teleological and normative characteristics of living things. Since it will be difficult to overcome the Mechanistic Consensus on the strength of these negative arguments alone, I now turn to a positive account of biological value, based on some promising, albeit speculative, lines of contemporary scientific research.

First, we must view our problem against the backdrop of a general picture of cosmic evolution (Denbigh 1975; Layzer 1990). The key concept here is spontaneous symmetry breaking, which is the framework within which the origin of all novelty and all complex structures and processes in the universe must ultimately be understood (Icke 1995). In order to explain this phenomenon, physicists have developed a variety of mathematical tools (above all, the renormalization group) for extracting certain universal properties shared by systems across length scales by abstracting away from physically irrelevant details (Cao 1997; Batterman, 2002). Such techniques work extremely well and seem to reveal a layered world of hierarchical levels, each with its own intrinsic stability and characteristic physical properties (Georgi 1989). The idea is that over the course of its history, the universe has repeatedly produced qualitatively new forms of matter with distinctive causal powers. Anderson (1994) famously encapsulated this insight in the slogan "more is different" (see also Schweber, 1997; Cao, 1998). Thirring (1995) has even gone so far as to speak of the evolution of the laws of nature themselves.

It is true that the asymptotic methods used to model these empirical phenomena have often been interpreted as a gimmick employed to circumvent our own cognitive limitations. But this epistemic interpretation of physical theory is based on little more than reductionist faith (Laughlin and Pines 2000; Laughlin et al. 2001). It is inconsistent with the principle that the best explanation for the success of a theory is that it has a purchase on reality. On the other hand, if we take the success of modern field theoretic methods in physics at face value, then we begin to see the possibility of a new conception of emergence, one that is directly linked to the properties of matter itself in its various guises. Let us call this notion *material emergence*, in order to distinguish it from the more usual idea that emergence is a purely formal property of organization per se.

What reason do we have to believe that biological value is emergent in the material sense? Batterman (2002, 135) notes that "[i]n the physics literature one often finds claims to the effect that [emergent] phenomena constitute 'new physics' requiring novel theoretical work – new asymptotic theories – for their understanding." In other words, wherever novel kinds of material systems are to be found, we can expect to find qualitatively distinctive causal powers, and hence to need "new physics" to describe those powers. For example, condensed matter in general required the development of many new physical concepts and remains imperfectly understood to this day. Why should this same principle not apply to life in particular? Given these considerations, it is unsurprising that physicists are beginning to articulate the need to tackle the expected "new physics" intrinsic to the living state of matter head-on (Laughlin et al. 2000).

Of course, this understanding of the general principle of material emergence still leaves us with one very pressing question: how can we make scientific sense out of biological value as a physical phenomenon? There are two lines of research that seem to me to bear directly on this question. The first of these is nonlinear dynamics (Auyang 1998; Walleczek 2000). Nonlinear dynamical systems are interesting in this context because their behavior possesses a number of properties that seem to be of potential significance for biology. One of these is robustness, meaning that the system will spontaneously damp perturbations to its dynamical regime, within limits. Such robust dynamical equilibria may be modeled mathematically as "attractors." Another important property of nonlinear dynamical systems is metastability, which means that, within the abstract landscape of possible dynamical regimes accessible to the system, other attractors exist in the vicinity of the original one. If a metastable system is pushed past the boundaries of its original attractor, it will not necessarily cease its dynamical activity altogether. Instead, it may be pulled onto a new attractor. Such a shift to a somewhat different dynamical regime constitutes a bifurcation event. This phenomenon is of the highest interest for understanding the directed or selective switching between different dynamical regimes in metabolism (Jackson 1993; Petty and Kindzelskii 2001) and other forms of robust short-term (ontogenetic) adaptive behavior in cells (Barkai and Leibler 1997; Alon et al. 1999; Jeong et al. 2000; Yi et al. 2000). Ravasz and colleagues (2002, 1555) point out that "[t]he organization of metabolic networks is likely to combine a capacity for rapid flux reorganization with a dynamic integration with all other cellular function." Nonlinear dynamics gives us a way of conceptualizing and modeling this cascading functional reorganization of relationships among the components of living systems. By showing how new functional states may be found through the operation of physical principles, it may also serve some day as the basis for a genuine understanding of long-term (phylogenetic) adaptive shifts in molecular structures and dynamical regimes – that is, evolution (Kauffman 1993; Flyvbjerg et al. 1995; Gordon 1999; New and Pohorille 2000; Segré, Ben-Eli, and Lancet 2000; Jain and Krishna 2001; Zhou, Carlson, and Doyle 2002).

Another interesting property intrinsic to nonlinear dynamical systems is the lack of proportionality between causes and effects in their interactions with the wider world around them. This disproportionate response to events impinging upon them from their surroundings is a hallmark of all living things. If organisms are conceived of, not as machines made up of rigidly connected parts, but as a dense network of loosely coupled, non-linear oscillators, each sensitive to a range of specific low-energy inputs from its surround, then we begin to see how information in the semantic sense just discussed is possible. On this "homeodyamic" view of the organism (Yates 1994), information is anything that acts as a trigger for the action of such an oscillator (Barham 1996). The role of such a trigger in the functional action of an organism is to coordinate the timing of actions in such a way that they become correlated with favorable environmental conditions, where "favorable" means tending to support the continued homeodynamic stability of the oscillator. On this view, then, the meaning of information consists in the prediction of the success of functional action, where "success" likewise means the continued homeodynamic stability of the oscillator. This dynamical interpretation of semantic information provides us with a new physical picture of the cognitive component of adaptive functional action.

Most, if not all, of the authors of the studies just cited would probably contend that they are working squarely within the Mechanistic Consensus. So why do I interpret their work as contributing to the overthrow of that worldview? Because nonlinear dynamics cannot be the whole story. After all, inorganic dynamical systems such as hurricanes and candle flames are not alive. They do not utilize information in the dynamical sense just described, nor do they draw on internal energy stores to do work against local thermodynamic gradients in order to preserve themselves in existence – all of which are hallmarks of living things. Rather, they are thermodynamically tightly coupled to their surrounds and are merely minimizing energy under a given set of constraints. Furthermore, dynamical networks with many of the properties just discussed can be constructed out of inorganic materials, as in neural networks. And yet a neural network is just fulfilling our functions, not its own. It has no internal tendency to prefer one energetically equivalent configuration over another. It is we who choose which configuration counts as a correct solution to a given problem. Once the boundary conditions have been set by us, everything else is just minimizing energy.

With nonlinear dynamics, we have still not reached the heart of the matter, where the leap from passive energy minimization to the active directing of energy in accordance with preferred goal states occurs. Still lacking is an understanding of how it is possible for dynamical networks to strive to preserve themselves – that is, to value their own continued existence. Philosophical mechanism posits a contingent link between function and matter. Therefore, in order to transcend mechanism, we must penetrate the mystery of the essential link between biological value and the living state. To do this, we must look beyond nonlinear dynamics, which is necessary but not sufficient for this task.

One of the most interesting lines of contemporary research that holds some promise of providing the missing piece of the puzzle is the investigation of the cell as a sui generis condensed phase of matter – the living state. There are several different approaches here that will eventually need to be integrated. One is work on the global properties of the protein–phosphate–ordered water gel that constitutes the main phase of the cytoplasm in all living cells (Ho et al. 1996; Watterson 1997; Pollack 2001). Another is work stressing the direct link between the physical structure of the cell components and the coordination of cellular functioning (Hochachka 1999; Kirschner, Gerhart, and Mitchison 2000; Surrey et al. 2001; Whitesides and Grzybowski 2002). A third is work on the intrinsic dynamical properties of proteins arising from their energy degeneracy and manifold competing self-interactions ("frustration") (Frauenfelder, Wolynes, and Austin 1999). Finally, there is the highly suggestive, if speculative, work on adapting the formalism of quantum field theory for use in describing the directed transfer of energy along macromolecular chains via coherent resonances within a hypothetical electric dipole field (Li 1992; Wu 1994; Fröhlich and Hyland 1995; Ho 1997, 1998; Vitiello 2001). All of these approaches share the assumption that there is more to the coordination of functional action *in vivo* than can be explained by mechanistic interactions observed to date *in vitro* or even *in silico* (Srere 1994, 2000). It is becoming clear that new, nondestructive experimental techniques for probing the real-time dynamics of macromolecular interactions are needed if we are ever to achieve a genuine theoretical biology (Laughlin et al. 2000).

How can such research programs help us to understand the ground of normativity in nature? By showing how life is "an expression of the self-constraining nature of matter" (Moreno Bergareche and Ruiz-Mirazo 1999, 60). Ultimately, this means showing how living systems function as integrated wholes, using information in the dynamical sense and doing work in order to maintain themselves in existence. It means showing how a mere physical system acquires the capacity for striving and preferring, how it becomes a self existing *pour soi* (Jonas 1982). And it means showing how all of this occurs through a process of material emergence.

It is impossible to say exactly how this happens in advance of the scientific breakthrough that will provide the eventual explanation. Whether any of the specific lines of research alluded to here are on the correct path is perhaps doubtful. But it is important to see that there are already ideas on the table that seem to be moving us in the right direction. Today, the emergence of objective biological value is no longer scientifically unthinkable.[2]

5. CONCLUSION

In the end, the idea that life is special is just plain common sense. After all, it is a matter of everyday observation that animate systems are fundamentally different from inanimate ones. A broken bone heals; a broken stone doesn't. If this homely truth has been lost sight of, it is undoubtedly because the mysteriousness of teleology and normativity have made them ripe for exploitation by irrationalist opponents of science. I think it is mainly for this reason that those sympathetic to science have seized upon every advance in biology since Friedrich Wöhler synthesized urea in 1828 to proclaim that "organic macromolecules do not differ in principle from other molecules" (Mayr 1982, 54). But this essentially defensive maneuver holds water only against the backdrop of reductionism. Against the backdrop of material emergence, it makes no sense at all. Why should organic macromolecules not be very different in principle from small molecules, when liquids are very different from gases, and solid matter is very different from both? If more really is different, then why should those behemoths, proteins, not have special causal powers that small molecules do not possess? The time has come for naturalists to rethink their metaphysical commitments in light of the Mechanist's Dilemma. After all, why should biologists and philosophers feel that they must be *plus mécanistes que les physiciens?*

As a scientific methodology, the machine metaphor has been extraordinarily fruitful. No doubt it will remain so for a long time to come, although there are many signs that we are beginning to bump up against the limits of its usefulness. But however that may be, as a metaphysics, mechanism has always been incoherent. The idea that a machine could occur naturally at all, much less that it might have its own intrinsic purposes and values, is simply an article of faith for which there is no rational support. Nevertheless, I want to emphasize that the attack mounted here against the Mechanistic Consensus should not be construed as an attack on science itself. We must carefully distinguish between the operationally verifiable results of science and philosophical extrapolations from those results. As Unger (2002, 10) has recently reminded us, the reductionist outlook is "a particular *philosophical approach to* science, rather than something science itself actually delivers."

Whether any of the alternative approaches too briefly surveyed here will prove to be of lasting value, only time will tell. But one thing is certain: it is not necessary to choose between the Mechanistic Consensus and Intelligent Design. The emergence of objective biological value as an intrinsic property of living matter is a coherent alternative that warrants further investigation.

ACKNOWLEDGMENTS

I am very grateful to William Dembski, Lenny Moss, and F. Eugene Yates for critical comments, and to Ellen F. Hall and James McAdams for editorial assistance. Many other friends, too numerous to name here, have given me the benefit of their criticism and encouragement over the years. I would like to express my heartfelt thanks to all of them – Darwinian, Christian, and "left-Aristotelian" alike – and to absolve them all of responsibility. Philosophy makes strange bedfellows.

Notes

1. Chance may still play a more modest role, of course. The role that chance plays in evolution is analogous to that of trial-and-error search in individual organisms. All learning, whether ontogenetic or phylogenetic, involves groping for new ways of functioning. Life is intelligent, not clairvoyant. But the crucial point is that even trial-and-error search is still essentially teleological and normative in character. Searches are aimed at particular preferred states, and trials are evaluated accordingly.
2. For further references, as well as discussion of the philosophical significance of this literature, see Barham (2000, 2002).

References

Albrecht-Buehler, G. 1990. In defense of "nonmolecular" cell biology. *International Review of Cytology* 120: 191–241.

Alon, U., M. G. Surette, N. Barkai, and S. Leibler. (1999). Robustness in bacterial chemotaxis. *Nature* 397: 168–171.

Alt, W. 1994. Cell motion and orientation: Theories of elementary behavior between environmental stimulation and autopoietic regulation. In *Frontiers in Mathematical Biology*, ed. S. A. Levin. Berlin: Springer-Verlag, pp. 79–101.

Anderson, P. W. (1994). More is different. In P. W. Anderson, *A Career in Theoretical Physics*, pp. 1–4. Singapore: World Scientific. (Originally published in *Science* 177 (1972): 393–6.)

Auyang, S. Y. 1998. *Foundations of Complex-System Theories in Economics, Evolutionary Biology, and Statistical Physics*. Cambridge: Cambridge University Press.

Barham, J. 1996. A dynamical model of the meaning of information. *BioSystems* 38: 235–41.

   2000. Biofunctional realism and the problem of teleology. *Evolution and Cognition* 6: 2–34.

   2002. Theses on Darwin. *Rivista di Biologia/Biology Forum* 95: 115–47.

Barkai, N., and S. Leibler. 1997. Robustness in simple biochemical networks. *Nature* 387: 913–17.

Batterman, R. W. 2002. *The Devil in the Details: Asymptotic Reasoning in Explanation, Reduction, and Emergence*. New York: Oxford University Press.

Cao, T. Y. 1997. *Conceptual Developments of 20th Century Field Theories*. Cambridge: Cambridge University Press.

   1998. Monism, but not through reductionism. In *Philosophies of Nature: The Human Dimension*, ed. R. S. Cohen and A. I. Tauber, Dordrecht: Kluwer, pp. 39–51.

Dembski, W. A. 1998. *The Design Inference: Eliminating Chance through Small Probabilities*. Cambridge: Cambridge University Press.

Denbigh, K. G. 1975. *An Inventive Universe*. New York: Braziller.

Depew, D. J., and B. H. Weber. 1998. What does natural selection have to be like in order to work with self-organization? *Cybernetics and Human Knowing* 5: 18–31.

Eden, M. 1967. Inadequacies of neo-Darwinian evolution as a scientific theory. In *Mathematical Challenges to the Neo-Darwinian Interpretation of Evolution*, ed. P. S. Moorhead and M. M. Kaplan. Philadelphia: Wistar Institute Press, pp. 5–19.

Elsasser, W. M. 1998. *Reflections on a Theory of Organisms*. Baltimore: Johns Hopkins University Press. (Originally published in 1987.)

Flyvbjerg, H., P. Bak, M. H. Jensen, and K. Sneppen. 1995. A self-organized critical model for evolution. In *Modelling the Dynamics of Biological Systems*, ed. E. Mosekilde and O. G. Mouritsen. Berlin: Springer, pp. 269–88.

Frauenfelder, H., P. G. Wolynes, and R. H. Austin. 1999. Biological physics. *Reviews of Modern Physics* 71: S419–S430.

Fröhlich, F., and G. J. Hyland. 1995. Fröhlich coherence at the mind-brain interface. In *Scale in Conscious Experience*, ed. J. King and K. H. Pribram. Mahwah, NJ: Erlbaum, pp. 407–38.

Georgi, H. 1989. Effective quantum field theories. In *The New Physics*, ed. P. Davies. Cambridge: Cambridge University Press, pp. 446–57.

Gordon, R. 1999. *The Hierarchical Genome and Differentiation Waves*. 2 vols. Singapore: World Scientific.

Ho, M.-W. 1997. Towards a theory of the organism. *Integrative Physiological and Behavioral Science* 32: 343–63.

    1998. *The Rainbow and the Worm: The Physics of Organisms*, 2nd ed. Singapore: World Scientific. (Originally published in 1993.)

Ho, M.-W., J. Haffegee, R. Newton, Y.-M. Zhou, J. S. Bolton, and S. Ross. 1996. Organisms as polyphasic liquid crystals. *Bioelectrochemistry and Bioenergetics* 41: 81–91.

Hochachka, P. W. 1999. The metabolic implications of intracellular circulation. *Proceedings of the National Academy of Sciences (USA)* 96: 12233–9.

Icke, V. 1995. *The Force of Symmetry*. Cambridge: Cambridge University Press.

Jackson, R. C. 1993. The kinetic properties of switch antimetabolites. *Journal of the National Cancer Institute* 85: 539–45.

Jain, S., and S. Krishna. 2001. A model for the emergence of cooperation, interdependence, and structure in evolving networks. *Proceedings of the National Academy of Science (USA)* 98: 543–7.

Jeong, H., B. Tombor, R. Albert, Z. N. Oltvai, and A.-L. Barabási. 2000. The large-scale organization of metabolic networks. *Nature* 407: 651–4.

Jonas, H. 1982. *The Phenomenon of Life: Toward a Philosophical Biology*. Chicago: Phoenix Books/University of Chicago Press. (Originally published in 1966.)

Jonker, C. M., J. L. Snoep, J. Treur, H. V. Westerhoff, and W. C. A. Wijngaards. 2002. Putting intentions into cell biochemistry: An artificial intelligence perspective. *Journal of Theoretical Biology* 214: 105–34.

Kauffman, S. A. 1993. *The Origins of Order: Self-Organization and Selection in Evolution*. New York: Oxford University Press.

Kirschner, M., J. Gerhart, and T. Mitchison. 2000. Molecular "vitalism." *Cell* 100: 17–88.

Lauffenburger, D. A., and A. F. Horwitz. 1996. Cell migration: a physically integrated molecular process. *Cell* 84: 359–69.

Laughlin, R. B., G. G. Lonzarich, P. Monthoux, and D. Pines. 2001. The quantum criticality conundrum. *Advances in Physics* 50: 361–5.

Laughlin, R. B., and D. Pines. 2000. The theory of everything. *Proceedings of the National Academy of Sciences (USA)* 97: 28–31.

Laughlin, R. B., D. Pines, J. Schmalian, B. P. Stojkovic, and P. Wolynes. 2000. The middle way. *Proceedings of the National Academy of Sciences (USA)* 97: 32–7.

Layzer, D. 1990. *Cosmogenesis: The Growth of Order in the Universe.* New York: Oxford University Press.

Lecomte du Noüy, P. 1948. *The Road to Reason.* New York: Longmans, Green. (Originally published as *L'homme devant la science.* Paris: Flammarion, 1939.)

Li, K.-H. 1992. Coherence in physics and biology. In *Recent Advances in Biophoton Research and Its Applications,* ed. F.-A. Popp, K.-H. Li, and Q. Gu. Singapore: World Scientific, pp. 113–55.

Loewenstein, W. R. 1999. *The Touchstone of Life: Molecular Information, Cell Communication, and the Foundations of Life.* New York: Oxford University Press.

Mayr, E. 1982. *The Growth of Biological Thought.* Cambridge, MA: Harvard University Press.

Millikan, R. G. 1998. In defense of proper functions. In *Nature's Purposes: Analyses of Function and Design in Biology,* ed. C. Allen, M. Bekoff, and G. Lauder. Cambridge, MA: Bradford Books/MIT Press, pp. 295–312. (Originally published in *Philosophy of Science,* 56 (1989): 288–302.)

Monod, J. 1972. *Chance and Necessity.* New York: Vintage. (Originally published as *Le hasard et la nécessité.* Paris: Editions du Seuil, 1970.)

Moreno Bergareche, A., and K. Ruiz-Mirazo. 1999. Metabolism and the problem of its universalization. *BioSystems* 49: 45–61.

Nagel, E. 1998. Teleology revisited. In *Nature's Purposes: Analyses of Function and Design in Biology,* ed. C. Allen, M. Bekoff, and G. Lauder. Cambridge, MA: Bradford Books/MIT Press, pp. 197–240. (Originally published in *Journal of Philosophy* 76 (1977): 261–301.)

New, M. H., and A. Pohorille. 2000. An inherited efficiencies model of non-genomic evolution. *Simulation Practice and Theory* 8: 99–108.

Pattee, H. H. 1982. Cell psychology: An evolutionary approach to the symbol-matter problem. *Cognition and Brain Theory* 5: 325–41.

2001. The physics of symbols: bridging the epistemic cut. *BioSystems* 60: 5–21.

Petty, H. R., and A. L. Kindzelskii. 2001. Dissipative metabolic patterns respond during neutrophil transmembrane signaling. *Proceedings of the National Academy of Sciences USA* 98: 3145–9.

Polanyi, M. 1969. Life's irreducible structure. In his *Knowing and Being.* Chicago: University of Chicago Press, pp. 225–39. (Originally published in *Science* 160 (1968): 1308–12.)

Pollack, G. H. 2001. *Cells, Gels and the Engines of Life.* Seattle, WA: Ebner and Sons.

Ravasz, E., A. L. Somera, D. A. Mongru, Z. N. Oltvai, and A.-L. Barabási. 2002 Hierarchical organization of modularity in metabolic networks. *Science* 297: 1551–5.

Rosen, R. 1991. *Life Itself: A Comprehensive Inquiry into the Nature, Origin, and Fabrication of Life.* New York: Columbia University Press.

Schoffeniels, E. 1976. *Anti-Chance.* Oxford: Pergamon. (Originally published as *L'anti-hasard,* 2nd ed. Paris: Gauthier-Villars, 1975.)

Schrödinger, E. 1992. What is life? In his *What Is Life? with Mind and Matter and Autobiographical Sketches*. Cambridge: Cambridge University Press, pp. 1–90. (Originally published in 1944.)

Schweber, S. S. 1997. The metaphysics of science at the end of a heroic age. In *Experimental Metaphysics*, ed. R. S. Cohen, M. Horne, and J. Stachel. Dordrecht: Kluwer Academic, pp. 171–98.

Segré, D., D. Ben-Eli, and D. Lancet. 2000. Compositional genomes: Prebiotic information transfer in mutually catalytic noncovalent assemblies. *Proceedings of the National Academy of Sciences (USA)* 97: 4112–17.

Srere, P. A. 1994. Complexities of metabolic regulation. *Trends in Biochemical Sciences* 19: 519–20.

2000. Macromolecular interactions: Tracing the roots. *Trends in Biochemical Sciences* 25: 150–3.

Surrey, T., F. Nédélec, S. Liebler, and E. Karsenti. 2001. Physical properties determining self-organization of motors and microtubules. *Science* 292: 1167–71.

Thirring, W. 1995. Do the laws of nature evolve? In *What Is Life? The Next Fifty Years*, ed. M. P. Murphy and L. A. J. O'Neill. Cambridge: Cambridge University Press, pp. 131–6.

Unger, P. 2002. Free will and scientiphicalism. *Philosophy and Phenomenological Research* 65: 1–25.

Vitiello, G. 2001. *My Double Unveiled: The Dissipative Quantum Model of Brain*. Amsterdam and Philadelphia: John Benjamins.

Walleczek, J. (ed.) 2000. *Self-Organized Biological Dynamics and Nonlinear Control*. Cambridge: Cambridge University Press.

Watterson, J. G. 1997. The pressure pixel – unit of life? *BioSystems* 41: 141–52.

Whitesides, G. M., and B. Grzybowski. 2002. Self-assembly at all scales. *Science* 295: 2418–21.

Wright, L. 1998. Functions. In *Nature's Purposes: Analyses of Function and Design in Biology*, ed. C. Allen, M. Bekoff, and G. Lauder, Cambridge, MA: Bradford Books/MIT Press, pp. 51–78. (Originally published in *Philosophical Review* 82 (1973): 139–68.)

Wu, T.-M. 1994. Fröhlich's theory of coherent excitation – a retrospective. In *Bioelectrodynamics and Biocommunication*, ed. M.-W. Ho, F.-A. Popp, and U. Warnke. Singapore: World Scientific, pp. 387–409.

Yates, F. E. 1994. Order and complexity in dynamical systems: Homeodynamics as a generalized mechanics for biology. *Mathematical and Computer Modelling* 19: 49–74.

Yi, T.-M., Y. Huang, M. I. Simon, and J. Doyle. 2000. Robust perfect adaptation in bacterial chemotaxis through integral feedback control. *Proceedings of the National Academy of Sciences (USA)* 97: 4649–53.

Yockey, H. P. 1992. *Information Theory and Molecular Biology*. Cambridge: Cambridge University Press.

Zhou, T., J. M. Carlson, and J. Doyle. 2002. Mutation, specialization, and hypersensitivity in highly optimized tolerance. *Proceedings of the National Academy of Sciences (USA)* 99: 2049–54.

PART III

THEISTIC EVOLUTION

12

Darwin, Design, and Divine Providence

John F. Haught

To the theist, a central question after Darwin is whether evolution renders implausible the notion of divine Providence. Do the rough and ragged features of the new story of life place in question the idea of a personal God who cares for the world? Most theologians today would say no, but the more intimately the idea of Providence is tied to that of "intelligent design," the more difficult becomes the task of reconciling theology with evolutionary biology. I suspect that much of the energy underlying so-called Intelligent Design (ID) theory, in spite of explicit denials by some of its advocates, is an achingly religious need to protect the classical theistic belief in divine Providence from potential ruination by ideas associated with Darwinian science. It is impossible not to notice that the advocates of IDT are themselves almost always devout Christian, Muslim, and occasionally Jewish theists. It is difficult, therefore, for most scientists and theologians to accept the claim that no theological agenda is at work in the ID movement.

It is highly significant, moreover, that scientific proponents of ID, although often themselves experts in mathematics and specific areas of science, are generally hostile to evolutionary theory (Behe 1996; Dembski, 1998, 1999; Johnson, 1991, 1995). The justification they usually give for rejecting what most scientists take as central to biology is that Darwinism, or neo-Darwinism, is simply a naturalist belief system and not science at all. Evolution, they claim, is so permeated with materialist metaphysics that it does not qualify as legitimate science in the first place (Johnson 1991, 1999). This protest is indicative of a religious sensitivity that recognizes materialism to be inherently incompatible with theism.

Although, as I shall illustrate more extensively, it is clearly the case that contemporary presentations of evolution are often interlaced with a heavy dose of materialist ideology, it is not likely that a concern for science's methodological purity is the driving force behind ID's energetic protests against Darwinism. Rather, I would suggest, the flight from Darwin is rooted quite simply in an anxiety that his evolutionary ideas may be incompatible with any coherent notion of God or divine Providence. And the suspicion that Darwinism conflicts with the doctrine of Providence is ultimately rooted in the ID judgment that Darwinism, if true, would render the notion of intelligent design unbelievable. Hence the way to defend Providence – by which I mean here the "general" doctrine that God cares or "provides" for the universe – is to defend design.[1] Quite candidly, it seems to me that beneath all of the complex logical and mathematical argumentation generated by the ID movement there lies a deeply human and passionately religious concern about whether the universe resides in the bosom of a loving, caring God or is instead perched over an abyss of ultimate meaninglessness.

What may add some credibility to the ID preoccupations, rendering them less specious than they might at first seem, is the fact that many evolutionary biologists (and philosophers of biology) agree that Darwin's "dangerous idea" does indeed destroy the classical argument from design and that in so doing it exorcizes from scientifically enlightened consciousness the last remaining traces of cosmic teleology and supernaturalism. Since religion, as the renowned American philosopher W. T. Stace pointed out long ago, stands or falls with the question of cosmic purpose (Stace 1948, 54), the Darwinian debunking of design – and with it the apparent undoing of cosmic teleology as well – strikes right at the heart of the most prized religious intuitions of humans, now and always. Darwinism seems to many Darwinians – and not just to IDT advocates such as Phillip Johnson, Michael Behe, and William Dembski – to entail a materialist and even anti-theistic philosophy of nature. Michael Ruse even refers to Darwinism as "the apotheosis of a materialist theory" (Ruse 2001, 77). Consequently, it seems to many theists as well as to many scientists that we must choose *between* Darwinism and divine Providence.

A straightforward example of this either/or thinking is Gary Cziko's book *Without Miracles* (1995), a work that from beginning to end explicitly places "providential" in opposition to "selectionist" explanations for all the various features of life. For Cziko, as for many other Darwinians, there isn't enough room in the same human mind to hold both scientific and theological explanations simultaneously, so we must choose one over the other. In her discussion of sociobiology, Ullica Segerstråle (2000, 399–400) insightfully comments that Richard Dawkins likewise assumes that in accounting for living phenomena there can be only one "explanatory slot." And so, if Darwinism now completely fills that single aperture, there can be no room for any theological explanation to exist alongside it.

In my own reading of contemporary works on evolution, I have observed time and again a tacitly monocausal or univalent logic (laced with curt appeals to Occam's razor) that inevitably puts biological and providential arguments into a competitive relationship. To give just one of many possible examples, in *Darwin's Spectre*, Michael R. Rose illustrates the widespread belief that accounting for life must be the job *either* of theology *or* of Darwinism, but not of both. "Without Darwinism," he claims, "biological science would need one or more deities to explain the marvelous contrivances of life. Physics and chemistry are not enough. And so without Darwinism science would remain theistic, in whole or in part" (Rose 1998, 211). Clearly, the assumption here is that evolutionary science has now assumed occupancy of the *same* explanatory alcove that was formerly the dwelling place of the gods. And now that Darwin has expelled the deities from this niche, there is no longer any plausible explanatory place left for religion or theology.

Historically, it is true, religious ideas have often played a quasi-scientific or prescientific explanatory role, even while also providing ultimate explanations. But science has now – providentially, we may say – liberated theology from the work of satisfying the more mundane forms of inquiry. Yet even today, scriptural literalists want religious ideas to fill explanatory spaces that have been assigned more appropriately to science. Not everyone embraces the distinction that mainstream Western theology has made between scientific and theological levels of explanation. Cziko's and Rose's books, along with the better-known works of Richard Dawkins (1986, 1995, 1996), Daniel Dennett (1995), and E. O. Wilson (1998), demonstrate that today's biblical literalists are not alone in assuming that religious and theological readings of the world lie at essentially the same explanatory level as natural science.

Only this assumption could have led to the forced option between Providence, on the one hand, and natural selection, on the other. Thus, for many evolutionists there is no legitimate cognitive role left for religion or theology after Darwin, only (at best) an emotive or evaluative one. For them, as Rose's book lushly exemplifies, Darwinism goes best with materialism (Rose 1998, 211). It is not entirely surprising, then, that religiously sensitive souls would balk at evolution if they were persuaded by the words of evolutionists themselves that Darwinism is indeed inseparable from "materialism" – a philosophy that is logically irreconcilable not only with intelligent design but also with each and every religious interpretation of reality.

If the appeal by biologists to "materialism" were simply methodological, then the ID community would have no cause for complaint. By its very nature, science is obliged to leave out any appeal to the supernatural, and so its explanations will always sound naturalistic and purely physicalist. In many cases, I believe that ID advocates unnecessarily mistake methodological naturalism/materialism for metaphysical explanation. Alvin Plantinga (1997) even argues that there can be no sharp distinction between methodological and metaphysical naturalism. Practicing the latter, he thinks, is a slippery slope to the former. But even aside from Plantinga's questionable proposal, the ID intuition that Darwinians often illegitimately conflate science with materialist ideology is completely accurate. The problem is that, like their Darwinian opponents, ID theorists typically accept the assumption that only one "explanatory slot" is available and that if we fill it up completely with naturalistic explanations, there will be no room left anywhere for theological explanations.

Consequently, if the contemporary discussion of the question of Darwinism and design is ever going to penetrate beneath surface accusations, it must consider two questions. First, is Darwinian biology unintelligible apart from a philosophical commitment to materialism – a philosophy of nature that theists everywhere and of all stripes will take to be inherently atheistic? That is, does the information gathered by the various sciences tributary to evolutionary theory (geology, paleontology, comparative anatomy, radiometric dating, biogeography, genetics, etc.) remain unintelligible unless it is contextualized within a materialist philosophy of nature? And, second, would the elimination of the notion of "intelligent design" in scientific explanations of life's organized complexity logically entail the downfall of a credible doctrine of divine Providence, as both ID theorists and their evolutionary antagonists generally seem to agree would be the case? In the interest of fairness, we owe the IDT advocates a careful consideration of their suspicion that Darwinism is materialist atheism in disguise. But for the sake of giving a fair hearing to the full spectrum of theological reflection after Darwin, we should also look at the question of just how vital the notion of "intelligent design" is to a religiously robust notion of Providence. I will now consider each of these two questions in turn.

I. IS DARWINISM INHERENTLY MATERIALISTIC?

It is not without interest to our inquiry that in the intellectual world today, critics of theism are increasingly turning for support to Charles Darwin. Many skeptics who seek to ground their suspicions about the existence of God in science no longer look as fervently to Freud, Marx, Nietzsche, Sartre, or Derrida as they do to Darwin. Especially for those already convinced that science is essentially ruinous to religion, Darwin has become more appealing than ever. His portrait of nature's apparent indifference seems to offer more compelling reasons than ever for scientific atheism. In fact, for some critics today natural selection provides much more secure grounds for atheism than do the impersonal laws of physics, which had already rendered the idea of divine action apparently superfluous several centuries ago. Even the renowned physicist Steven Weinberg considers Darwinism to be a much more potent challenge to theism than his own discipline (Weinberg 1992, 246). He singles out the ID enthusiast Phillip Johnson as the most sophisticated example of a theological alternative to Darwin and then proceeds to shred theism by destroying the arguments of one who, at least to Weinberg, speaks most eloquently for belief in God after Darwin (247–8). For many others among the scientific elite today, the ways of evolution are so coarse that even if the universe appears on the surface to be an expression of design, beneath this deceptive veneer there lurks a long and tortuous process in which an intelligent Deity could not conceivably have played any role.

It is not only the waste, struggle, suffering, and indifference of the evolutionary process that place in question the idea of a benevolent providential Deity. The three main evolutionary ingredients – randomness, the impersonal law of selection, and the immensity of cosmic time – seem to be enough to account causally for all the phenomena we associate with life, including design. The apparent completeness of the evolutionary recipe makes us wonder whether the universe requires any additional explanatory elements, including the creativity of a truly "interested" God. We may easily wonder, then, whether we can reconcile the ragged new picture of life not only with the idea of an intelligent Designer but also with any broader notion of divine Providence. Darwin himself, reflecting on the randomness, pain, and impersonality of evolution, abandoned the idea that nature could have been ordered in its particulars by a designing Deity. It is doubtful that he ever completely renounced the idea of God, since he often seems to have settled for a very distant divine law maker. But he gradually became convinced that the design in living beings could be accounted for in a purely naturalistic way. After Darwin, many others, including a number of the most prominent neo-Darwinian biologists writing today, have come close to equating Darwin's science with atheism. Sensitive to the conflation of Darwin's science with philosophical materialism that prominent biologists often make, ID proponents have drawn the conclusion that Darwinian biology, as evidenced in the publications of Darwinians themselves, is simply incapable of being reconciled with theistic belief. In order to save theism, then, Darwin must be directly refuted.

Not only scientific skeptics but also other intellectuals are now making the figurative pilgrimage to Down House in order to nourish their materialist leanings. A good example is the noted critic Frederick Crews, who recently published a titillating two-part essay, "Saving God from Darwin," in *The New York Review of Books* (October 4 and 18, 2001). Crews is best known for his constant pummelling of Sigmund Freud, whose ideas he considers blatantly unscientific. But in all of his blasting of psychoanalysis he has never challenged Freud's materialist metaphysics. Crews clearly shares with Freud the unshakable belief that beneath life, consciousness, and culture there lies *ultimately* only mindless and meaningless material stuff.

In Crews's opinion, Darwin has uncovered the ultimate truth to which all intelligent and courageous humans must now resign themselves. Referring to Daniel Dennett's radically materialist interpretation of Darwin, Crews is convinced that Dennett has "trenchantly shown" that Darwin's ideas lead logically to "a satisfyingly materialistic reduction of mind and soul" and that evolutionary theory entails a "naturalistic account of life's beginning" (Crews, October 4, p. 24). Even though the materialism and naturalism he is referring to are really examples of metaphysics and not pure science, for

Crews they have become part and parcel of biology itself. Crews, of course, is not a biologist, but he could easily point to many ideological associates in the scientific community who share his view that the ultimate "truth" of Darwinism is a materialist and Godless cosmos.

This, of course, is exactly the same not-so-subtle message that proponents of ID have detected in contemporary evolutionary thought. Crews upbraids the ID literature for sneaking theology into an explanatory slot that science alone should inhabit. But interestingly, his own comprehension of Darwinism – a conflation of science with materialist metaphysics – is identical to that of his ID opponents. In both instances, the idea of evolution is understood to be inseparable from the nonscientific *belief* that matter is all there is and that the universe is inherently pointless – an assumption that is inherently antithetical to theism of any kind. For Crews, as well as for numerous biologists and philosophers today, evolution and materialism come as a package deal (see also Dennett 1995). And so the only difference between them and ID disciples is that the latter throw the package away, whereas the former hold onto it tightly. Both evolutionary materialists and ID advocates discern at the bottom of Darwinism a fundamentally pointless universe.

We may have good reason to wonder, then, whether the evolutionist alloy of scientific information and philosophically materialist belief is any closer to pure science than the conflation of biology with Intelligent Design that is now the object of so much scientific scorn. If ID is advised to keep theological explanation (under the guise of an abstract notion of "Intelligent Design") from intruding into biology, are not Darwinians also obliged to keep whatever philosophical biases they may have from invading their public presentations of evolutionary science?

Strictly speaking, after all, it is no more appropriate to say that Darwinism is a materialist theory than it is to say that the theory of relativity is. All science *must* be methodologically materialist – in the sense that it is not permissible when doing science to invoke nonphysical causes. It is one thing to hold that evolutionary science provides a picture of nature that supports a purely materialist philosophy of life, if you happen to have one. But it is quite another to claim – as Rose, Cziko, Dawkins, Dennett, and many others do – that the facts of evolution *do not make sense* outside of a materialist philosophical landscape. How would we ever know for sure that this is the case? Such a claim is based as much on belief as on research, and it is one that will forever remain logically unsupportable by scientific evidence as such. Moreover, there is always the possibility that alternative metaphysical frameworks may turn out to be no less illuminating settings for interpreting evolutionary information (Haught 2000).

For now, however, it is sufficient to note that, strictly speaking, neither Darwinians nor their ID adversaries can logically claim that evolutionary biology is an expression of materialism. Like all other applications of scientific method, evolutionary biology remains methodologically natural-istic. *As such*, it makes no formal appeal to the idea of God, purpose, or intelligence in its own self-restricting mode of explanation. But likewise, any inferences that a scientist might make from doing the work of pure science to materialist conclusions about that work is not itself an exercise intrinsic to science. The energizing force behind scientism and materialism is never the purely scientific desire to know, but something quite extrinsic to science. The slippage from methodological naturalism into metaphysical materialism is not justifiable by scientific method itself. Logically speaking, therefore, we must conclude that it is not at all evident that evolution nec-essarily entails philosophical materialism.

II. IS DESIGN ESSENTIAL TO THE IDEA OF PROVIDENCE?

The more intimately the idea of God or "Providence" is associated with "In-telligent Design," as is implicit in the theological assumptions underlying most of the ID movement, the more it seems that the most efficient way to oppose materialist "Darwinism" is to shore up arguments from design. But just how closely do we have to connect Providence with Intelligent Design in the first place? I shall propose here that the two ideas are quite distinct and that evolutionary biology may cohere quite nicely with a theologically grounded notion of Providence, even if it does not fit a simplistic under-standing of divine design. If such a case can be made, then there should be no reason for theists to oppose evolutionary biology, even if they must oppose Darwinian materialism.

There is no denying, however, that to many scientists and philosophers, as well as to devotees of ID, Darwinian biology connotes a universe empty of any conceivable divine governance, compassion, or care. In view of the obvious challenges that so many sincere skeptics and religiously devout peo-ple perceive to be inherent in evolution, can the idea of Providence now have any plausibility at all? Responses to this question fall roughly into three distinct classes. Evolutionary materialism and ID fall together as one, since they both view Darwinian accounts of evolution as incompatible with Provi-dence. But there are two distinct kinds of theological response that have no difficulty embracing both conventional biological science and, at the same time, a biblically grounded notion of divine Providence. Let us consider each of these in turn.

A. Theological Response I

Evolutionary science, though perhaps disturbing to a superficial theism, is no more threatening to theistic faith than is any other development in modern science. Science and religion, after all, are radically distinct ways of understanding, and they should be kept completely apart from each other.[2]

Science answers one set of questions, religion an entirely different one. Science asks about physical causes, while religion looks for ultimate explanations and meanings. If we keep science and religion separate, there can be no conflict. The ugly disputes between Galileo and the Roman Catholic Church, and later between Darwin and Christianity, could have been avoided if theologians had never intruded into the world of science and if certain highly visible evolutionists had refrained from making sweeping metaphysical claims about evolution as though they were scientific statements.

Thus Darwin's ideas – which may be quite accurate, scientifically speaking – carry not even the slightest threat to theism. The apparent contradiction arises not from the scientific theory of evolution itself, but from the confusion of the biblical accounts of creation with "science" in the case of biblical literalists, the confusion of Providence with intelligent design in the case of ID theorists, and the equally misbegotten confusion of evolutionary data with metaphysical materialism in the case of some evolutionary scientists and philosophers. There is no squabble here with the purely *scientific* aspects of evolution. What is objectionable is the uncritical mixing of evolutionary science with nonscientific beliefs, whatever these beliefs may be. The "danger" of Darwinism to theism, then, is not so much Darwin's own ideas but the way in which they get captured by materialist ideologies that are indeed incompatible with theism but that have nothing inherently to do with *scientific* truth.

At some point, of course, if we dig toward the deepest roots of life's designs, we will have to yield to metaphysical explanations. But both evolutionary materialism and ID move prematurely into metaphysical discourse. They reach for ultimate explanations at a point when there is still plenty of room left for more subtle scientific inquiry into the proximate causes of the complex patterns evident in living phenomena. Darwinian materialists, therefore, cannot credibly object that ID theorists turn prematurely to metaphysics, since materialism – as a worldview and not just as a method – permeates their own inquiry from the outset, even tacitly helping them to decide what are and are not worthwhile research projects. Their materialist metaphysics consists of the controlling belief that mindless "matter" is the ultimately real stuff underlying everything – even if contemporary physics has shown this "stuff" to be much more subtle than was previously thought. (One may use the term "physicalist" here if the term "materialist" seems too harsh.) In any case, many evolutionists commit themselves to a physicalist or materialist creed, and not just to methodological naturalism, long before they ever embark on their "purely scientific" explorations of life. So the fact that ID would want to propose a metaphysical framework of its own as the setting for explaining living design does not, *as such*, make it any more objectionable than evolutionary materialism.

The real problem, however, is that *both* ID and evolutionary materialism take flight into ultimate metaphysical explanations too early in their explanations of life. One of the lessons that a more seasoned theology has learned from modern science is that we must all postpone metaphysical gratification. To introduce ideas about God or intelligence as the direct "cause" of design would be theologically as well as scientifically ruinous. A mature theology allows natural science to carry its own methods and explanations as far as they can possibly go. This reserve does not entail, however, that theology is irrelevant at every level of a rich explanation of life. Theology is now freed from moonlighting in the explanatory domain that science now occupies, so that it may now gravitate toward its more natural setting – at levels of depth to which science cannot reach. Theology can now devote its full attention to the truly big questions that constitute its proper domain. Theology, after all, assumes that there is more than one level of explanation for everything. It endorses the idea of a plurality of explanations, perhaps hierarchically arranged, such that no discipline can give an exhaustive account of anything whatsoever. Any particular explanation, including the Darwinian explanation, is inevitably an abstraction and needs to be complemented by a luxuriant explanatory pluralism. When it comes to living beings, for example, there is more than one explanatory slot available, though it is entirely appropriate to push scientific explanations (physical, chemical, biological) as far as they can possibly go at their own proper levels within the many explanatory layers. The main problem with ID is that, ironically, it shares with evolutionary materialism the unfounded belief that only one authoritative kind of explanation is available to us today – namely, the scientific – and so feels compelled to push impatiently a metaphysically and theologically loaded notion of "intelligent design" into a logical space that is entirely too small for it.

If there is anything like a providential significance to the historical arrival of the scientific method, it may lie in the fact that science has now distanced the divine from any immediate grasp or human cognitional control. Darwin's science, in particular, has removed easy religious access to an *ultimate* explanation of design that formerly seemed to lurk just beneath the surface of living complexity. By allowing for purely scientific inquiries into living design, theology can now function at a deeper level of explanation, addressing questions such as why there is any order at all, rather than just chaos; or why there is anything at all, rather than nothing; or why the universe is intelligible; or why we should bother to do science at all. Today, as a result of science, the long path from surface design down to nature's ultimate depths turns out to be much less direct than ID theorists seem to crave.

Postponing metaphysics, however, calls for an asceticism that neither ID nor evolutionary materialism is disciplined enough to practice. They both try to arrive at the ultimate foundations of design too soon, pretending to have reached the basement level before even commencing the long journey down the stairs. One way of manifesting this metaphysical impatience is to fasten the phenomenon of living complexity directly onto the cozy idea of divine intelligent design without first looking into nature's own self-organizing, emergent spontaneity. But no less impatient, and prematurely metaphysical, are assertions that design is "nothing but" the outcome of blind natural selection of inheritable variation. ID is a "science stopper," since it appeals to a God-of-the-gaps explanation at a point in inquiry when there is still plenty of room for further scientific elucidation. But invoking the idea of "natural selection" as though it were an incomparably *deep* explanation of life's design could be called a "depth suppressor." Evolutionist materialism, not unlike ID, capitulates to the craving for ultimate explanations of life at a point when the journey into the depth of design may have just barely started.

It seems to me that "dreams of a final theory" are as conspicuous among Darwinians today as they are among physicist-philosophers. The fantasies of categorical finality among some evolutionary thinkers exhibit a dogmatism as rigid as that of any creationist. Science can have a future, however, only if its devotees retain a tacit sense of the unfathomable depth beneath nature's surface. And this is why any premature appeal to either theological or naturalistic metaphysics blunts our native intuition of nature's endless depths, supposing as it does that human inquiry has already arrived at the bottom of it all. The deadening thud of metaphysical finality is audible on both sides of the ID versus evolution debate. An appropriate theology of divine Providence and its relation to nature, on the other hand, abhors such premature metaphysical gratification. It sets forth a vision of the world in which science has an interminable future, since nature has an inexhaustible depth. Accordingly, the unwillingness of either ID or evolutionary materialism to dig very deep into the explanatory roots of life's organized complexity is an insult to the human mind's need for an endless horizon of intelligibility.

If there are still any doubts about the conflation of materialist ideology with biological science, then the following words of Stephen Jay Gould, one of the most eloquent interpreters of Darwin, should dispel them:

I believe that the stumbling block to [the acceptance of Darwin's theory] does not lie in any scientific difficulty, but rather in the *philosophical content* of Darwin's message – in its challenge to a set of entrenched Western attitudes that we are not yet ready to abandon. First, Darwin argues that evolution has no purpose. Individuals struggle to increase the representation of their genes in future generations, and that is all.... Second, Darwin maintained that evolution has no direction; it does not lead inevitably to higher things. Organisms become better adapted to their local environments, and that is all. The "degeneracy" of a parasite is as perfect as the gait of a gazelle. Third, Darwin applied a consistent philosophy of materialism to his interpretation of nature. Matter is the ground of all existence; mind, spirit and God as well, are just words that express the wondrous results of neuronal complexity. (Gould 1977, 12–13, emphasis added)

Gould would argue that Darwin himself had already begun the process of mixing materialism and evolution and that the current materialism among biologists is not a departure from a trend initiated by the master himself. In any case, today what is so threatening to ID about Darwinism is not just the scientific information it gathers, but even more the ideology of materialism that has taken hold of this data, twisting it into a tangle of science and faith assumptions – all in the name of science. The only solution to the ID versus Darwinism debate, therefore, is for everyone to distinguish science from all belief systems, whether religious or materialist. This would mean that the idea of divine Providence after Darwin remains pretty much the same as it was before. Evolutionary science cannot tell us anything significant about God that we did not already know from revelation, nor can our experience of God add much to our understanding of evolution. Evolution is a purely scientific theory that should be taken hostage neither by theism nor by materialism.

Of course, the inevitable objection will arise, from both ID and evolutionary materialism, that evolution cannot really be separated from the materialist beliefs that have dogged it ever since Darwin. For example, don't the elements of chance, suffering, and impersonal natural selection, operative over the course of a wasteful immensity of time, entail a materialist and therefore Godless universe? Aren't Dennett, Dawkins, Rose, Cziko, Crews, and the rest fully justified in reading evolution as the direct refutation of any plausible notion of divine Providence?

Such a conclusion is not at all necessary. For example, what appears to the scientist to be contingent, random, or accidental in natural history or genetic processes may have these apparent attributes because of our general human ignorance. After all, any purely human angle of vision, as the great religions have always taught, is always limited and narrow. As for complaints about the struggle, cruelty, waste, and pain inherent in evolution – realities that seem at first to place the universe beyond the pale of any conceivable Providence – these are already familiar issues to religion and theology. Darwin has contributed nothing qualitatively new to the perennial problem of suffering and evil, the harsh realities to which religions have always been sensitive. Robust religious faith has never cleared up these mysteries, but mystery is a stimulus and not a defeat for the endless religious adventure. Faith, after all, means unflagging trust *in spite of* the cruelties of life. Too much certitude actually renders faith impossible. So our new knowledge of evolution's indifference is no more an impediment to trust in divine Providence than suffering and evil have always been.

B. Theological Response II

The response just summarized, Theological Response I, is one that scientifically educated believers have often made in defense of the notion of

Providence in the post-Darwinian period. A good number of scientists and theologians have reacted to Darwin's allegedly "dangerous" ideas at least roughly along the lines of this response. However, let us now consider another approach, one that goes far beyond just affirming the logical compatibility of evolutionary biology and theism. The following proposal argues that theology can fruitfully and enthusiastically embrace the Darwinian portrayal of life once the latter has been purged of its excess philosophical baggage.[3]

Although the sharp distinctions made by Theological Response I are essential, it is too willing to let science and religion go their separate ways. It is true that science and religion are distinct, but they both flow forth from a single human quest for truth. They are oriented toward reality in different ways, but the unity of truth demands that we not keep such scientific facts as evolution separate from our reflections on the notion of Providence. Once we accept evolutionary science in an intellectually serious way, we cannot have exactly the same thoughts about Providence as we had before Darwin. And so another kind of theological response than the one just given is necessary.

Like the first approach, this second one is critical of the implicitly theological maneuvers of ID as well as of the usually unacknowledged ideological agenda of evolutionary materialism. However, it candidly admits that contemporary evolutionary biology demands a deepening of the theological sense of what divine care or Providence could possibly mean in the light of the disturbing picture of life that Darwin's science has now set before us.

Unfortunately, most contemporary theology has still not undergone the transformation that evolution demands. This delay is not entirely surprising, since the Darwinian shock is still ringing, and religions often take centuries to react to such an earthquake. The sobering fact, however, is that the world's religions have still barely begun to respond to evolutionary biology. A few religious thinkers have made evolution the backbone of their understanding of God and the world, but they are the exception rather than the rule. Modern Western theology began to lose interest in the natural world soon after the historical emergence of science, and this is one reason why the topic of evolution still remains outside the general concerns of theology. Such a slighting of evolution, however, is a lost opportunity for theological growth and renewal. A serious encounter with Darwinism might not only deepen the theology of Providence but also help to expose both the theological and the scientific evasiveness of ID.

In fact, it is now conceivable that theology's understanding of divine Providence can be restated in evolutionary terms without any loss in its power to provide a basis for religious hope. How, though, can the idea of divine providential wisdom be rendered compatible with the obvious randomness or contingency in life's evolution? This is a question that both ID proponents and Darwinian materialists will inevitably ask. The first theological response, summarized earlier, would respond that we humans resort to words like "randomness," "accident," and "chance" simply because of our abysmal human ignorance of the larger divine plan for the universe. In other words, randomness and contingency are illusions beneath which the properly initiated may be able to discern God's deeper designs. There is admittedly something deeply religious about trusting in a vision wider than meager human awareness can command, but it is entirely unnecessary – and indeed wrong theologically – to deny that chance is a real aspect of the world. Contingency, to use the general philosophical term, must be accepted as a fact of nature – as Stephen Jay Gould, among others, has rightly emphasized. But instead of logically refuting Providence or rendering the universe absurd, the fact of contingency in natural history and in genetic processes allows us to correlate the facts of nature more intimately than ever with actual religious experience and belief.

The real issue here is not whether evolution rules out a divine designer, but whether the Darwinian picture of life refutes the notion that divine Providence is essentially self-giving love. The approach of evolutionists such as Dawkins and Dennett is first to reduce the idea of God to that of a designer, then to argue that Darwinism explains design adequately, and finally to conclude that Darwinism has thus made God superfluous. It does not help things theologically, of course, that ID also – at least in its formal argumentation – implicitly reduces ultimate explanation to that of intelligent design. However, in any serious discussion of evolution and theism, there is little point in abstract references to emaciated philosophical ideas of deity, especially those that picture this ultimate reality as essentially an engineer, mechanic, or designer. Instead, scientists and scientifically educated philosophers must converse with thoughts about God that arise from *actual* religious symbols and teachings. This advice will make the conversations regarding evolution and theology more complex, but also more pertinent, than they are when the only idea being debated is that of design or a designer. Most theists, after all, are not willing to spend every ounce of their theological energy defending such simplistic philosophical abstractions as "divine design." Many of them will be much more concerned with the question of whether and how Darwinism fits the vision of God as humble, self-giving love.

To be more specific, if there is anything to the religious intuition that the universe is the consequence of selfless divine love (a belief held by countless theists), then the fact of contingency in the created universe and life's evolution may easily be understood as a manifestation rather than a refutation of that love. An infinite love, if we think about it seriously, would manifest itself in the creation of a universe free of any rigid determinism (either natural or divine) that would keep it from arriving at its own independence, autonomy, and self-coherence. Contingency, it is true, can lead life down circuitous pathways, but that is precisely the risk that love takes when it allows genuine otherness to emerge as the object of that love. In any wholesome human experience of being loved, we already intuitively know that true concern or care for the other does not resort to control or compulsion, but rather in some way "lets the other be." Can we expect less than this respect for the otherness and eventual freedom of created reality on the part of what religious faith confesses to be an infinite love?

Love, at the very minimum, allows others sufficient scope to become themselves. So if there is any substance at all to the belief that God really loves and *cares* for the integrity of the created universe as something truly other than God, then this universe must possess a degree of autonomy. Relative independence must be inherent in the universe from the outset. If the universe were not percolating with contingency, it would be nothing more than an extension of God's being, an appendage of the Deity, rather than something genuinely other than God. A central tenet of theism, as distinct from pantheism, is that the world is not God. But if the world is distinct from and *other* than God, then its existence is not necessary (as is God's existence), nor are its specific characteristics. In other words, the world is contingent, and wherever there is contingency, there has to be room for the undirected events we call accidental or random. Chance is an inevitable part of any universe held to be both distinct from and simultaneously loved by God (Johnson, 1996). Even the medieval philosopher and theologian Saint Thomas Aquinas observed that the idea of a universe devoid of accidents would be theologically incoherent (Mooney 1996, 162).

Moreover, once we allow that God's creative and providential activity is essentially a liberation of the world to "be itself," then it comes as no surprise to the informed theist that the creation would have its own autonomously operative "laws" (such as natural selection), and that the universe would probably not be finished in one magical moment but would instead unfold stepwise – perhaps consuming billions and trillions of human years. The point is, there could be no genuine self-giving of God to any universe that is not first allowed to become something radically distinct from God. This means that in some sense the created world must be self-actualizing, and even self-creative, though within the limits of relevant possibilities held out to it by its Creator. Theologically speaking, therefore, the vastness of evolutionary duration, the spontaneity of random variations or mutations, and even the automatic machinations of natural selection could be thought of as essential ingredients in the emergence of cosmic independence. Perhaps only in some such matrix as this could life, mind, and eventually human freedom come into being. Maybe the entirety of evolution, including all the suffering and contingency that seem to render it absurd, are quite consistent, after all, with the idea of a Providence that cares for the internal growth and emergent independence of the world.

To some evolutionary materialists and followers of ID, the "wasteful" multi-millennial journey of evolution counts against a plausible trust in divine Providence. For both schools of thought the idea of Providence has been implicitly reduced to design, so evolution is logically inconsistent with theism. If God were intelligent and all-powerful, both sides assume, then the creative process would not be as rough and ragged as it is. Nontheistic evolutionists and ID proponents alike insist that a competent creator would have to be an intelligent designer. But their blueprints for an acceptable deity seem, at least to a theology in touch with actual religious experience, to resemble a conjurer or a distant architect rather than the infinitely humble and suffering love that theologians such as Wolfhart Pannenberg (1991), Elizabeth Johnson (1996), and Karl Rahner (1978) have set before us. Unfortunately, these and many other highly respected interpreters of theism are seldom consulted by those involved in the so-called Darwin wars.

A theology after Darwin also argues that divine Providence influences the world in a persuasive rather than a coercive way. Since God is love and not domineering force, the world must be endowed with an inner spontaneity and self-creativity that allows it to "become itself" and thus to participate in the adventure of its own creation (Haught 1995, 2000). Any other kind of world, in fact, is theologically inconceivable. If God were a directive dictator rather than persuasive love, the universe could never arrive at the point of being able to emerge into freedom. If God were the engineering agency that ID and evolutionary materialism generally project onto the divine, evolution could not have occurred. The narrative grandeur that Darwin observed in life's evolution would have given way to a monotonously perfect design devoid of an open future.

If one still wonders why Providence would wait so long for the world to be finished, some words of theologian Jürgen Moltmann are worth hearing:

God acts in the history of nature and human beings through his patient and silent presence, by way of which he gives those he has created space to unfold, time to develop, and power for their own movement. We look in vain for God in the history of nature or in human history if what we are looking for are special divine interventions. Is it not much more that God waits and awaits, that – as process theology rightly says – he 'experiences' the history of the world and human beings, that he is "patient and of great goodness" as Psalm 103:8 puts it? . . . "Waiting" is never disinterested passivity, but the highest form of interest in the other. Waiting means expecting, expecting means inviting, inviting means attracting, alluring and enticing. By doing this, the waiting and awaiting keeps an open space for the other, gives the other time, and creates possibilities of life for the other. (Moltmann 2001, 149)

In the well-known words of Saint Paul, such a picture of God's hiddenness, patience, and vulnerability sounds like "foolishness" (1 Cor 1:25) when compared to our typical preference for Providence in the form of a designing divine magician. But a magician could never provide what a self-effacing love can provide – namely, the space within which something truly unpredictable, and therefore truly other than God, might emerge into being. Ever since Darwin, the most profound theologies have had no difficulty reconciling evolution with a God who does not compel but rather awaits the world so as not to detract from its independence and integrity. Today, of course, many will still find Darwinian evolution too troubling a notion to be allied so closely with theology. But the notion of a God whose essence is self-giving love, a God who opens up an endlessly new future for the world, actually anticipates the idea of evolution. Thus it should not be difficult for theology to understand Providence in terms of a picture of life approximately like the one that Darwin and contemporary evolutionary science have given us, rather than in terms of the abstract – and rather lifeless – notion of a divine designer.

CONCLUSION

I hope to have demonstrated in this chapter, first, that Darwinian evolution does not inevitably entail a materialist metaphysics, and second, that the theological notion of Providence is quite distinct from the idea of Intelligent Design. I have summarized two quite different theological positions, both of which argue for the compatibility of evolution and divine Providence without conceiving of Providence as essentially design. The special strength of the first approach is that it distinguishes clearly between metaphysics and science, a point lost on both evolutionary materialism and ID theory. Sooner or later, the appeal to metaphysics is unavoidable and necessary for deep explanation. Both materialism and ID theory implicitly realize this point, but they lurch too quickly toward ultimate explanations. The advantage of the second theological approach, then, is that it both allows science full scope to do its own work and simultaneously sets forth a metaphysical alternative to evolutionary materialism that can fully contextualize Darwinian discovery without jeopardizing either scientific explanation or the religious doctrine of Providence. In doing so, it also responds compassionately to the religious anxiety that turns ID theorists away from good biology and shows how unnecessary it is to import theological ideas into the work of science.

Notes

1. Theologians often distinguish between general Providence, God's care for the universe as a whole, and particular Providence, God's involvement in a specific event or in an individual's life. Here the focus will be on general Providence.
2. Ian Barbour (1990) refers to this theological way of relating science to religion as the "independence" model, and I refer to it (1995) as the "contrast" approach.
3. I have developed this approach more fully in *God after Darwin* (Haught 2000).

References

Barbour, Ian G. 1990. *Religion in an Age of Science.* San Francisco: Harper and Row.

Behe, Michael J. 1996. *Darwin's Black Box: The Biochemical Challenge to Evolution.* New York: The Free Press.

Crews, Frederick. 2001. Saving us from Darwin. *The New York Review of Books* 48 (October 4 and 18).

Cziko, Gary. 1995. *Without Miracles: Universal Selection Theory and the Second Darwinian Revolution.* Cambridge, MA: MIT Press.

Dawkins, Richard. 1986. *The Blind Watchmaker.* New York: Norton.

1995. *River Out of Eden.* New York: Basic Books.

1996. *Climbing Mount Improbable.* New York: Norton.

Dembski, William A. 1999. *Intelligent Design: The Bridge between Science and Theology.* Downers Grove, IL: InterVarsity Press.

Dembski, William A. (ed.) 1998. *Mere Creation: Science, Faith and Intelligent Design.* Downers Grove, IL: InterVarsity Press.

Dennett, Daniel C. 1995. *Darwin's Dangerous Idea: Evolution and the Meaning of Life.* New York: Simon and Schuster.

Gould, Stephen Jay. 1977. *Ever since Darwin.* New York: Norton.

Haught, John F. 1995. *Science and Religion: From Conflict to Conversation.* Mahwah, NJ: Paulist Press.

2000. *God after Darwin: A Theology of Evolution.* Boulder, CO: Westview Press.

Johnson, Elizabeth. 1996. Does God play dice? Divine Providence and chance. *Theological Studies* 57 (March): 3–18.

Johnson, Phillip E. 1991. *Darwin on Trial.* Washington, DC: Regnery Gateway.

1999. *The Wedge of Truth: Splitting the Foundations of Naturalism.* Downers Grove, IL: InterVarsity Press.

Moltmann, Jürgen. 2001. God's kenosis in the creation and consummation of the world. In *The Work of Love: Creation as Kenosis,* ed. John Polkinghorne. Grand Rapids, MI: Eerdmans.

Mooney, Christopher, S. J. 1996. *Theology and Scientific Knowledge.* Notre Dame and London: University of Notre Dame Press.

Pannenberg, Wolfhart. 1991. *Systematic Theology,* Vol. I, trans. Geoffrey W. Bromiley. Grand Rapids, MI: Eerdmans.

Plantinga, Alvin. 1997. Methodological naturalism. *Perspectives on Science and Christian Faith* 49: 143–54.

Rahner, Karl. 1978. *Foundations of Christian Faith: An Introduction to the Idea of Christianity,* trans. William V. Dych. New York: Seabury Press.

Rose, Michael R. 1998. *Darwin's Spectre: Evolutionary Biology in the Modern World.* Princeton: Princeton University Press.

Ruse, Michael. 2001. *Can a Darwinian Be a Christian?* Cambridge: Cambridge University Press.

Segerstrale, Ullica. 2000. *Defenders of the Truth: The Battle for Science in the Sociology Debate and Beyond.* New York: Oxford University Press.

Stace, W. T. 1948. Man against darkness. *The Atlantic Monthly* 182 (September 1948).

Weinberg, Steven. 1992. *Dreams of a Final Theory.* New York: Pantheon Books.

Wilson, E. O. 1998. *Consilience: The Unity of Knowledge.* New York: Knopf.

13

The Inbuilt Potentiality of Creation

John Polkinghorne

Our understanding of the very early universe tells us that we live in a world that seems to have originated some fourteen billion years ago from a very simple state. The small, hot, almost uniform expanding ball of energy that is the cosmologist's picture of the universe a fraction of a second after the Big Bang has turned into a world of rich and diversified complexity – the home of saints and scientists. Although, as far as we know, carbon-based life appeared only after about ten billion years of cosmic history, and self-conscious life after fourteen billion years, there is a real sense in which the universe was pregnant with life from the earliest epoch.

ANTHROPIC FINE-TUNING

One can say this because, although the actual realisation of life has pro-ceeded through an evolutionary process with many contingent features (the role of "chance"), it has also unfolded in an environment of lawful regularity of a very particular kind (the role of "necessity"). The so-called Anthropic Principle (Barrow and Tipler 1986; Leslie 1989) refers to a collection of scientific insights that indicate that necessity had to take a very specific form if carbon-based life were ever to be a cosmic possibility. In other words, it would not have been enough to have rolled the evolutionary dice a suffi-cient number of times for life to have developed somewhere in the universe. The physical rules of the cosmic game being played also had to take a very precise form if biology were to be a realisable possibility. The given physi-cal fabric of the world had to be endowed with anthropic potentiality from the start.

It is worth recalling some of the many considerations that have led to this conclusion. A good place to begin is by asking where carbon itself, along with the nearly thirty other elements also necessary for life, comes from. Because the very early universe was so simple, it only made very simple things. Three minutes after the hectic events of the immediate post–Big Bang era, the universe settled down into a state in which its matter was three-quarters hydrogen and one-quarter helium. These are the simplest of the chemical elements; on their own, they have too boring a chemistry to be the basis for complex and interesting developments. It was only when the nuclear furnaces of the first generation of stars started up, a billion years or so after the Big Bang, that richer possibilities began to be realised. Every atom of carbon in our bodies was once inside a star – we are people of stardust. One of the great successes of twentieth-century astrophysics was to unravel the chain of processes by which the chemical elements were made, within stars and in the death throes of supernova explosions.

As an example, consider how carbon was formed. The first stars contained α-particles, the nuclei of helium. To make carbon, three α-particles must combine to yield carbon 12. One might suppose that the natural way to achieve this would be via the intermediate state of a berylium nucleus (made of two αs), to which a third α might subsequently become attached; but this possibility is made problematic by the extreme instability of berylium 8. The only way in which carbon could be made would be if the attachment of that third α was a process that went anomalously fast. At first, the astrophysicists were completely baffled. Then Fred Hoyle realised that this route to carbon would be possible only if there were a resonance (a greatly enhanced effect) in carbon 12 at exactly the right energy to facilitate the process. Such a resonance was not then known, but so precise was the pinpointing of what its energy would need to be that Hoyle could tell the experimentalists exactly where to look to see if it were in fact there. They did so, and they found it. Any change in the strengths of the basic nuclear forces would have displaced the resonance and so frustrated any possibility of carbon production by stars. When Hoyle saw the remarkable degree of fine-tuning that was necessary if the process of nucleogenesis was to get off the ground, he is reported to have said that the universe is a "put-up job." Hoyle could not believe that the fulfilment of so precise a condition was just a happy accident.

Further investigation revealed that not only the first link but the whole chain of processes by which the elements are made in viable abundance is beautifully and delicately balanced. After carbon, the next element to be formed is oxygen, requiring the addition of yet another α to carbon. Here it is important that there is not a resonance in the oxygen nucleus, enhancing the effect, for if the process were too efficient, all the carbon would be lost as it was transmuted rapidly into oxygen. And so on. If the basic nuclear forces that control the processes of stellar nucleogenesis had been in the slightest degree different, this would have broken links in this chain and so frustrated the possibility of life's developing anywhere in the universe.

Many more examples could be given of necessary anthropic "fine-tunings" in the laws of nature. For instance, if complex life is to develop on their encircling planets, stars must burn reasonably steadily and for the several billion years that such a process requires. The behaviour of stars in our universe is well understood, and it is found to depend upon a delicate balance between the intrinsic strengths of two basic forces of nature, gravity and electromagnetism. If that balance were out of kilter, stars would either burn so furiously that they would quickly burn themselves out, lasting only for a few million years, or burn so feebly as not to be able to function as effective energy sources to fuel the development of life.

Turning to a terrestrial example, one might consider the many remarkable properties of water that have made it indispensable to living beings (Denton 1998, Chapter 2). For instance, the density of water behaves anomalously near freezing point, decreasing rather than showing the normal liquid behaviour of increasing as the temperature falls. This property has been of vital significance for the development of life, for it means that lakes and pond freeze from the top down, rather than from the bottom up. This prevents such life-breeding locations from freezing solid, with the lethal consequences that would follow. Ultimately, aqueous properties of this kind must stem from the exact form of electromagnetism (the force that holds matter together); if this force were different, presumably water would not behave in these ways that are so hospitable to life.

It would be possible extensively to multiply examples of anthropic coincidences, but it will be enough for the present finally to focus on three cosmological considerations that are of anthropic significance. One is simply the vast size of the observable universe, with its 10^{22} stars. Rather than being daunted by such cosmic immensity, we should be thankful for it. Only a universe at least as big as ours could have lasted the fourteen billion years that must elapse between the initial Big Bang and the possibility of evolved self-conscious life. It is not a process that can be hurried – for example, it takes about ten billion years to get enough carbon, as the first step.

A second point relates to the fact that the very early universe was smooth and homogeneous and also that there was a very precise balance (of the order of one part in 10^{60}) between the explosive effects of expansion and the contractive effects of gravity. Both of these conditions must be satisfied in a life-evolving universe. Large inhomogeneities would have produced destructive turbulence, and a greater degree of inbalance between expansion and contraction would have either induced rapid cosmic collapse or blown matter apart too quickly for it to condense into galaxies and stars. However, there is a scientific explanation for these particular fine-tunings. A speculative, but very plausible, process called "inflation" is thought to have operated when the universe was about 10^{-35} seconds old, in which, for a short period, space expanded with extreme rapidity. This process would have resulted in a smoothing and balancing effect, even if still-earlier cosmic circumstances had been different. Yet the possibility of inflation itself requires that the laws of nature take a particular form, so this insight deepens, but does not dispose of, the question of anthropic specificity.

The third point relates to what is called the "cosmological constant." Essentially, this is a kind of energy associated with space itself, for which the phrase "dark energy" has been coined. Until recently, cosmologists believed that this constant was in fact zero. They now tend to think that it is nonvanishing but extremely small, amounting to about 10^{-120} of what one would have expected to be its natural value. If this constant had been in any degree larger, life would again have been impossible, for the universe would have either been blown apart very quickly or collapsed extremely rapidly, depending on the sign of the constant. The minute value of the cosmological constant is the most exacting of all the conditions of anthropic fine-tuning.

Let us return to the general question of the strengths of the basic forces that we observe today to be controlling physical processes. We have noted that they are tightly constrained by anthropic considerations: nuclear forces capable of producing the elements, gravity and electromagnetism capable of suitably regulating the burning of the stars, and so on. Often the same force is subject to multiple conditions of this kind (electromagnetism in relation to both stars and water, for example), which nevertheless clearly, and remarkably, turn out to be mutually compatible. Many physicists believe that these observable force strengths are a consequence of processes taking place in the very early universe, by which, as the cosmos cooled, a highly symmetric ur-force (described by a Grand Unified Theory) was broken down into the less symmetrical set of forces that we observe today. This reduction would have been triggered by a process that is called spontaneous symmetry breaking. It is induced by contingent circumstances, and it need not have happened in a literally universal way. It is possible that symmetry breaking took different forms in different cosmic domains. (The process is rather like the way in which a vertical pencil, balanced on its point, may fall asymmetrically in a particular horizontal direction, as the result of a very small disturbance. These disturbances may vary from place to place – a set of such pencils would not all fall in the same direction.) Thus there could be different parts of the universe in which the resulting balance of observable forces takes different forms. The whole of our observable universe must lie within one of these domains, for we see no sign of the variation of natural force strengths within our field of observation. Yet, there might also be other domains, blown away from our view by inflation. Of course, human observers must find themselves located within an appropriately fine-tuned cosmic domain, because we could not have evolved anywhere else. As with the discussion of inflation itself, this consideration does not remove anthropic particularity altogether, but it serves to deepen its discussion. It still remains necessary to assume that the ur-force of the original Grand Unified Theory takes some constrained form, if its broken symmetry consequences are to be capable anywhere of generating forces of the kind that would yield a domain able to evolve carbon-based life

Although the exploration of possibility through evolving process is certainly part of the story of the universe's fruitful history, by itself evolution would have been powerless to bring about carbon-based life if the physical fabric of the world had not taken a very specific form. Realisation of this fact came not only as a surprise to almost all scientists, but also as an unwelcome shock to many. The scientific inclination is to prefer the general to the unique, and so it had seemed natural to assume that our world was just a fairly typical specimen of what a world might be like. The Anthropic Principle makes it clear that this is not the case at all.

This realisation has evoked a number of different responses. Since science (physics) takes the laws of nature as its given and unexplained starting point, on which it grounds all of its explanations of cosmic process, its own discourse cannot take it behind those laws to explain their ultimate character. Thus all responses to the Anthropic Principle are essentially metaphysical in their character, whether this is acknowledged by their proponents or not.

One way of responding simply says that if the universe were not the way it is, then we humans would not be here to worry about it. Any world of which we are aware just has to be consistent with our being its inhabitants. (This is sometimes called the Weak Anthropic Principle.) Of course, this is true, but simply resting on this truism misses the real point. What is truly surprising, and what surely must in some way be significant and call for further understanding, is that satisfying this harmless-looking criterion turns out to impose such stringent conditions on the physical character of the universe. One might have supposed that any "reasonable" sort of cosmos might be expected to have had the possibility of its own kind of fruitful history. Consider, for instance, a world whose physical fabric is exactly the same as ours but with the sole difference that its gravity is somewhat stronger (say, for the sake of definiteness, three times stronger) than ours. One might well have thought that so similar a universe would in due course evolve its own form of life – obviously not *Homo sapiens*, but little green men, maybe. (And they would be little, for the stronger gravity of that world would make it more difficult to grow tall, so that its inhabitants would be expected to be rather squat.) In fact, there would be no living beings of any kind in that world, for its stars would burn themselves out in just a few million years, before life would have time to develop.

Simply to shrug off this remarkable specificity seems an intellectually lazy response. I have proposed the Moderate Anthropic Principle, "which notes the contingent fruitfulness of the universe as being a fact of interest calling for an explanation" (Polkinghorne 1991, 78). John Leslie makes the same point in a more picturesque fashion (Leslie 1989, 13–15). He tells the story of a person facing a firing squad composed of fifty highly trained marksmen. After the shots ring out, the prisoner finds that he has survived. Will he not ask himself why he has been so fortunate? Of course, he could not do so if he were not still alive, but simply to shrug one's shoulders and say "That was a close one" would be to carry incuriosity to ridiculous extremes. Equally, we should not be indifferent to the questions that relate to anthropic fine-tuning. Leslie suggests that there are two rational kinds of explanation possible for the prisoner's escape from death. One is that the marksmen were on his side and missed by design. The other is that even trained marksmen occasionally miss, and there were so many executions taking place that day that this was one instance in which they all missed.

Clearly, theism can provide a coherent response to the anthropic question, corresponding to the first of the explanations that Leslie suggested for his parabolic story. Those who believe in God do not regard the universe as being just "any old world," but they understand it to be a creation whose Creator may be expected to have endowed it with just those finely tuned laws and circumstances that will enable it to have the fruitful history that would be a fulfilment of the divine will. Two important aspects of this theological response need to be noted. One is that its metaphysical character does not put it in conflict with what an honest science has to say, but rather complements the story that the latter can tell. One should welcome the possibility of extended understanding that theism offers, since the laws of nature, in their anthropic specificity, are not so intellectually self-contained that they can fittingly be treated as requiring no further explanation, in the way that David Hume recommended that they be considered. Rather, these laws may be held to point beyond themselves in what may credibly be conceived to be a theistic direction. A kind of natural theology of this kind must be sharply distinguished from one, say, that tries to argue that some form of direct divine "intervention" is needed to bring about life. The latter is in contention with science in accounting for the unfolding process of the world; the former is concerned with matters that lie beyond the grasp of science, as theology seeks to make intelligible the form of those laws of nature that are the assumed ground of all scientific discussion of physical/biological processes. Putting the matter in Hebraic terms, science is concerned with *'asah* (ordinary making, of the kind that Paley's watchmaker might also have been supposed to be engaged in), but theology's proper concern is *bara* (the word reserved in the Hebrew scriptures uniquely for divine creative activity, which can be understood as being the sustaining of created reality).

The second aspect of the theistic response that needs to be noted also follows from the last remark. Hume had criticised the natural theology of his day as being too anthropomorphic, as if the Creator were to be compared to a carpenter making a ship. We are presenting a picture of creation made fruitful through inbuilt potentiality, an activity that is intrinsically divine, having no corresponding human analogue. (The point at issue relates precisely to the distinction between creation "out of nothing" and the human artist's creative manipulation of an already existing medium.)

There is an alternative conceivable response to the anthropic question that looks, not to God, but to a greatly expanded concept of the scope of physical reality. It is the analogue of Leslie's second explanation of his parable. Those who wish to defuse the threat of theism can turn instead to the supposition that our universe is part of a much more extensive multiverse, a vast collection of component universes. If there are indeed a very great variety of different universes of all kinds (and there would have to be a truly vast cosmic portfolio for this argument to work), then this universe in which we live is just the one where, by chance, carbon-based life is a possibility, since, of course, we could not have appeared in the history of any other. According to this many-worlds approach, all that is significant about this universe is that it represents a winning ticket in the grand multiversal lottery.

An immediate comment to make on this response is that, if it is really to do the work required of it, it is as much a metaphysical proposal as is the response of theism. In respectable scientific terms, the only well-motivated concept remotely like a multiverse would be the set of cosmic domains of spontaneous symmetry breaking, as discussed earlier. But we have already seen that they do not dispose of anthropic particularity, since their Grand Unified Theory is itself subject to anthropic constraints. An appeal to the many-worlds interpretation of quantum theory (see Polkinghorne 1990, 67–8) – even if one accepted its ontological prodigality – would not help, since its parallel universes differ only in the outcomes of quantum events and not in their underlying physical laws. If a many-worlds response to the anthropic question is really to work, it has to go beyond these limited possibilities, invoking the existence of universes with laws of nature having radically different forms. Such a proposal is manifestly a metaphysical guess – and one that might be considered to be of a very ontologically prodigal kind.

One might make a similar kind of comment on an ingenious but highly speculative proposal by Lee Smolin (see Rees 1997, 261–4), which seeks to assimilate anthropic explanation to quasi-Darwinian principles. Smolin assumes that black holes generate new universes with slightly changed natural laws. This entirely ad hoc metaphysical assumption then provides variation. Selection is imposed by the suggestion that anthropic universes are particularly apt to generate many black holes and so come to dominate the supercosmic evolutionary process. This latter claim is in principle scientific in its character, but there is considerable doubt as to its validity.

How then are we to conclude the discussion? Shall it be theism or many worlds? In his even-handed philosophical way, Leslie simply presents us with the choice:

Now much evidence suggests that Life's prerequisites could only amazingly have been fulfilled anywhere unless this is a truth: *that God is real and/or there are vastly many, very varied universes.* (Leslie 1989, 204)

If one is only considering the anthropic question, so balanced a conclusion seems reasonable. However, once one looks beyond the immediate point at issue, the scales begin to tilt. There are many other reasons for belief in God – the deep intelligibility of the world (a kind of cosmological argument), the existence of moral and aesthetic values (a kind of axiological argument), the human encounter with the sacred (an argument from religious experience), and so on (see Polkinghorne 1998, Chapter 1). Together, these constitute a cumulative case for theism in which the God hypothesis does a number of pieces of explanatory work. On the other hand, the conjecture that there are many, very varied universes seems to do only one piece of explanatory work. It is simply invoked as an alternative to theism in the metaphysical response to anthropic fine-tuning. An important form of argument in defence of any metaphysical position is the explanatory scope that it offers. In this respect, theism seems to score over the multiverse.

Metaphysical questions do not lend themselves to categorical knockdown answers. There will always be some room for more tacit considerations to come into play in determining a personal conclusion (room for the commitment of faith, a theologian might say). It would be wrong to accuse atheism of irrationality, but there are grounds for claiming that theism offers a greater breadth of understanding of the nature of the rich reality within which we live.

Two further points need to be made before we pass on to a brief final discussion of some related topics. The first point relates to the fact that there are some physicists who believe that, rather than there being a vast range of conceivable laws of nature (and so a great variety of universes in which they might hold), there is, in fact, ultimately only one totally consistent possibility. In their view, when eventually we discover this "theory of everything," we shall find that, in fact, all of the forces of nature could not be other than they actually are. This universe is simply the way it is because it could not be any other way.

As a matter of scientific judgement, I personally very much doubt that this ambitious hope is justified. It seems highly likely that any physical theory will contain within it some adjustable scale parameters that have to be determined empirically and are not fixed by logic alone. One of the motivations that lie behind this conjecture of uniqueness is the difficulty that has been found in combining quantum theory and general relativity in a wholly consistent manner. In fact, that problem has not yet fully been solved, even after decades of strenuous effort. Some think that it will turn out that there is only one way in which the synthesis can be accomplished. Even were this to turn out to be the case, there are further points to be made.

One is that it would be highly surprising, and surely significant, if a theory defined by such abstract – essentially mathematical – conditions of consistency should also prove to be just the one that permitted the fulfilment of all the intricate conditions that alone have allowed the evolution of self-conscious beings. To my mind, that would be the most astonishing anthropic coincidence of all.

Another comment is that quantum theory and general relativity are by no means self-evidently necessary properties of universes in general, though they are certainly critical for a life-evolving universe. The anthropic specificity of the cosmos extends not only to the parameters that specify force strengths, but also to the *form* that its physical laws take. A universe composed of Newtonian billiard ball atoms with hooks would be perfectly consistent, though boringly sterile.

The issue of the form of the laws of nature is one of particular interest. The coming to be of complex novelty seems to require a combination of flexibility and stability. It is precisely this combination that enables the evolutionary interplay of chance (historical contingency) and necessity (lawful regularity) to be so fruitfully effective. As we have learnt to say, the emergence of the new takes place "at the edge of chaos." Too far on the deterministic side of that border, and things are so rigid that only rearrangements are possible, so that there is no true novelty. Too far on the haphazard side of that border, and things are so fitful that nothing of novelty can ever persist. In biological evolution, one needs both a degree of genetic mutability to yield new forms of life, and also a degree of genetic stability to preserve the species on which selection acts.

Quantum theory, in contrast to the rigidity of Newtonian mechanics, has just that kind of controlled flexibility – it explains the stability of some atoms and the probabilistic decay of others. A similar interplay between openness and order seems to characterise the insights beginning to arise from the infant science of complexity studies. At present, that discipline is at the natural history stage of the investigation of the behaviour of computerised models. These are found to display quite astonishing powers of spontaneous self-organisation. Such phenomena of autopoesis are currently not well understood, but they seem to point in the direction of there being pattern-generating propensities present in the whole that are not discernible from consideration simply of the properties of the parts. I venture to guess that by the end of the twenty-first century, science will be as much concerned with the information-bearing behaviour of totalities as it has traditionally been with energetic exchanges between constituents. It is entirely possible that there are holistic laws of nature of an information-generating kind that are unknown to us today. Stuart Kauffman has suggested that concepts of this kind may be especially significant for biology (Kauffman 1995). He thinks that the homologies between species that comparative anatomists study may not always, as conventional Darwinism supposes, derive from a remote common ancestor but may arise from the convergent propensity for natural processes to generate certain specific forms of complex structure (see also Conway-Morris 1998).

These are interesting possibilities that await further investigation and evaluation. They certainly raise important *scientific* questions, and they should not be dismissed because of some neo-Darwinian "certainty" that there is nothing more to be found out about the basis of biological evolution. What then might be the metaphysical significance of the possibility that a full understanding of the fertile potentiality of the laws of nature will require a recognition of the role of their holistic, informational aspects? Once again, one encounters the inescapable degree of ambiguity present in metaphysical construction that seeks to build on a given physical base. If there are indeed holistic, pattern-generating laws of the kind just discussed, and if they are relevant to the coming to be and development of life, then the atheist can say, "Well, it's all nature, after all," while the theist can say, "Such very remarkable lawful potentiality calls for an explanation beyond the brute fact that things are this way." If, on the other hand, there seems to be an irreducible degree of lucky happenstance in life's having developed, then the theist can say, "Such fertile serendipity requires the providential action of the Creator to make it intelligible," while the atheist can say, "It is all simply a meaningless, if rather happy, accident."

The second point to make concerns what is meant by "life." The name of the "Anthropic Principle" would have been most infortunately chosen if it suggested that the issues centred on the coming to be of humans as such (literally, *anthropoi*). It is, of course, the generality of carbon-based life, with its potentiality to develop self-conscious beings, rather than *Homo sapiens* as such, that is the focus of concern. There are no doubt many specific features of humanity that are the contingent deposits of past evolutionary history. It has been suggested, however, that even a concern with carbon-based life is too parochial. Maybe other universes, of a character different from ours, would evolve their own versions of intelligent complexity that are beyond the power of us carbon-creatures to imagine. Maybe, or maybe not. This is the point in the discussion where the holders of all metaphysical positions have difficulty knowing what to say. One can only remark that the rich, structured complexity that self-consciousness seems to require – think of the human brain, with its 10^{11} neurons and their 10^{14} connections – is so immense that it doesn't encourage the thought that there would be many alternative ways of generating it.

As this section comes to a close, we should recall that ambiguities in how to respond to the interpretation of anthropic coincidences can be resolved only by a willingness to incorporate the discussion into a wider review. A key principle of metaphysical choice remains that of scope, the attainment of the widest possible understanding of the human encounter with reality. We have already indicated that it is along these lines that the religious believer can best mount a rational defence of theism.

PROVIDENTIAL INTERACTION

If one were to grant the theistic arguments so far set out their maximum persuasiveness, they would still lead only to a picture of the Creator that is as consistent with the spectator God of deism as with the God of the Abrahamic faiths, interacting providentially with the course of unfolding history. Putting the point in different terminology, so far we have considered only the case for considering the divine will and purpose as lying behind the necessity half of the evolutionary dialectic. A strong account of "Intelligent Design" (Behe 1996; Dembski 1999; for a critique, see Polkinghorne 2000, Chapter 4), however discreetly expressed, presumably has behind it an account of a Creator who intervenes in the developing history of life. The more widely theologically supported concept of "continuous creation" (Barbour 1966, Chapter 12; Peacocke 1979, Chapters 2 and 3), if it is to amount to more than a religious gloss on natural process, will itself need some account of a Creator who interacts with the developing history of life. The latter supposition need not put theology on a collision course with science if physics remains sufficiently open to permit within its account the acts of agents, whether human or divine. Whether this possibility of seeing God as active also within the chance half of the evolutionary dialectic is a rational one, is the issue that this section will seek briefly to explore. While full treatment would need to be very extensive (Russell, Stoeger, and Ayala 1998), here it will be sufficient to focus on three specific points relating to scientific insights and the theological reflections to which they give rise.

(1) *Evolving Process.* Whether one is considering cosmic history, with the coming to be of galaxies and stars, or terrestrial history, with the coming to be and complexification of life, a common feature is that unfolding process is driven by the evolutionary interplay of chance and necessity. "Chance" is simply the recognition of historical contingency, that not everything that might happen does happen. The Anglican clergyman Charles Kingsley, writing soon after the publication of the *Origin of Species* in 1859, had already grasped the theological significance of this. Although the Creator could, no doubt, have brought into being a ready-made world, God has in fact done something cleverer than that, bringing into being a creation that could "make itself." In that pregnant phrase, Kingsley had encapsulated an important insight into the way in which the God of love has chosen to order creation. The universe is not a divine puppet theatre in which everything dances to God's tune alone. The Creator is not the Cosmic Tyrant whose unrelenting grip holds on tightly to all. Such an enslaved world could not be the creation of a loving God. Rather, creation is allowed to be itself and to make itself, realising the inbuilt potentiality with which the Creator has endowed it, but in its own time and in its own way. Creatures live at some epistemic and ontological distance from their Creator, as they enter into the liberty that God has given them. Just as we may understand the reliability of "necessity" as being a pale reflection of the Creator's faithfulness, so we may understand the role of "chance" as being an expression of God's loving gift of freedom. The history of the universe is not the performance of a

fixed score, determined from all eternity, but rather an improvisatory performance, led by the Creator who is, in Arthur Peacocke's striking phrase, "an Improviser of unsurpassed ingenuity" (Peacocke 1986, 98), developing the grand fugue of creation, whose themes are provided by anthropic potentiality. An extremely important aspect of twentieth-century theology has been the recognition that creation is an act of *divine kenosis* (Polkinghorne 2001), God's self-limitation in allowing the creaturely other to be and to make itself.

(2) *Open Process*. Twentieth-century physical science identified the unexpected presence of widespread *intrinsic* unpredictabilities in the processes of the universe. Both at the subatomic level (through quantum theory) and at the everyday level of macroscopic phenomena (through chaos theory), many limitations on our ability to predict future behaviour have been discovered. It is important to recognise that these inherent limitations cannot be overcome by refining observational techniques or by improving theoretical calculations. The old mechanical picture of the physical world as tame, controllable, and predictable is certainly dead.

Unpredictability is an epistemological property, and it is a matter for further metaphysical decision to propose how these discoveries should affect our ontological thinking. Are we simply confronted by unavoidable patches of ignorance, or is unpredictability a sign of an actual ontological openness? In the case of quantum theory, the majority verdict has been in favour of the latter conclusion. Almost all physicists regard Heisenberg's uncertainty principle (which initially was concerned with the epistemological question of what could be measured) as being a principle of actual indeterminacy, rather than just a principle of ignorance. Yet the choice is not forced, for there is an alternative – and equally empirically adequate – interpretation of quantum theory, due to David Bohm, that is entirely deterministic in its character. One of the motivations for physicists' opting for indeterminacy in quantum theory may be supposed to lie in an instinctive realist stance, aligning epistemology and ontology as closely as possible with each other. Realism encourages the thought that unpredictability is the sign of a lack of total determinacy – a degree of physical openness – for it implies that what we know or cannot know is a guide to what is actually the case.

A similar choice would be possible in the case of chaos theory, though it has been much less popular. People seem to have been bewitched by the familiar determinsitic Newtonian equations from which chaos theory was originally derived. Yet affirming their unquestioned validity is, in fact, a metaphysical decision, and I am among those who have chosen a different option by seeking an open interpretation of chaotic dynamics (Polkinghorne 1998, Chapter 3).

This is not the place to rehearse the various arguments about these conflicting metaphysical strategies (see Russell, Murphy, and Peacocke 1995).

If the openness option is to make much sense, it has to amount to more than the mere presence of randomness. Its implication is not that chance rules, but that the familiar principles of the exchange of energy between constituents need to be complemented by new causal principles of a holistic and pattern-forming kind, for which the phrase "active information" has been coined. One can see here a *glimmer* (no more than that, as yet) of how one might begin to see how it is that agency, whether involving the execution of human willed intentions or involving divine providential interaction, might function within the open grain of physical process.

While these matters need much more development, enough has been done to indicate that science does not preclude the possibility of *divine interaction within unfolding evolutionary process*. This is the proposal contained in the concept of "theistic evolution." It pictures God as acting within the open grain of nature and not against that grain. Exactly how the balance between divine activity and creaturely activity is struck is the theological problem of grace and free will, now writ cosmically large. If the "causal joint" by which special Providence is exercised does indeed lie within the cloudy unpredictabilities of physical process, then there is an important consequence for theological thinking. There will be an inescapably apophantic aspect to such veiled action. The unfolding process of the world cannot be disentangled and itemised. One cannot say "nature did this, human will did that, and God did the third thing." The eye of faith may be able to discern God at work, but the particularity of that fact cannot be demonstrated by empirical analysis. (A sudden wind blows aside the waters of the Reed Sea and a bunch of fleeing slaves get across before the wind drops and the waters return to drown their pursuers. A member of the Israelite community of faith can see this as divine deliverance from slavery in Egypt, but a neutral observer cannot be forced to see more than a remarkably fortunate coincidence.)

(3) *Precarious Process.* The evolutionary exploration of potentiality is inevitably a process with ragged edges and blind alleys. Its programme will be more like the spreading of a bush than the straight growth of an upright tree. A creation making itself may rightly be held to be a great good, but it also has a necessary cost. Exactly the same biochemical processes that enable some cells to mutate and produce new forms of life – the very engine that has driven the three-and-a-half-billion-year history of life on Earth – will necessarily also allow other cells to mutate and become malignant. The presence of cancer in the world is the necessary cost of the evolution of complexity.

This insight is of some help to theology as it wrestles with its greatest problem, *the existence of evil and suffering*. The more we understand the process of the world scientifically, the more it seems to be a package deal in which processes interrelate in mutual entanglement. The idea that it would be possible to create a world with all the nice features of this one and none of the nasty ones seems more and more implausible (Polkinghorne 1989, Chapter 5; Ruse 2001, Chapter 7). This insight does not by any means remove all the perplexities that theodicy faces – one can argue whether the cost of creation's making itself is a cost worth paying – but it does offer some modest help. The occurrence of malignancy is deeply anguishing, but it is not gratuitous, something that a Creator who was less callous or more competent could easily have remedied.

The greatest difficulty for any theological argument that seeks to discern a form of divine design in the universe has always been the ambiguity of the evidence in a world both fruitful and destructive, both beautiful and terrifying. The point was made with characteristic trenchancy and bluntness by Hume, when he questioned whether it did not all look like a rather botched attempt by an "infant deity," still struggling to learn how to create and not yet making a very good job of it. At least modern science has helped us to see that the issue is not so simple a matter as Hume, in the eighteenth century, could suppose it to be.

A Christian writer such as Holmes Rolston must still acknowledge that, when one moves from the rather abstract beauty of fundamental physics to the much more messy story told by biology, one encounters what he calls a "cruciform creation" (Rolston 1999, 303–7). That phrase encapsulates both the problem of theodicy and also its most profound Christian answer. The Christian God is the Crucified God (Moltmann 1974), not just a compassionate spectator of the travail of creation, but also truly a 'fellow sufferer who understands', for Christians believe that in the cross of Jesus it is God in Christ whom we see stretched out on the gibbet, embracing and redeeming the pain and suffering of creation.

References

Barbour, I. G. 1966. *Issues in Science and Religion*. London: SCM Press.

Barrow, J. D., and Tipler, F. J. 1986. *The Anthropic Cosmological Principle*. Oxford: Oxford University Press.

Behe, M. 1996. *Darwin's Black Box*. New York: The Free Press.

Conway-Morris, S. 1998. *The Crucible of Creation*. Oxford: Oxford University Press.

Dembski, W. M. 1999. *Intelligent Design*. Downers Grove, IL: InterVarsity Press.

Denton, M. 1998. *Nature's Destiny*. New York: The Free Press.

Kauffman, S. 1995. *At Home in the Universe*. Oxford: Oxford University Press.

Leslie, J. 1989. *Universes*. London: Routledge.

Moltmann, J. 1974. *The Crucified God*. London: SCM Press.

Peacocke, A. R. 1979. *Creation and the World of Science*. Oxford: Oxford University Press.

  1986. *God and the New Biology*. London: Dent.

Polkinghorne, J. C. 1989. *Science and Providence*. London: SPCK.

  1990. *The Quantum World*. London: Penguin.

1991. *Reason and Reality*. London: SPCK.

1998. *Belief in God in an Age of Science*. New Haven, CT: Yale University Press.

2000. *Faith, Science and Understanding*. London: SPCK and New Haven, CT: Yale University Press.

(ed.) 2001. *The Work of Love*. Grand Rapids, MI: Eerdmans.

Rees, M. 1997. *Before the Beginning*. London: Simon and Schuster.

Rolston, H. 1999. *Genes, Genesis and God*. Cambridge: Cambridge University Press.

Ruse, M. 2001. *Can a Darwinian Be a Christian?* Cambridge: Cambridge University Press.

Russell R. J., N. Murphy, and A. R. Peacocke. (eds.) 1995. *Chaos and Complexity*. Vatican City: Vatican Observatory.

Russell, R. J., W. R. Stoeger, and J. Ayala. (eds.) 1998. *Evolution and Molecular Biology*. Vatican City: Vatican Observatory.

14

Theistic Evolution

Keith Ward

As a theologian, I renounce all rights to make any authoritative statements about matters of natural science. To some, that may mean that I renounce any claim to speak about matters of fact at all. But there are many matters of fact that are not matters of natural science. It is a fact that Napoleon lost the Battle of Waterloo, but that, and many other particularities of history, will not find a place in any record of scientific discoveries or hypotheses. It is a fact that I am now thinking about what facts are, but again, natural scientists will not be professionally concerned with that fact. And it is a fact that either there is a God or there is not, though now that theology is not usually claimed to be a science (and it is definitely not a natural science), it is not a scientific fact.

I take it that it is an established fact of science that evolution occurs, and that human beings have descended by a process of mutation and adaptation from other and simpler forms of organic life over millions of years. The evidence for that does seem to be overwhelming. It seems to be equally firmly established that natural selection is the main driving force of evolutionary change; but my colleagues in evolutionary biology inform me that it is not agreed that natural selection is the only force, and that the mechanisms of evolution remain to some extent open to diverse interpretations.

I cannot decide on an undecided issue in natural science. So my reflections will not depend upon any particular view of what the physical mechanisms of evolution are, and they will not rely upon the insufficiency of natural selection as a scientific explanation of evolution. They will, however, touch on the question of what can be expected of a scientific explanation, and on what answer to that question an acceptance of the existence of God might suggest.

Central to intellectual reflection about God is the nature of consciousness. Those who believe there is a God claim that consciousness is ontologically distinct from material existence, and that consciousness is ontologically prior to material existence. For God, they argue, is an immaterial being distinct from the universe, and the universe depends solely upon God for its existence. Moreover, God is usually said to have some reason for having created the universe. Creation is an intentional act, it is a bringing about in order to realise some purpose. This purpose need not consist in some end state toward which the universe strives. It may consist simply in the existence of the universe as a total process. But it follows that evolution – now shown to be a major feature of the universe as we know it – must either itself be chosen by God for some reason, or lead to some state that is so chosen. In other words, evolution must in some sense be purposive – either good in itself, or good as leading to some state that is good in itself.

This entails that evolution cannot simply be random or the result of blind necessity. It must be choosable, and must have been chosen, by a rational agent for the sake of some good that it, and perhaps it alone, makes possible. This may not say anything about the precise mechanisms of evolution, but it does say something about its value, and it suggests that it contains some goal or goals that the process is intended to realise, and that humans may help to realise. Whatever the mechanisms are, they will not be wholly fortuitous or accidental, and what they produce must be a possible and actual object of rational choice.

It is difficult to deduce specific implications of the thesis that the universe was created by God. But some observable states of affairs are more compatible with that thesis than others, and some will be incompatible with it. To this extent, at least indirect verification or falsification of the creation hypothesis is possible. As is well known, some specific forms of the creation hypothesis are falsified by acceptance of evolution. That the world was created only about six thousand years ago, as a literal reading of the Old Testament suggests, is the clearest example.

More generally, any theory that God directly creates every event for the sake of its goodness is falsified by the occurrence of events that are not in themselves good, and that also may be harmful to complexes of surrounding events. If natural science shows that many genetic mutations are fatally harmful to organisms, that is a strong indication that any theory of creation that attributes every event to the directly intended action of a good and omnipotent God is mistaken. The evolutionary process is one in which not all mutations are directed to the flourishing of organisms (that is largely, I am told, what is meant by calling such mutations "random"). Therefore, if one is to think of a Creator, one must envisage the creation of a process that is morally random in many of its details, though it will be morally good overall. That will affect one's conception of the way in which God relates to the cosmos, and it will affect ideas of Providence and divine action. One must think of God as not undermining the generally stochastic processes of genetic modification. God will be a God who chooses chance, as part of the process of creation – and in order to support that hypothesis, one will have to be able to suggest some reason why this should be so.

The most obvious suggestion is that chance, or indeterminacy – that is, lack of sufficient causality – is a necessary, though by no means a sufficient, condition for the emergence of free and responsible choice on the part of agents generated by the evolutionary process. At this point, the argument from what is sometimes called "libertarian freedom" reinforces the argument from the observed stochastic nature of genetic mutation. Libertarian freedom is that property of a rational agent by which, on at least some occasions, no prior physical state, even with the addition of a set of general laws of nature, entails one specific outcome of a given situation. There are real alternative possible futures, and only the agent's uncompelled decision decides which future is realised.

A God who creates every specific event for the best is a God who must sufficiently determine every event, leaving nothing to the causality of finite agents. To give real causality to finite agents is to allow for the possibility that they will not choose the best. Why should they not? A clue is given by our ordinary human experience of consciousness and intentionality. As conscious beings, we have a natural orientation to truth and understanding. When consciousness functions properly, it generates an understanding of truth. We also have a natural orientation to sensitivity toward and appreciation of beauty. To become fully conscious is to appreciate the form and structure of the objects of consciousness. And we have a natural orientation to happiness. Consciousness naturally seeks pleasure and avoids pain.

From reflection on consciousness, we can come to recognise a natural orientation toward truth, beauty, and happiness. Insofar as we do so, we are already postulating some sort of teleology in the existence of consciousness. We are supposing that there is an innate, inbuilt goal of conscious activity, which is to realise its nature by generating and attending to truth, beauty, and the happiness of sentient beings. There is a practical imperative that consciousness should realise itself in these ways.

In this way, if one begins by stressing the fundamental and irreducible importance of consciousness, one is already employing concepts of purpose and value as central to an understanding of how things are. For consciousness aims, so far as it can, to realise states of value – states of knowledge and understanding, of creativity and appreciation, of happiness and contentment. And such realisation, in conceiving of valued states and seeking to bring them about by consciously directed activity, is purposive. Moreover, that purpose is not just an option that may or may not be chosen. Insofar as conscious agents distinguish between what is truly of value and what only seems to be so, the pursuit of true values comes to have the character of obligation, of something that ought to be realised, even in opposition to forces that lead one not to do so. Where value, purpose, and belief in the freedom of the agent have a place, the concept of obligation is naturally generated as describing the course of action that the agent ought, if fully rational, to pursue.

Now, this is a very specific way of seeing human consciousness, and it may be far from obvious to many people. It may perhaps be described as a fundamental stance that arises from a basic way of seeing one's own experienced existence. In saying that way is "basic," I mean that it is not based on any simpler set of observations or on any further justifying arguments. It is just how things appear, upon reflection, to a being that seeks to comprehend its own nature as existing.

In particular, it is not based on any scientific findings or on any previously known general philosophical or revealed principles. It is what has been called a "phenomenological" claim – a claim to reflective insight into what the nature of consciousness, as an experienced phenomenon, shows itself to be. My claim is that, to a properly reflective agent, consciousness shows itself to be a distinctive form of existence that has an inbuilt disposition toward self-realisation by attending to objects that manifest truth, beauty, and happiness. It has a disposition toward the good.

However, it is at once clear that this disposition is an ideal or obligation rather than a universally manifest fact. Human agents very often live by the acceptance of conventional opinion or wish-fulfilling fantasies. They often seek passive entertainment and diversion rather than creative understanding. They often choose easy sensual gratification rather than the welfare of all persons and sentient beings. Falsehood, diversion, and gratification are the easier options. They are forms of inauthentic existence. They contract, rather than expand, consciousness, but they also offer direct and immediate rewards. It is not hard to understand why finite sensuous agents should choose such things, rather than the good or the best, to which they are naturally – but not actually – disposed.

On this understanding, humans find themselves existing in a world that calls them imperatively to realise their disposition to the good, while at the same time placing before them a host of inclinations to self-gratification. In such a world, freedom is the physically undetermined choice between imperative and inclination. Only in the moment of decision is the future determined. Personal agents can be seen as points at which the universe itself determines its own nature, either toward good – that is, toward greater creativity, comprehension, and community – or toward a passivity, ignorance, and egoism that in the end denies any distinctive value to personal existence itself.

All this may seem heavily evaluative, and it is. It is an evaluation of personal existence that springs from a sustained attempt at reflexive understanding – an understanding based not on experimental observation and hypothesis but on the effort to understand from within, from one's personal experience, one's own distinctive form of existence as a human being. If this is admitted as a source of knowledge and understanding, then it must stand alongside the experimental observations of the natural sciences as a way of providing an adequate account of human nature and of the nature of the universe of which humans are an integral part. But these two sources of understanding – the experimental and the reflexive – must cohere with one another. They must be compatible, and more than that, they must be mutually reinforcing, if a plausible view of human nature is to be achieved.

It is the difficulty of attaining such coherence that has largely occasioned the division between the sciences and the humanities, between a positivistic view of science and a broadly idealist view of the humanities, that has marked Western culture over the last two hundred years. With the division of European philosophy in the early twentieth century into logical positivism and existentialism, leading on to postmodernism, the gap became a chasm.

The theory of evolution has become the location at which this chasm has emerged most sharply. But it is also perhaps the place at which it may be bridged. The chasm seems to be a yawning gulf when neo-Darwinian philosophers claim that everything about human life and consciousness can be explained wholly in terms of natural selection and a few basic laws of physics, and when postmodernist philosophers claim that scientists are merely producing social constructions of their own imagination, that there is no such thing as objective truth at all, and that evolutionary theory is a way at first of supporting and later of subverting traditional values of human, intellectual – and probably Western and male – superiority.

The chasm between the view that "real" explanations need to be proposed in terms of physical, publicly observable (or inferable) atomistic elements and a few basic laws connecting them, and the view that the personal conscious experience of subjectivity is an ineliminable and inexponable constituent of reality – perhaps even its deepest nature – seems vast. Yet evolutionary theory contains the possibility of bridging it, insofar as it offers a view of the emergence of consciousness as a natural and continuous development from simple material elements into the complex integrated forms – the neural networks – that make consciousness possible. Physical explanations are real and vital, and they are truly explanatory of the genesis of physical complexity. But such complexity issues in conscious states, which then require a different form of explanation, one in terms of intentions and purposes, in order to account for their behaviour. One form of explanation shades into the other by a continuous developmental process, which is nonetheless significantly different at each end of the spectrum of development. In such a context, the basic physical processes themselves might be seen as purposively oriented from the start toward the development of consciousness, purpose, and value.

That is in fact how evolutionary theory began, in the works of cosmically optimistic philosophers such as Hegel, who saw history as the self-realisation of consciously experienced values, implicit in the very structure of the cosmos itself. From the phenomenological point of view, we might see ourselves as beings poised in freedom between the realisation of many forms of goodness potential in the cosmos, on the one hand, and the actualisation of the destructive possibilities of conflict and negation that are also implicit in the structure of physical reality, on the other. This may well suggest a view of the cosmos as a process of emergence through a polarisation of creative and destructive forces (the dialectic of history, in Hegel's terms) – a process oriented toward the ultimate realisation of goodness but driven also (and essentially so) by destructive forces that impede, but also act as catalysts of, that realisation, a process in which we play a pivotal, if minor, role. For Hegel, the universe evolves from brute unconscious objectivity toward a perfect self-conscious subjectivity, through a dialectical process in which human freedom expresses the decisions of finite beings to realise or to impede the self-realisation of the potencies of being, which is the history of the cosmos.

The subsequent history of philosophy has not dealt kindly with Hegel. His cosmic vision was felt to be too grandiose, and too much like armchair speculation, to satisfy the increasing demand for precise measurement and testable observation. His postulation of an Absolute Spirit that realises itself in an evolutionary cosmos was felt to be too redolent of the all-determining God of theism, whose very existence compromises the possibility of human freedom. So philosophers divided into those – the linguistic analysts – who ceased to speak about the universe at all, leaving the field clear for scientists, and those – the existentialists – who insisted on absolute freedom and the particularity of personal experience to such an extent that no objectivity was left for a general view of the universe to latch onto.

Where philosophers feared to tread, scientists rushed in to fill the void, and the result has been the creation of a new grand metaphysics of the cosmos on an allegedly purely scientific basis – a metaphysics of the genesis of complexity by a combination of interminable random shuffling and recursive elimination of the unfit. But in this metaphysics, subjectivity and consciousness have no place, except as mysterious and useless byproducts of an intrinsically blind and automatic process without direction or goal.

Perhaps there is still something to be said for Hegel, however, from a theistic point of view. One basic problem of traditional theism has been the question of why the creator should have apparently failed to choose the best possible world – only Leibniz was brave enough to suggest that the creator had done so, and he was memorably mocked for his pains by Voltaire. If theism is approached from a phenomenological perspective, and enriched by the addition of an evolutionary view of the universe, the problem might not appear in that form. One would begin from the experienced facts – or necessarily postulated facts – of freedom, of orientation to the good, and of inclination to the bad. Then one might be led to postulate a partly indeterminate universe as the necessary condition for such freedom – a universe containing both creative and destructive possibilities, as a condition of its dialectically emergent character. One can see the cosmos as oriented toward goodness, but only through struggle and the exercise of finite freedoms. The history of the cosmos would be an increasingly self-shaping emergent process oriented toward the realisation of new states of value, of truth, beauty, and community – toward what the Jesuit visionary Teilhard de Chardin called the "Omega Point," where goodness is perfectly realised as the natural culmination of cosmic history. The whole process would be purposive, tending toward a supremely valuable goal, and the nature of that goal would be defined in part by the nature of the process itself, so that the goal could not exist without the process by which it had been generated.

Admittedly, this vision is rather grandiose and speculative. It is a bold hypothesis. But if it were entertained as a possibility, it would enable one to speak of the universe as purposively shaped toward the realisation of a unique and distinctive state of value. This might not suggest a picture of God as planning everything for the best, out of an infinite range of possible worlds among which God could freely choose. But it might suggest a picture of God as creating a relatively autonomous, purposive universe for the sake of a goal of great value that such a universe alone could realise. This picture could be strengthened by supposing that the creation of such a universe is part of a necessary process of self-realisation by God, even though many features of the created universe may be contingent or even fortuitous. If one could add that such a God would know and remember all the values that the universe would ever realise, and could give finite creatures the possibility of sharing in that knowledge to an appropriate degree, one might not hesitate to say that such a God was a worthy object of awe and desire, and that such a universe was grounded in a mindlike reality with intention and purpose, rather than in chance or blind necessity.

It should immediately be said that theistic faith does not rest simply on such a bold hypothesis. Its main basis is undoubtedly the sense of experiential encounter with a transcendent reality. That sense is adumbrated in many ways – in the beauty of nature and art, the courage and compassion of devoted human beings, the order and elegance of the laws of physics, the ecstasy of love, the desolation of loss, and the terror of storm and earthquake. In all these experiences, many people believe that they can sense a reality that makes itself known in and through them – powerful, mysterious, beautiful but dangerous, attractive yet austere, always present yet infinitely removed and always beyond our grasp. That is the reality that theistic traditions name "God."

Such experiences of transcendence are often refined by disciplines of overcoming attachment and attending to supreme goodness. They are then shaped by traditions of revelation, of what are felt to be initiatives of the Transcendent itself, revealing its nature more fully in paradigmatic personal experiences, inspired interpretations, and historical events. For Christians, the life of Jesus, the experience of his presence after the Crucifixion, and of the power of the Spirit that flowed from that experience can become a disclosure of the nature of the Transcendent and of the final transfiguration of the finite into God that will come through the experience of suffering and self-renunciation.

All of these strands – personal experience of the divine, revelation, the sense of responsible freedom and moral obligation, and speculative conjecture about the nature and destiny of the cosmos in which consciousness has a central role – come together to form a cumulative case for theism. Their source is not experimental observation of physical and measurable processes. Theism is not primarily a theory meant to explain why the universe is the way it is. It is primarily a commitment to a more-than-physical reality experienced as morally demanding, yet also as forgiving and sustaining. It is incumbent upon the theist, however, to provide some account of how such a reality relates to the observed nature of the universe.

The theistic postulate clearly must not contradict well-established observations and hypotheses, and such observations may suggest that some forms of theism are more plausible than others. Thus a form of theism that allows for a partly indeterministic yet purposively oriented emergent cosmos is, for the reasons given, more plausible than a theism that supposes that a perfectly good God directly chooses every particular state for a good reason.

Is the theory of evolution, as it is accepted by the consensus of competent scientists – however provisional that consensus may be – compatible with a theistic interpretation of evolution? Atheists have offered a number of reasons why it is not so compatible. One is that the process of evolution is too random, that it shows no overall improvement, and that it has no discernable direction. Another is that the process is too cruel, too indifferent to the welfare of the life forms it produces, to be seen as providentially created or directed. A third is that it is virtually certain that all life will become extinct in the long run, as the Second Law of Thermodynamics inexorably takes effect. So the whole thing will end in futility. The Omega Point will be the same nothingness as that from which the universe began – by some random quantum fluctuation, perhaps – and any talk of a final goal is thereby shown to be no more than wishful fantasy. A fourth reason is that God is simply superfluous as an explanation. One can explain what happens in the universe most simply and elegantly without any reference to a God. We have no need of that hypothesis.

These reasons do not seem to me to be very strong. When mutation is said to be random, the point is that most mutations do not seem to benefit either the individual or the species. This might suggest that there is no inherent tendency of mutations to proceed in any specific direction. However, mutations are not random in the sense of being free from the laws of physics. Since God would design both the laws of physics and the environment that selects some mutations over others, it is easy to see the mutational process as a whole as directional, as leading to the selection of cumulatively complex systems and eventually to the existence of conscious agents. The system as a whole may have a tendency toward a goal, while individual events within it have a large degree of variability. One might see evolution as akin to the shuffling of a pack of cards, which, in a suitable set of environmental conditions, will produce a well-sorted pack in a finite time.

It is undeniable that organisms have developed from unicellular organisms to beings with central nervous systems and brains. If a necessary condition of the actual existence of values is the existence of consciousness, understanding, and feeling, then there is here a development of values from a cosmos in which there were no actual values (except, perhaps, those experienced by God in contemplating the beauty of nature). That is a strong enough sense of direction, development, and progress for any theist.

To the objection that there is too much suffering in the evolutionary cosmos for it to have been designed by a good God, there are three main responses. First, God's goodness lies primarily in the perfection or supreme value of the divine nature itself, and it is unaffected by what happens in creation. Second, the suffering involved in evolution may be a necessary or ineliminable part of the structure of the cosmos, which itself may arise by necessity from the divine nature – for instance, as an actualisation of potencies within the divine nature that inevitably express themselves in some form. And third, creaturely suffering may in large part be sublimated by being taken into a postcosmic experience in which all creatures that are capable of it will share in a divine experience and happiness incomparably greater than any finite suffering.

Of course, these suggestions are disputable, even among theists. But they are all to be found in some form within most religious traditions, and in my view they are sufficient to demonstrate the compatibility of the existence of suffering with a good Creator. All that needs to be shown here is that there is some set of conditions that would render divine goodness compatible with suffering in creation. We need not know what they are. But these points do suggest, I think, that there could well be some.

As to the suggestion that the universe will come to an end, that is wholly irrelevant to the question of whether its existence, at least for large periods of time, has been of great value. In addition, most theists will suppose, as part of their initial hypothesis, that God will be able, and will probably wish, to give creatures a postcosmic existence (in Paradise, or in the presence of God) that will not come to an end. After all, God will not come to an end with the death of the universe, and so there is reason to think that God would give creatures continued existence in another form. So the questions of how the physical universe began, of how long it will last, and of how it will end are not directly relevant to the issue of whether God chooses to create great and enduring values that originate in and are importantly shaped by, but not necessarily confined to, an emergent evolutionary physical cosmos.

The most challenging objection to theistic evolution is probably that the postulate of God is superflous to scientific explanation. The paradigm of scientific explanation is one for which an observable initial state gives rise, by the operation only of general quantifiable laws, to a subsequent observable state. But there is another form of explanation, one that was one of Aristotle's four forms of causality, though it does not form part of modern science. That is explanation in terms of value, the value of that for the sake of which some occurrence happens. This has been called "personal explanation" by Richard Swinburne, and it is well established in ordinary usage.

I explain my writing these words by stating that for the sake of which I am acting – the production of a chapter, something that I want to produce, something that is of value to me, at least. This is a value, a state of affairs envisaged in my mind, that I am trying to realise. So a clear case of personal explanation is the formation of intentions in human minds, and the undertaking of actions in order to realise those intentions.

Whereas the basic elements of scientific, causal explanation are "initial conditions" and "law," the basic elements of personal explanation are "intention" and "action." No account of initial physical states and laws will ever mention an intention or a purposive action. These are different categories of explanation, using different basic terms. We ordinarily assume that, while they do not conflict, they are irreducible to one another. The relationship between them, like the relationship between physical brain states and mental occurrences, remains deeply mysterious in the present state of knowledge.

Some scientifically minded philosophers, however, would say that the postulate of mind, and of personal explanation, is superfluous to a proper explanation of how things occur, which must be a scientific explanation. That is, we could in principle give a complete account of all the processes of nature just in terms of brain states and the laws of physics. Of course, no one has come near to doing such a thing. It would be far too cumbersome and complicated to be of any practical use, in any case. So personal explanations, in terms of intentions and actions, are useful as a sort of shorthand for referring to the total states produced by the interactions of millions of fundamental particles. Personal explanations, in short, are useful currency for human beings, but they are strictly eliminable and superfluous, and they do not require the postulation of any immaterial or mental entities or states, in addition to physical states.

Such a hypothesis is verifiable in principle, if one could successfully predict all occurrences in terms of laws governing the interactions of fundamental particles. But any such verification is vastly unlikely, if only because the calculations required for reliable prediction would require a virtually infinite amount of information. One would need to know the total state of the whole universe, and all the laws that would ever apply to it. Moreover, the present state of quantum theory suggests that access to a precisely specified set of initial conditions as they are in themselves is barred to any means of observation we can conceive.

All that is actually available to the natural sciences is the well-established fact that general laws do apply, other things being equal, and that complex entities are built out of smaller component parts, which account for the behaviour of the complex entities, in general if not in precise detail. (There seem to be indeterminacies in physical systems that can cause macrocosmic changes that are unpredictable in detail by finite observers.) In this situation, personal explanations are not in fact superfluous or eliminable. Whatever their ultimate status, they are nonscientific forms of explanation that are invaluable for understanding human life and conduct.

The basic theistic hypothesis is that there is a personal explanation for the universe as a whole. The universe exists in order to realise values that are envisaged by something akin to a cosmic mind. If one asks why that mind exists, the answer is because it realises value in a uniquely preeminent way. God exists because God realises the highest conceivable value, and the universe exists because it realises values conceived by God that can exist only in such a universe, and which perhaps, in their general nature, are necessarily emergent from the reality of God itself.

If the theistic hypothesis is true, not only will personal explanation be nonreducible to scientific explanation, it will be the more basic form of explanation. The fact that the natural sciences do not mention or refer to it does not indicate that it does not or cannot exist. Since there are no observable initial states in God, and since God does not act in accordance with general causal laws, it is not surprising that there can be no scientific explanation of the creation of the universe. Nevertheless, theistic explanation is not superfluous, for it explains why the universe exists by referring to the nature of a divine mind. If there is a God who has purposes for the universe, and if one of our roles is to implement those purposes, theistic explanation will add substantial knowledge to what natural science tells us about the nature of the universe. If one of those purposes is that finite beings should come to know and love the Creator in some appropriate way, this will be knowledge that the natural sciences are in principle unable to provide.

Evolutionary biology can tell us, with quite a large component of speculation but also with a good deal of positive evidence, the mechanisms by which organic life forms developed on Earth. It does not, as a natural science, even deal with the question of whether evolution has purpose or value. If it has, then the theistic hypothesis, far from being superfluous, will be one of the most important things we could know about human existence.

It is vital to insist that God is not here being postulated as the simplest explanation of design. Despite what early anthropologists such as James Frazer and Edward Tylor have said, there is no evidence that any religion began with attempts to explain why events happen as they do. On the contrary, it is fairly obvious that religious beliefs rarely offer satisfactory explanations of specific events. They most often pose problems about why God should act in the way God does, at which point religions often resort to appeal to mystery. The sources of religious belief lie elsewhere, in experience of a transcendent reality, of an overwhelming moral demand, and of awareness of a personal presence. Explanation comes in only when one seeks to integrate such experiences with knowledge of the natural environment, in all its ambiguity and complexity.

It is only if there is believed to be a reality of supreme value, which is knowable in and through the natural world, that one will be tempted to speak of Intelligent Design or of creation. An analogous case is our knowledge of other persons, other centres of conscious experience and agency. We do not postulate other persons because that is the simplest explanation for why bodies move as they do. We believe there are other persons because we see and react to parts of the world as mediating personhood. This is a matter not of inference but of encountering objects as mediators of personal forms of existence. It is a basic, noninferred, epistemic attitude. Once we have the concept of persons, we will naturally explain the movements of their bodies as being at least in part, the result of conscious states. But those states may remain largely mysterious, and we may often only guess at what the detailed explanation for them might be.

So it is with the idea of God as supreme objectively existing value. Presumably there is some relation between God and the physical universe, and religions offer various accounts of just what that relation is. Theistic accounts generally suppose that God is at least a designing intelligence, and perhaps the generator of physical reality for some good purpose. For such an account to be plausible, the observed processes of physical reality must be adapted to a good purpose. In other words, if theism is a plausible way of integrating reliable experience of God with observations of the physical universe, and if Darwinian explanations give a good account of that universe, then Darwinian explanation must be compataible with theistic explanation.

The Darwinian says that apparent design could have come about by the cumulative repetition of simpler, unconscious, and unintelligent elements. We therefore do not need to postulate any other causal influence. The theist says that the universe was created by God in order to realise otherwise unobtainable states of great intrinsic value. There is a goal for which the universe was created, and it was envisaged by a cosmic intelligence. There is no incompatibility between these views. One asks about the causal processes by which states of affairs have come into being. The other asks about the purpose for which they have come into being. As Michael Ruse has argued, they are complementary forms of explanation.

Yet would a theist not wish for some more positive and particular interaction of God and creation? Already, in speaking of awareness of God, some causal interaction between God and the world has been posited. For, on most accounts of knowledge, the objects of knowledge play a causal role in the perception of them. Our knowledge of a chair is caused in part by the existence of the chair. So if we know God, then God plays some causal role in that knowledge, and therefore in the genesis of the physical states of the brain in which that knowledge is embodied. It seems that anyone who speaks realistically of knowledge of God is committed to saying that God causally interacts with at least some physical states of affairs.

The case here seems not dissimilar to the case of mind-body interaction in general. If it makes sense to speak of intentions causing actions, in the human case, then it probably makes as much sense to speak of divine intentions causing physical states of affairs. We do not speak of the laws of nature being broken in the one case, so why should we do so in the other? There are areas here where human knowledge is as yet very meagre. But my own predilection is for an account of laws of nature along the lines proposed by Rom Harre – not as inviolable and inflexible rules that all events must obey, but as pointing to propensities of objects, variously realised by the sorts of interactions and forms of integration into complex structures that those objects exhibit. Laws are general principles of the interaction of objects, but those laws become more complex as objects relate to one another in more complex and structured ways.

If the whole physical universe is generated and continuously sustained by a spiritual intelligence, it is to be expected that the presence of that intelligence will have particular causal effects on the universe. Just as we consider the forming of intentions by humans as having causal effects in history, so we may assume that God's intentions will have causal effects in the cosmos, even if we do not have any satisfactory account of how such causal links "work." Believers in God have no difficulty in thinking that sometimes the general laws of nature will permit exceptions. For though the general uniformity of nature is a condition of reliable science (and that may be one good reason for such uniformity), it is by no means necessary to science that there will never be, much less that there can never be, exceptions to such uniformity.

In most cases, however, the theist will not wish to rely on exceptions to the laws of nature. Rather, the theist may wish to affirm that natural causal processes are influenced or modified (not violated or "broken") by specific divine intentions, even if we cannot pinpoint where such influence occurs. It is a statement of faith, not of science, that everything that happens, even on the physical level, must do so in accordance with exceptionless laws of nature. It will probably never be possible to trace in detail all of the causal factors that go to determine the evolutionary process. So if theistic evolution posits that God causally influences evolution in order to ensure, for example, that it results in the existence of free moral agents, it is virtually certain that the natural sciences could not falsify the claim. If this were a strictly scientific dispute, the victor would be the most economical hypothesis, the one without God. But it is not. It is a dispute about the ultimate nature of reality, and about whether the natural sciences alone give an adequate account of that nature.

Purely naturalistic accounts of evolution have, as their main strength, the fact that they offer apparently simple and elegant accounts of evolution that do not need to refer to non-natural entities or forces. They strongly motivate research programmes, and they have been heuristically fruitful. Theistic accounts of evolution have, as their main strength, the fact that they seek to take into account the facts of consciousness and experience in an integrated, purposive explanation. They seek to integrate subjectivity with the objective world investigated by the natural sciences. They seek ultimately to assign good reasons why the universe should exist as it does.

There may be no such reason; but the nature of science itself, which seeks reasons for as many things as possible, prompts the mind to seek one. In this sense, the postulate of God is a nonscientific hypothesis that nevertheless has an intellectual affinity with the natural sciences. Both seek an adequate understanding of what the universe is like. One looks to objective, dispassionate, quantifiable, and publicly accessible evidence. The other looks to the data of subjective experience, of feeling, evaluation, intention, and obligation, which require a more engaged, intuitive approach. The true consilience of humane and experimental understanding, which is a proper goal of intellectual endeavour, is perhaps most likely to be found, not in reducing one form of understanding to the other, but in an integrated worldview that can coherently relate one to the other. One main argument for belief in theistic evolution is that it offers at least the prospect of framing such a view, which takes the findings of modern science and the testimony of the world's ancient religious and philosophical traditions with equal seriousness.

15

Intelligent Design

Some Geological, Historical, and Theological Questions

Michael Roberts

The design argument evokes William Paley walking on a Cumbrian moorland and discovering a watch. In the windswept silence, he developed his watchmaker analogy of an intelligent designer, and thus Intelligent Design may be considered as the restatement of the old argument refuted by Charles Darwin. There are similarities but also important differences between the old design arguments of Paley, William Buckland, and even John Ray and those of Behe, Dembski, and other proponents of Intelligent Design. In order to make a valid comparison, it is essential to consider the content and context of design, old and new, and the relationship of both to geological time, biological evolution, naturalism (or secondary causes), and a "theological approach" to science. My first major concern is the refusal of design theorists to take sufficient cognisance of the vastness of geological time. Second, I show that, historically, scientists cannot simply be typecast as "theistic" or "naturalist," as both some design theorists and some critics imply. Third, I show that Intelligent Design is more an argument from rhetoric than from science, and finally I seek to demonstrate the difference between Intelligent Design and the nineteenth-century design arguments and to show how Intelligent Design is very different from Paley's argument.

Design arguments came to prominence in the seventeenth-century, evolving from theological arguments of "nature leading to nature's God" in a culture dominated by mechanistic science. There are roots in Calvin, who wrote in Book I of *The Institutes*, "Hence, the author of . . . Hebrews elegantly describes the visible worlds as images of the invisible (Heb. 11. 3), the elegant structure of the world serving as a kind of mirror, in which we may behold God, though otherwise invisible."[1] He then writes "innumerable proofs, not only those more recondite proofs which astronomy, medicine, and all the natural sciences, are designed to illustrate, but proofs which force themselves on the notice of the most illiterate peasant, who cannot open his eyes without beholding them."[2] Calvin made clear the *general* appeal of his argument, including both the scientific and the popular. *Proof* is not rational demonstration but rather the sense of awe and beauty "demonstrating" "the admirable wisdom of its maker." The "recondite" side of Calvin's "innumerable proofs" was taken up a century later by members of the Royal Society, as in the *Physico–theology* of William Derham and other works. Robert Hooke in *Micrographia* (1665) provides a fine example when he compares the perfect design of living things to the blemishes of man's artefacts. Brooke comments, "Compared with the filigree precision of nature, human artefacts made a very sorry sight: 'the more we see of their shape', Hooke observed, 'the less appearance will there be of their beauty.'"[3]

The development of the design argument in the eighteenth century culminated in William Paley's *Natural Theology* (1802) and William Buckland's Bridgewater Treatise in 1836. Paley and Buckland emphasised the perfection of natural structures, but Hugh Miller, writing in the 1850s, focussed on the beauty of natural structures, indicating a shift in the design argument.[4] After Darwin, the detailed appeal to design went out of vogue, though in 1884 the liberal Anglican Frederick Temple could write, "The fact is that the doctrine of Evolution does not affect the substance of Paley's argument at all."[5] Clearly, Temple's "substance" excludes the detailed design argument of a Paley or a Dembski. The detailed design argument has resurfaced in recent years in both Intelligent Design and in the more general arguments of both Old Earth and Young Earth creationists. The focus here is on Intelligent Design.

THE IMPLICATIONS OF GEOLOGICAL TIME AND THE FOSSIL
SUCCESSION FOR INTELLIGENT DESIGN

Of most concern to a geologist is the near-absence of reference to geological time in studies on Intelligent Design. It is as if the origin of species, whether by direct intervention or by evolution, can be discussed without reference to deep time, or to the succession of life. As Nancy Pearcey wrote, "For too long, opponents of naturalistic evolution have let themselves be divided and conquered over subsidiary issues like the age of the earth."[6] Like Pearcey, who is a Young Earth creationist, most intelligent designers simply ignore issues of age as irrelevant. The issue of the succession of life through the 4.6 billion years of time clearly has an effect on how one conceives of how life forms have come into being. If the idea of aeons of geological time is correct – and Pearcy, Nelson, and Wise consider that idea to be wrong – then life forms have appeared during time and have gradually changed, either through an outside force or naturally. If the earth is only 10,000 years old, then there is insufficient time for changes through natural means, and thus it is reasonable to hold the abrupt appearance of species so poetically expressed by Milton:

> The grassy clods now calved, now half appeared
> The tawny lion, pawing to get free
> His hinder parts, then springs as broke from bonds,
> And rampant shakes his brinded mane . . .
> *Paradise Lost*, Book VII, 1463-6

The Problem of Geological Time

In his *Natural Theology*, William Paley discussed the design of biological structures. In 1800, however, little was known of the succession of life because the geological column had not been worked out, so Paley could not have attempted to consider "creation" over geological time. By 1820, as the geological column was elucidated, a progressive creation over millions of years was seen as the most reasonable explanation – and one inevitable from the fossil record – though Uniformitarians such as Lyell rejected progressivism. This meant that instead of a few creative acts in the Six Days of Reconstitution,[7] there had been innumerable creative acts during the vastness of geological time. Thus in the 1850s the French geologist Alcide d'Orbigny "recognised 27 successive fossil faunas in one part of the geological column (part of the Jurassic at Arromanches in Normandy) each of which he believed became entirely extinct as the next was created."[8] This was used to justify his concept of a *geological stage*, which is still accepted, though shorn of its creationist roots. If d'Orbigny was correct and that part of the Jurassic was 10 million years, then at the same rate of creation there would some have been some 1,500 creations since the beginning of the Cambrian.[9]

This raised some difficult questions. Why did God create/design a succession of forms differing only slightly from previous forms? Why was extinction allowed? Assuming evolution has not occurred, then the Designer returned at regular intervals to modify a previous creation as motor manufacturers give an annual revamp to their models. In England, such questions were put aside for a time after the formation of the Geological Society of London in 1807, as the most important task was stratigraphy – that is, elucidating the historical succession of strata, rather than providing any interpretative framework, thus avoiding the problem of design over time. From 1800 to 1850, geologists worked out the geological column from the Cambrian to the postglacial and the fossils embedded in them, without acceptance of evolution. This demonstrated the succession of life, which is derived from *the principle of superposition* rather than based on any hypothesis on the origin of life. Thus by 1850, the general order was the same as what we have today, though there was a marked absence of human fossils. However, this avoided the question of change over time, which would not go away.

A fine early example of a study on the succession of life can be found in John Phillips' *Treatise of Geology* of 1838. He dealt with the subject again that year for Baden Powell.[10] After giving "[t]he order of development of life," he wrote: "Is the present creation of life a continuation of the previous ones;...? I answer, Yes; but not as the offspring is a continuation of its parent." His meaning is clear – there *has* been a succession of similar species, each separately created and differing only slightly from its predecessor, but no descent. Phillips thereby allowed the direct creation of each species and thus retained the argument from design almost intact. This meant that any possibility of evolution could be sidestepped.

The Young Darwin on a Nonevolutionary Succession of Life

Phillips was a lifelong opponent of evolution, but Darwin made a fascinating use of Phillips's ideas, when toying with evolution in his B Notebook of 1837–38.[11] This was nine months before he read Malthus and thus predates natural selection. Darwin agreed with Phillips's historical ordering of fossils, but not with his successive creations. In the B Notebook we see Darwin the *geologist* arguing historically and abductively for evolution. From page 167, he was using Phillips for *historical* information on fossils, "fish approaching to reptiles at Silurian age" (B 170), and asking, "How long back have insects been known?" (B 171) Having asked the *when* questions, he then asked the *why*. Crucial is his earlier statement, "Absolute knowledge that species die & others replace them," but "two hypotheses [individual creation and common descent] fresh creation mere assumption, it explains nothing further, points gained if any facts are connected" (B 104). Here Darwin appears to dismiss the view of Phillips cited earlier. Later he asked, "Has the creator since the Cambrian formations gone on creating animals with same general structure. – miserable limited view" (B 216). And he argued, "My theory will make me deny the creation of any new quadruped since days of Didelphus[12] in Stone[s]field" (B 219). This is in contrast to the *Origin of Species*, where Darwin argues by analogy from artificial selection and then from the fossil record and biogeography. In the B Notebook he was arguing for the inference from the best explanation to explain the succession of life, but in 1859 he argued for the mechanism first and then gave a minor abductive argument from biogeography and the fossil record. However, the original basis of his "one, long argument" was abduction from the fossil record. In fact, Darwin was more successful in convincing others that evolution was the best historical interpretation of the fossil record than he was in arguing for natural selection.[13] This is contrary to Johnson's alleged *materialist* model of evolution, where "a materialistic evolutionary process that is at least roughly like neo-Darwinism follows as a matter of deductive logic, regardless of the evidence."[14] Darwin had argued abductively and inductively from the historical evidence, and then by analogy. He had taken the long chronology of "creationist" geologists and then, **and only then**, argued for evolution and the virtual absence of creative acts to explain the progression of life forms. This was a bold step, as there were few detailed sequences such as the elephant, the horse, and *Triceratops* and allied species.

Miller, in *Finding Darwin's God*,[15] mischievously considers design in relation to elephants, with twenty-two species in the last six million years and many more going back to the Eocene. If all were "formed" at about the same time, ca. 8000 B.C., then the only reasonable explanation is some kind of intelligent intervention, which designed each to be different, rather like cars made by Chrysler and GM over several decades.

If the geological time scale is correct, then these different fossil elephants appeared consecutively and, despite "gaps," form a graded sequence. They indicate only "annual model upgrades." Assuming that this is a fairly complete sequence, the Intelligent Designer seemed to have adopted the same sequence of modifications as would be expected by evolution. This is exactly the point Darwin made in his 1844 draft:

I must premise that, according to the view ordinarily received, the myriads of organisms, which have during past and present times peopled this world, have been created by so many distinct acts of creation. . . . That all the organisms of this world have been produced on a scheme is certain from their general affinities; and if this scheme can be shown to be the same with that which would result from allied organic beings descending from common stocks, it becomes highly improbable that they have been separately created by individual acts of the will of a Creator. For as well might it be said that, although the planets move in courses conformably to the law of gravity, yet we ought to attribute the course of each planet to the individual act of the will of the Creator.[16]

The Playing Down of Geological Time in Intelligent Design

The example from Miller highlights why the avoidance of geological time results in problems. Behe focuses entirely on biochemistry and Dembski on detecting design. Both accept a long time scale but do not consider the implications for their understanding of design. Thus the formation of biological complexity is considered without any reference to the history of life and its time scale in a way that is reminiscent of Lessing's ditch, in that "*accidental truths of history can never become the proof of necessary truths of reason.*"[17] The accidental truths of geology are simply ignored for the purposes of demonstrating *Intelligent Design*. In the volume *The Creation Hypothesis*, Stephen Meyer argued cogently for what he called *The Methodological Equivalence of Design and Descent*, but swung the argument in favour of design by omitting any reference to geological time. If geological time is accepted, then the choice is between Phillips (design or multiple abrupt appearance) and Darwin (descent), as discussed earlier. If geological time is not accepted, then design is the only choice. Kurt Wise likewise avoided the issue of age in his essay

"The Origins of Life's Major Groups," failing to see that the awareness of the change in organisms over time came through detailed stratigraphy rather than by interpreting them though the theories of "macroevolution, progressive creation, global deluge."[18] The early geologists tediously recorded the order of strata without asking questions of origins, though their vast age was common knowledge.[19] (Wise's idea that the fossil record is explained by rising flood waters is simply absurd. This type of approach justifies critics such as Pennock and Eldredge in dismissing ID as a variant of Young Earth creationism.)

Perhaps the demonstration of evolution from the fossil record falls short of "rational compulsion," as the *geological* argument for evolution is *abduction* or *inference of the best fit*. Considering the fossil record within a four-billion-year time scale abductively, the best fit is gradual change over time (with or without interference). But within a short timescale of 10,000 years, the best and only fit is *abrupt appearance*. To avoid citing the evidence of the fossil record and vast time (and merely to mention, or even to parody, the Cambrian Explosion)[20] may be good practice for a defence lawyer, but not for a scientist.

Unless one rejects geological time, the fossil record points either to progressive creation with regular interventions (the common pre-Darwinian view) or to evolution, possibly with occasional "interventions." The starting point has to be an ancient Earth and the "absolute knowledge that species die & others replace them." To regard geological time as a subsidiary issue would deny that.

DESIGN, THEISTIC SCIENCE, AND NATURALISM: A HISTORICAL PERSPECTIVE FROM 1690 TO 1900

Whereas Johnson, Behe, and Dembski often present the case for Intelligent Design without reference to theology, Plantinga and Moreland stress the need for *theistic science*, whereby theology almost becomes part of science. *Theistic science* is open to the direct activity of God, whereby these acts are demonstrated on theological grounds. Thus J. P. Moreland itemises "*libertarian, miraculous acts of God*" as being "the beginning of the universe, the direct creation of first life and the various kinds of life, the direct creation of human beings in the Middle East, the flood of Noah," and "for some, the geological column"[21] and the crossing of the Red Sea.[22] This has great appeal to those who wish to stress the supernatural nature of Christian belief.

DILUVIAL OR FLOOD GEOLOGY; THEISTIC OR NATURALISTIC?

John Ray and Edward Lhwyd

With the apparently Christian origin of science in the seventeenth century, it is tempting to see science as moving from a *theistic* base to a *naturalistic* one over two centuries. This appears to be so in considering the formation of life and also in seeing the Flood as the cause of strata. Thus the Flood may be seen as an example of divine intervention, invoked from the seventeenth to the early nineteenth century. Because the seventeenth-century theorists of the Earth wrote so biblically of Creation, Flood, and Conflagration, their espousal of a kind of naturalism or theory of "secondary causes" is overlooked. S. J. Gould expounds this view, writing, "Burnet's primary concern was to render earth history not by miracles, but by natural physical processes."[23] Gould described Burnet as a "rationalist"; Johnson would call him a *naturalist*. A similar willingness to explain geological features by natural processes is found in Ray's discussion of erratic blocks[24] in the second edition of *Miscellaneous Discourses Concerning the Dissolution of the World.*[25] Ray was writing in response to a letter from Edward Lhwyd, who wrote to Ray on February 30, 1691, "Upon the reading on your discourse of the rains continually washing away and carrying down earth from the mountains, it puts me in mind [of something that] I observed," and then described what he had observed in Snowdonia. He described innumerable boulders that had "fallen" into the Llanberis and Nant Ffrancon valleys, which are two U-shaped glacial valleys. (Most of these rocks are erratics deposited by retreating glaciers.) As "there are but two or three that have fallen in the memory of any man now living, in the ordinary course of nature we shall be compelled to allow the rest many thousands of years more than the age of the world."[26] Lhwyd was reluctant to ascribe them to the Deluge, and Ray commented evasively on Lhwyd's findings in order to avoid facing the logic of Lhwyd's comments.[27] On the issue of geological time, Ray nailed his colours firmly to the fence, without explicitly rejecting an Ussher chronology. His evasiveness to Lhwyd shows that he was reluctant to posit a divine intervention at the Deluge. Ray equivocated between a *naturalistic* and a supernatural explanation.

Early Nineteenth-Century Geologists

Moving on to the 1820s, let us consider the clerical geologists Henslow, Sedgwick, Buckland, and Conybeare, who contributed so much to geology: Henslow on Anglesey,[28] who influenced Darwin's geology far more than Lyell; Sedgwick, who elucidated the Cambrian and taught Darwin; Buckland for the first Mesozoic mammal and for introducing the Ice Ages to Britain; and Conybeare on the ichthyosaur. Cannon, in a classic article, claimed that they were Broadchurchmen,[29] that is, proto-modernist; but Pennock portrayed Henslow and Sedgwick as ardent adherents of "the detailed hypotheses of catastrophist flood geology,"[30] which they allegedly taught Darwin. Thus Henslow and Sedgwick were *supernaturalists*. Both cannot be right. Pennock has presented a simplistic ***either/or*** – either to be totally naturalistic, as Darwin (possibly) became, or to misguidedly base one's science on theology, as did(n't) Sedgwick and Henslow. It is a simple thesis, which misunderstands Sedgwick's and Henslow's geological method. Ironically, Pennock's polarisation is almost identical to Johnson's, which he so effectively demolished in *The Tower of Babel*, where he distinguished between ontological and methodological naturalism. As geologists, Henslow and Sedgwick were methodological naturalists, which becomes manifest when one studies their geology in the field in North Wales. They did not base their geology on "the detailed hypotheses of catastrophist flood geology." Sedgwick's geological work was straightforward stratigraphy of England from 1820 to 1831. After his so-called recantation in 1831, he started work in Wales without changing his practice or theory in his field notes or published papers. In 1825, he contributed a paper on the "Origins of Alluvial and Diluvial Formations,"[31] which contained much good information on the "drift," later seen as glacial. He realised that it had come from the north. He considered this "to demonstrate the reality of a great diluvian catastrophe during a comparatively recent period" and argued that "[i]t must... be rash and unphilosophical to look to the language of revelation for any direct proofs of the truths of physical science." In fact, this is a short step from the Ice Age. Henslow wrote a short paper in 1822[32] in which he proposed that a passing comet caused the Flood. Here Henslow is reiterating the *naturalistic* approach of Whiston 150 years earlier, where "Biblical Events" were explained by natural/secondary causes and thus not as *libertarian acts of God*. At about the same time, Henslow mapped and described the geology of Anglesey. His memoir, which had such a great and unrecognised influence on Darwin's geology,[33] is a superb pioneering work on pre–Cambrian geology and contains no theology.

By contrast, William Buckland at times called in divine agency to explain matters geological. However, in his reconciliation of Christianity and geology in both *Vindiciae Geologicae* (1820) and *Reliquiae Diluvianiae* (1823), he explained geological phenomena in a *naturalistic* way, without invoking God. Privately, he invoked God for some *libertarian* interventions. In some (barely legible) notes on the Deluge made during the early 1820s,[34] he grappled with the Deluge and geology and thought that God may have recreated all life throughout the world after the Flood waters, as there was no room in the Ark, and also contemplated a local Flood. Perhaps it is significant that Buckland's public face was *naturalistic*, but in private he considered divine intervention. That is the opposite of expectations, but does show that a leading geologist in the 1820s could be open to *supernatural* intervention. In the 1840s, Buckland still regarded the Flood as a geological event but as one caused by the Ice Age. In 1842, he wrote that icebergs had been carried from the North by "a great diluvial wave or current,"[35] words reminiscent of Sedgwick in 1825.

The Flood geology and diluvialism of the early nineteenth century had far more in common with Ryan and Pitman[36] on the Black Sea than with the divine hydraulics of Morris and Whitcomb.[37] Though diluvialists like Conybeare regarded uniformitarians, such as Scrope and Fleming (a

Calvinist evangelical), as geese and donkeys, they were equally *naturalistic* in their geology. (It is probably better to say *methodologically uniformitarian.*)[38]

From 1820 to 1860, one cannot divide geologists into *naturalists* and *theists.*[39] Many – Sedgwick, Buckland, Lyell (until 1864), and others – were *naturalist* and noninterventionist in geology and *theist* and interventionist in their interpretation of the fossil record. Geological revolutions were *natural*, but the creation of species was *supernatural.* This may be an inconsistent approach to the history of the Earth and of life. The problems were well known, as from 1820 to 1860 many scientists were questioning the fixity of species. The common view of progressive creationism was an unstable amalgam of *supernaturalism* and *naturalism.*

Darwin on "the Ordinary View of Creation"

Throughout the *Origin of Species*, Darwin referred to "the ordinary view of creation" and cited its weaknesses in order to make his ideas plausible. The rhetorical value of "the ordinary view of creation" will be discussed later, but its power was its lack of definition. Readers today will think of a six-day Creation, and that may have been Darwin's intention, though six-day creationism had virtually disappeared by 1855.[40] The "ordinary view of creation" was, in fact, progressive creation, which was emphatic on geological time and the succession of life but frankly confused about such issues as the fixity of species and how "vestigial organs" were designed. Darwin easily pointed out contradictions with devastating effect.

This he did by asking whether "species have been created at one or more points of the earth's surface" (352). He pointed out that geologists will find no difficulty in accounting for migration, as, for example, when Britain was joined to the European mainland some millennia ago. And then he asked, "But if the same species can be produced at two separate points, why do we not find a single mammal common to Europe and Australia or South America?" The implications he spelt out in detail in comparing the Cape Verde Islands flora and fauna with those of the Galapagos. The one flora and fauna were similar to those of Africa and the other to those of South America, yet their climates and landscape were almost identical. His conclusion was that "this grand fact can receive no sort of explanation on the ordinary view of independent creation" (398). He took this up again in the last chapter on naturalists; although they "admit variation as a *vera causa* in one case, they arbitrarily reject it in another." He then asked, with Miltonic overtones, "But do they really believe that at innumerable periods of the earth's history certain elemental atoms have been commanded suddenly to flash into living tissues?" (482) Dembski sees this as a concern that "the distinction of design and non-design cannot be reliably drawn,"[41] but this was not Darwin's point, as his concern was drawing the line between species and varieties – unless Dembski sees "species" as separately designed and not "varieties."

(Ultimately, Intelligent Design demands that one believe that atoms can flash into living tissue.) In the next paragraph, Dembski claims that there is "a rigorous criterion for distinguishing intelligently designed objects from unintelligently designed ones," referring to sciences such as "forensic science, . . . cryptography, archaeology and SETI." This is a rhetorical appeal and does not explain how it would work out in practice. Were Darwin alive today, I am sure he would direct his withering criticism to Dembski's argument from SETI and to Behe's partial acceptance of common descent *and* his biochemical mousetraps. To put this personally, as a conservative Christian, I feel the attraction of Intelligent Design both emotionally and religiously, but I cannot justify it rationally, scientifically, or theologically.

"Theism" and "Naturalism" from the Mid Nineteenth Century

Yet Darwin retained some of "the ordinary view of creation" for the initial Creation and the creation of life, which he saw virtually as *libertarian acts of God*. This enabled many Christians to accept his ideas, though often rejecting natural selection. Some added the creation of consciousness and of man as two more such acts, whether they were Christian or not – for example, A. R. Wallace, the Scottish theologian James Orr, and the American G. F. Wright.

Orr was a conservative Scottish Presbyterian whose Kerr Lectures for 1890–91 are significant. He discussed evolution in his lecture on "The Theistic Postulate of the Christian View." He said, "On the general hypothesis of evolution . . . , I have nothing to say, except that, within certain limits, it seems to me extremely probable, and supported by a large body of evidence." What comes next has a most contemporary ring: "On this subject two views may be held. The first is, that evolution results from development from within, in which case, obviously, the argument from design stands precisely where it did, except that the sphere of its application is enormously extended. The second view is, that evolution has resulted from fortuitous variations. . . . "[42] Clearly, Orr rejects pure chance. His discussion of evolution is highly informed, and he almost advocates a form of punctuated equilibrium, as "[t]he type persists through the ages practically unchanged. At other periods . . . there seems to be a breaking down of this fixity. The history of life is marked by a great inrush of new forms. . . . it in no way conflicts with design."

But Orr wishes to go beyond design: "The chief criticism . . . upon the design argument . . . , is that it is too narrow. It confines the argument to final causes – . . . it is not the marks of purpose alone which necessitate this inference (of God) but everything which bespeaks of order, plan, arrangement, harmony, beauty, rationality in the connection and system of things." We are now back to Calvin – "the elegant structure of the world serving as a kind of a mirror, in which we may behold God, though otherwise invisible" – and to Polkinghorne's "inbuilt potentiality of creation."

Orr seems to hover between van Till's "creatonomic" view and Behe's occasional interventions. Orr's criticism that design was too narrowly understood in the early nineteenth century ought to be recognised. In a sense, these writers were intermediate forms between *theistic scientists* and *methodological naturalists* and thus do not fall into Pennock's or Johnson's simplistic categories. They also give the lie to the claim that Darwin killed design.

Many years ago in his important study *The Principle of Uniformity*, Hooykaas[43] attempted to categorise the different approaches as atheism, deism, semi–deism, supranaturalism, and the biblical view. Deism allowed no divine involvement after the initial creation; supranaturalism could imply capricious divine involvement; semi–deism allowed the occasional interruption; and the biblical view stressed God's involvement in Creation at all times. By Hooykaas's delineation, "the ordinary view of creation" of Buckland and others, and of Intelligent Design, is semi–deism, or what van Till calls *punctuated naturalism*. It is an unstable position between deism (or even atheism) and supranaturalism. Hooykaas scarcely develops his "biblical view," of which he writes: "In his providence, God usually guides the world according to constant rules, but, as He is a free agent, He gives order as well as deviation from order." This opens up the question of miracles, as any creed that accepts the Virgin Birth or the Empty Tomb of the Resurrection must, in a sense, be super – or possibly supranaturalist in the eyes of thoroughgoing naturalists. As with Griffin's *supernaturalism* and *minimal naturalism*,[44] it is difficult to draw the line.

To summarise: it is not possible to make a neat distinction between *theistic* and *naturalistic* science, especially when we consider historical examples. Both Pennock and Johnson oversimplify matters to fit into their respective rhetorical schemes. One must give Darwin the last word. He was never totally consistent in his naturalism and always said of himself, "I am in a hopeless muddle."

A RESTATEMENT OF FACT: RHETORIC RATHER THAN ARGUMENT IN DESIGN

Rhetoric is an ambiguous term, but it has been an essential feature of design arguments in the nineteenth century and in Intelligent Design. It may be viewed as the persuasiveness of a lawyer where the argument is weak, but Phillip Johnson and Charles Lyell are not my concern! In their recent Gifford Lectures, Brooke and Cantor discuss *natural theology as rhetoric* and expound several examples from the eighteenth and nineteenth centuries, including Buckland on *Megatherium*. As they point out, "[I]t is important to re-emphasise that natural theologians did not deploy such evidence (from Design) to 'prove' (in the strong deductive sense) the existence and attributes of God." The design argument was an *inductive* argument, and its conclusion was deemed a "moral" truth. They cite Campbell, a contemporary writer: "In moral reasoning we ascend from possibility . . . to probability . . . to the summit of moral certainty." They observe that "the persuasiveness of arguments suggest a close similarity between natural theology and the proceedings of the courtroom." "Persuasion becomes the name of the game."[45]

Rhetoric in Buckland

Considered in this light, the design argument as employed by both Buckland and Intelligent Design creationists becomes a rhetorical argument with shades of a persuasive advocate. The rhetoric gives design both its strength and its fatal flaw. This highly charged courtroom atmosphere was present in the music rooms at Holywell in 1832, when Buckland gave his tour de force on *Megatherium*,[46] whom he christened *Old Scratch*. Buckland gave a superb scientific account of its peculiar anatomy, but throughout the lecture was the implicit message, "the adaptation of Old Scratch is so wonderful and demonstrates the skill of the Designer, who is none but the Father of our Lord Jesus Christ." Buckland began with the *possibility* that sloths were not as poorly designed as Buffon and Cuvier had claimed. As he described *Old Scratch* so favourably, he moved to *probability* and then to the *moral certainty* of his theistic conclusion. This worked, as Buckland gave an explanation of every part of its anatomy, but he did not describe vestigial organs. In his *Origin of Species*, Darwin picked up this flaw in design arguments and how this had been swept under the carpet by appeals to the divine plan. He wrote that "rudimentary organs are generally said to have been created 'for the sake of symmetry', or in order 'to complete the scheme of nature;' but this seems to me no explanation, merely a restatement of fact." The fact being that God is the Creator.

Rhetoric in Behe

Behe also makes great use of rhetoric. Having led the reader through many explainable and unexplainable biochemical functions and having used the rhetorical appeal of his mousetrap, he uses an inductive rhetorical argument and argues that the absence of an explanation, as in the case of blood clotting, indicates the direct activity of a Designer. He rapidly moves from possibility to probability to moral certainty, but that certainty is only certain until an explanation is found. Behe's mousetrap is a rhetorical flourish, and his conclusion of a Designer is only a "restatement of fact" based on his original argument.

Darwin on the Rhetoric in Design and Creation

At the end of the *Origin of Species*, Darwin wrote, "It is so easy to hide our ignorance under such expressions as the 'plan of creation', 'unity of design'

&c., and to think that we give an explanation when we only restate a fact." To argue rhetorically, surely Intelligent Design is a restatement of fact? We may also see argument by rhetoric in the work of Richard Dawkins, most notably with his computer-simulated evolving biomorphs in *The Blind Watchmaker*. Here the rhetoric is based on contemporary faith in computer simulation rather than in God, but it is ultimately no proof of evolution and likewise is "a restatement of fact." Hard proof would require an actual sequence of evolving plants or animals.

In the *Origin of Species*, Darwin used rhetoric to persuade readers of his case for "descent by modification." His persuasiveness is to be seen in the cumulative effect of his "one long argument" as he moves from the known fact of "artificial selection" to natural selection. Here he follows the rhetorical approach recommended by Campbell. In places, his rhetoric contains some ridicule. This ridicule is more pungent in his essays of the 1840s; in 1842, he wrote, "The creationist tells one, on a . . . spot the American spirit of creation makes . . . American doves."[47] His discussion of the three rhinoceroses of Java, Sumatra, and Malacca argues rhetorically for rejecting separate origins. As these three "scarcely differ more than breeds of cattle," how is it that "the creationist believes these three rhinoceroses were created (out of the dust of Java, Sumatra, . . .) with their deceptive appearance of true . . . relationship; as well can I believe the planets revolve . . . not from one law of gravity but from distinct volition of Creator."[48] These pachyderms did not make it to the *Origin of Species*, but Darwin developed this argument as part of his conclusion when he wrote, "It is so easy to hide our ignorance under such expressions as the 'plan of creation', 'unity of design', etc., and to think we give an explanation when we only restate a fact.' Darwin would undoubtedly say the same about ID today.

There is a parallel between Darwin's and Dembski's arguments. As Darwin moves from artificial to natural selection, Dembski moves from intelligent causes in forensic science, cryptography, and archaeology to intelligent causes in the natural world. He argues that if we accept intelligent causes in the former, we ought to consider them "a legitimate domain for scientific investigation" for the latter.[49] This has a considerable (rhetorical?) appeal, but we must be clear how we can determine the intentional design of an Intelligent Designer or outside cause, beyond citing our own of wonder and awe. That Dembski and others have failed to do.

THE PROBLEM OF *DESIGNED* AND *UNDESIGNED* CREATION IN INTELLIGENT DESIGN

Historical Understandings of Genesis, Creation, and Design

During the last half-millennium, the Genesis account of Creation has been variously understood. The dominant view during the sixteenth, seventeenth, and eighteenth centuries was the chaos–restitution interpretation, with an initial undefined period of chaos, followed by a re-creation with several creative acts spread over six days.[50] A minority, following Archbishop Ussher, reduced the period of chaos to twelve hours, thus confining the time for all creation to six days. This was later adopted by a minority during the early nineteenth century, and has also been adopted by Young Earth creationists today, though it would be truer to say that Young Earth creationism stems from the Seventh Day Adventists.[51]

After the discovery of deep time by geologists in the late eighteenth century, the period of chaos was extended indefinitely by Chalmers and his predecessors as the gap theory.[52] Here, as in the long day interpretation of G. S. Faber[53] and others, the number of creative acts was multiplied indefinitely. By 1900, the gap theory had become the preserve of "Fundamentalists," and many Christians had come to take a figurative or mythical view of Genesis, cutting the Gordian Knot on creative acts. However, a significant proportion still suggested three particular creative acts, those for initial life, consciousness, and humanity. A good example was the Scottish theologian James Orr; but here they were in agreement with A. R. Wallace. (Wallace's motivation was spiritualism rather than theism, which may be significant.) Whether there were a few creations during the Demiurgic Week, or three Creations throughout geological time, or the myriad creations of progressive creation, or just the one of theistic evolution, the common ground was that God was the Creator of all that is. To all Christians from 1650 to 1900, all of Creation was designed.

Split-level Creation in Intelligent Design

In their understanding of Creation, Intelligent Design theorists open themselves to having a split-level understanding of Creation, as being part designed and part not. Behe states that "[i]f a biological structure can be explained in terms of those natural laws, then we cannot conclude that it was designed."[54] To take one of Behe's examples, haemoglobin is not designed, but blood clotting is. This is in contrast to the design approach of Paley and Buckland, where all is designed. Buckland took this to its extreme in his lecture on *Megatherium*, the sloth that both Buffon and Cuvier thought was poorly designed. Buckland would have none of that and showed how such an apparently ill-designed beast was superbly designed "to dig potatoes"! I completed my paper on Buckland and Behe on an alpine holiday, and considered the implications of Behe's concept of design while ascending the Col du Lame at 3,040 meters, which is overshadowed by le Petit Combin and its glaciers. Despite the steepness of some immense lateral moraines, I kept a good pace, with the exhilaration of feeling fit with heart and lungs working well. Then I thought, "Behe argued that haemoglobin is not designed," and as I scrambled up the last few hundred feet of unstable scree, I thought, "Haemoglobin is not designed, thus my good aerobic condition is not God-given." I then realised that if I had slipped off the loose rock on to the glacier headwall below, I would have been shredded on the descent. And as I lay bleeding at the foot of the slope, design would come into action as my bleeding wounds began to clot! However, I did not slip, and at the summit I continued to think of design as I considered the panoramic view, with Mont Blanc to the west and the Great Saint Bernard Pass below me. The beauty was breathtaking, but to identify design in such a complex and chaotic landscape was impossible. Awesome and wonderful, yes! But designed?

Undoubtedly, glaciated scenery is an extreme case, but the question of design must be considered. However, the example of haemoglobin as undesigned and blood clotting as intelligently designed poses a problem. In this case, Intelligent Design will not satisfy any theist, but the perspective of Calvin and Orr would. Intelligent Design results in some of Creation being *designedly* created, whereas the rest is *undesignedly* created. We end up with a two-tier Creation, with some life systems, which are due to the process of natural laws, being undesigned, while the others are supernaturally designed. The further question is whether inanimate bodies are designed or not. This Van Till has aptly named *punctuated naturalism*,[55] whereby most processes are to be explained *naturalistically* (e.g., haemoglobin and glaciers) and only some *theistically* (e.g., blood clotting). There was no two-tier Creation for Buckland; God had created (and thus designed – whatever that means) "all things, visible and invisible," including Cuvier's woeful *Megatherium*. Viewed teleologically, apparently random and chaotic structures such as glaciers may reflect what Polkinghorne terms the inbuilt potentiality of Creation, or what Orr termed "the order, plan, arrangement, harmony, beauty, rationality in the connection and system of things."

It is essential to see what exponents of Intelligent Design are saying. They adopt *reverse engineering*,[56] and where this accomplishes a reduction to unintelligent causes – as in the case of haemoglobin, according to Behe – then that feature is *not* designed. **Design is reserved only for those features that cannot be explained by reverse engineering.** By using this strategy, they think they ensure a place for the creative activity of the Intelligent Designer – God. Our two advocates of *reverse engineering*, Buckland and Dennett, would concur that ultimately a reason for any structure will be found. If the Intelligent Design argument is followed consistently, the result is a two-tier Creation. To put matters as baldly as possible:

Paley saw the demonstration of Design in explaining.

ID sees the demonstration of Design in not-explaining.

Orr's comments apply even more to ID: "The chief criticism . . . upon the design argument, . . . is that it is too narrow. It confines the argument to final causes – . . . it is not the marks of purpose alone which necessitate this inference (of God) but everything which bespeaks of order, plan, arrangement, harmony, beauty, rationality in the connection and system of things." We are now back to Calvin, with "the elegant structure of the world serving as a kind of a mirror, in which we may behold God, though otherwise invisible."

Finally, one should ask whether design *on its own* is a biblical idea. I think not; I would argue that an overemphasis on design (Paleyan or ID) pushes the concept beyond the breaking point, as Orr stressed in 1890. The emphasis should be on God the Creator, not on God the Designer. If we follow the former and emphasise the Creator, we can say with Gerard Manley Hopkins;

> The World is charged with the grandeur of God.
> It will flame out, like shining from shook foil;
> It gathers to a greatness, like the ooze of oil
> Crushed. Why do men then now not reck his rod?

If we follow Intelligent Design, then we must parody Hopkins's poem;

> The clotting of blood is charged with the grandeur of God
> It will ooze out, like shining from shook foil.
> But haemoglobin is not charged with the grandeur of God.
> We know not when to reck his rod.[57]

CONCLUSION

Design is but one of the "natural" arguments for God; in its classic form, it evolves from the physico-theology of the seventeenth century. However, both in its classical form in Paley and in Intelligent Design, the design argument has tended to avoid the issue of geological time. In Paley's case, this was because geological time and the succession of species was scarcely known. In the case of Intelligent Design, there is a deliberate strategy to avoid "subsidiary issues like the age of the earth" – probably in order to retain the support of Young Earth creationists. It is essential to see the vast age of the Earth and universe as a matter of "rational compulsion," as opposed to evolution, which is an "inference to the best explanation."[58] It is because an evolutionary perspective on the succession of life provided a better "inference to the best explanation" in the 1860s that the progressive creationism of Sedgwick, Phillips, and others was rejected.

The appeal for a *theistic science* begs many questions. In the hands of Johnson, it tends to posit two polarised models of science – *materialist* and *empirical* – which does not allow for the diversity of opinion across science. This results in a game of Ping-Pong,[59] forcing an extreme **either/or**. These models do not survive historical scrutiny, whether we consider Ray, Sedgwick, or Darwin himself. This Ping-Pong over ID is often used as a (poor) rhetorical device.

Finally, here is Adam Sedgwick, after he supposedly gave up "the detailed hypotheses of catastrophist flood geology" in 1831:

> To the supreme Intelligence, indeed, all the complex and mutable combinations we behold, may be the necessary results of some simple law, regulating every material change, and involving within itself the very complications, which we, in our ignorance, regard as interruptions in the continuity of Nature's work.[60]

That is a far better statement of the relevance of the classical theological understanding of Creation to science than the *punctuated naturalism* of Intelligent Design, which inadvertently denies the involvement of the Creator in **all** creation.

Notes

1. J. Calvin, *Institutes*, Book 1, Chapter 5, section 1.
2. J. Calvin, *Institutes*, Book 1, Chapter 5, section 2.
3. J. H. Brooke and G. Cantor, *Reconstructing Nature* (Edinburgh: T. and T. Clark, 1998), p. 217.
4. H. Miller, *The Testimony of the Rocks* (London, 1858), pp. 238–44.
5. F. Temple, *The Relations between Science and Religion* (London: Longmans, 1884), p. 113.
6. N. Pearcey, "Design and the Discriminating Public," *Touchstone*, July/August 1999, p. 26. To my mind, the error at Kansas was to omit the teaching of geology and the age of the Earth in high school science. See Keith Miller, "The Controversy over the Kansas Science Standards," *Perspectives on Science and Faith* 51 (1999): 220–1.
7. Reconstitution and not Creation as the dominant understanding of Genesis was the idea that God first created chaos and then after an undefined interval "reconstituted" it. M. B. Roberts, "The Genesis of John Ray and His Successors," *Evangelical Quarterly* 74 (April 2002): 143–64.
8. John Thackray, *The Age of the Earth* (London: HMSO for the Institute of Geological Sciences, 1980), p. 8.
9. This is a crude calculation based on some 27 creations every 10 million years, i.e., 27 × 55 creations since the beginning of the Cambrian, i.e., a mere 1,485 creations.
10. J. Phillips, supplementary note in Baden Powell, *Connexion of Natural and Divine Truth* (London, 1838), p. 369, cited in J. P. Smith, *The Relation between the Holy Scripture and Some Parts of Geological Science* (London, 1848), p. 60.
11. C. Darwin, "B Notebook," in *Charles Darwin's Notebooks, 1836–1844*, ed. P. H. Barrett, P. J. Gautry, S. Herbert, D. Kohn, and S. Smith (Cambridge: Cambridge University Press, 1987).
12. A Jurassic marsupial first described by Buckland in 1824.
13. P. J. Bowler, *The Non–Darwinian Revolution* (Baltimore: Johns Hopkins University Press, 1988).
14. P. Johnson, "The Wedge," *Touchstone*, July/August 1999, pp. 19–20.
15. K. Miller, *Finding Darwin's God* (New York: HarperCollins, 1999), pp. 95–9.

16. C. Darwin, "The Essay of 1844," in *The Works of Charles Darwin*, vol. 10, ed. P. H. Barrett and R. B. Freedan (New York: New York University Press, 1986), pp. 133–4.

17. *Lessing's Theological Writings*, ed. H. Chadwick (London: A. and C. Black, 1956), p. 53.

18. Stephen Meyer, "The Methodological Equivalence of Design and Descent," and K. Wise, "The Origin of Life's Major Groups," in *The Creation Hypothesis*, ed. J. P. Moreland (Downers Grove: InterVarsity Press, 1994), pp. 67–112, pp. 211–34 at p. 226.

19. This is clear if one reads through the *Transactions of the Geological Society of London* and similar journals from 1810. Many of the papers are tedious stratigraphic and palaeontological descriptions substantiating Bragg's charge that geology is stamp collecting! But they show how undoctrinaire stratigraphy was, as it unravelled the chronology of the Earth.

20. N. Eldredge, *The Triumph of Evolution* (New York: W. H. Freeman, 2000), pp. 42–8.

21. Moreland, ed., *The Creation Hypothesis*, p. 51; J. P. Moreland in *Mere Creation*, ed. W. Dembski (Downers Grove, IL: InterVarsity Press, 1998), pp. 265–88.

22. Presumably Moreland would favour C. B. de Mille's interpretation of the crossing of the Red Sea.

23. S. J. Gould, "Burnet's Dirty Little Planet," in his *Ever Since Darwin* (Harmondsworth: Penguin, 1980), p. 144.

24. *Erratic blocks* are boulders of various sizes that were transported up to 100 miles or more by ice during the Ice Ages and are found in the northern United States, Britain, and Northern Europe.

25. John Ray, *Miscellaneous Discourses Concerning the Dissolution of the World* (London, 1691).

26. As there are at least 10,000 boulders in the Llanberis Pass and 60 years is the memory of any man, that gives an "age" of $10,000 \times 20 = 200,000$ years.

27. Ray, *Miscellaneous Discources*, pp. 285–9.

28. J. S. Henslow, "Geological Description of Anglesey," *Transactions of the Cambridge Philosophical Society* 1 (1822): 359–452.

29. W. S. Cannon, "Scientists and Broadchurchmen," *Journal of British Studies* 4 (1964): 65–88.

30. R. T. Pennock, *The Tower of Babel* (Cambridge, MA: MIT Press, 1999), p. 61.

31. A. Sedgwick, "Origins of Alluvial and Diluvial Formations," *Annals of Philosophy* (1825), cited from Clark and Hughes, *Life and Letters of Adam Sedgwick* (Cambridge: Cambridge University Press, 1890), pp. 292–3.

32. J. S. Henslow, "On the Deluge," *Annals of Philosophy* 6 (1823): 344–8.

33. See my "A Longer Look at the Darwin–Sedgwick Tour of North Wales 1831," forthcoming. I am preparing another paper on how Henslow, through his Anglesey memoir, influenced Darwin while he was on the Beagle – probably far more than Lyell.

34. In the *Deluge file* at University Museum, Oxford.

35. W. Buckland, "On the Glacia-diluvial Phaenomena in Snowdonia," *Proceedings of the Geological Society of London* 3 (1842): 579–84.

36. W. Ryan and W. Pitman, *Noah's Flood* (New York: Simon and Schuster, 1999).

37. H. M. Morris and J. C. Whitcomb, *The Genesis Flood* (Nutley: Presbyterian and Reformed, 1961).

38. S. J. Gould, *Time's Arrow, Time's Cycle* (Harmondsworth: Penguin, 1988); R. Hooykaas, *The Principle of Uniformity*, Leiden: E. J. Brill, 1957).

39. I use the term *theist* in the same way that Intelligent Designers use it today, to make my point.

40. In Britain, the only examples I can think of are Gosse and B. W. Newton. In the United States, there were Moses Stuart, Dabney, and a few others.

41. W. Dembski, *Intelligent Design* (Downers Grove, IL: InterVarsity Press, 1999).

42. J. Orr, *The Christian View of God and the World* (Edinburgh, 1897), pp. 98ff.

43. Hooykaas, "The Principle of Uniformity," pp. 170ff.

44. David Griffin, *Religion and Scientific Naturalism* (Albany State University of New York Press, 2000).

45. This is based very closely on J. Brooke and G. Cantor, *Reconstructing Nature* (Edinburgh: T. and T. Clark, 1998), pp. 181–2.

46. The manuscript of Buckland's lecture is cited with permission from Mrs. D. K. Harman and M. B. Roberts, "Design up to Scratch," *Perspectives on Science and Christian Faith* 51 (1999): 244–53.

47. Darwin, "The Essay of 1844," p. 31. N.B. This was a rough draft, hence incomplete sentences.

48. Ibid., p. 49.

49. Dembski, *Intelligent Design*, p. 105.

50. M. B. Roberts, "The Genesis of John Ray and His Successors," *Evangelical Quarterly* 74:2 (2002): 143–64.

51. R. Numbers, *The Creationists* (New York: Knopf, 1991).

52. This is a deliberate statement, as Chalmers was *not* the originator of the gap theory. R. Kirwan, *Geological Essays* (London, 1799), p. 47.

53. G. S. Faber, *A Treatise of the Three Dispensations* (London, 1823), pp. 111–65.

54. M. Behe, *Darwin's Black Box* (New York: The Free Press, 1996), p. 203.

55. H. van Till in *Darwinism Defeated?*, ed. D. Lamoureux (Vancouver: Regent College Publishers, 1999). "The intelligent design concept as it has been promoted by Johnson, with its intense emphasis on episodes of form-imposing intervention and its frequent association of material processes with naturalistic causes, could be more accurately called a theory of *punctuated naturalism*" (p. 88).

56. D. C. Dennett, *Darwin's Dangerous Idea* (Harmondsworth: Penguin, 1995).

57. The word "rod" is used in the Authorised Version to mean "sceptre," a symbol of God's power (see Psalm 23, verse 4).

58. Dembski, *Intelligent Design*, pp. 195–8.

59. B. Mitchell, *How to Play Theological Ping-Pong* (London: Hodder and Stoughton, 1987), pp. 166ff.

60. A. Sedgwick, "Anniversary Address to the Geological Society," February 18, 1831, *Proceedings of the Geological Society of London* 1 (1834), 281–316, 302.

16

The Argument from Laws of Nature Reassessed

Richard Swinburne

I have campaigned for many years for the view that most of the traditional arguments for the existence of God can be construed as inductive arguments from phenomena to the hypothesis of theism (that there is a God), which best explains them.[1] Each of these phenomena gives some probability to the hypothesis, and together they make it more probable than not. The phenomena can be arranged in decreasing order of generality. The cosmological argument argues from the existence of the universe; the argument from temporal order argues from its being governed by simple laws of nature; the argument from fine-tuning argues from the initial conditions and form and constants of the laws of nature being such as to lead (somewhere in the universe) to the evolution of animal and human bodies. Then we have arguments from those humans' being conscious, from various particular characteristics of humans and their enivronment (their free will, their capacity for causing limited good and harm to each other and especially for moulding their own characters for good or ill), from various historical events (including violations of natural laws), and finally from the religious experiences of so many millions of humans.

I assess these arguments as arguments to the existence of "God" in the traditional sense of a being essentially eternal, omnipotent, omniscient, perfectly free, and perfectly good; and I have argued that His perfect goodness follows from the other three properties.[2] God's omnipotence is His ability to do anything logically possible. God's perfect goodness is to be understood as His doing only what is good and doing the best, insofar as that is logically possible and insofar as He has the moral right to do so. So He will inevitably bring about a unique best possible world (if there is one) or one of a disjunction of equal best possible worlds (if there are such). But if for every good possible world there is a better one, all that God's perfect goodness can amount to is that He will bring about a good possible world.[3] So God will bring about any state of affairs that belongs to the best or all the equal best or all the good possible worlds. If there is some state of affairs such that any world is equally good for having it or not having it, then we can say that there is a probability of $^1/_2$ that He will make it. God will exercise this choice among worlds (and so among states of affairs), which it is logically possible for Him to bring about and which He has the moral right to bring about. There are some very good possible worlds and states thereof that God cannot, for logical reasons, guarantee to bring about – for example, worlds in which agents with a choice between good and evil always freely choose the good.[4] (When I write about "free choice," I mean libertarian free choice, that is, a choice that is not fully determined by causes that influence it.) Also, God can bring about a world only if He has the moral right to do so. There are, in my view, limits to His moral right to allow some to suffer (not by their own choice) for the benefit of others – limits of the length of time and intensity of suffering that He may allow.

The traditional arguments to the existence of such a God, which I have just listed, are, I claim, cumulative. In each case, the argument goes that the cited phenomena are unlikely to occur, given only the phenomena mentioned in the previous argument. That is, the existence of the universe is improbable a priori (i.e., if we assume nothing contingent at all); the universe being governed by laws of nature is improbable, given only the existence of the universe – and so on. The argument then claims that if there is a God, these phenomena are much more to be expected than if there is no God. For God, being omnipotent, has the power to bring about a universe and to endow it with the various listed characteristics, that is, to sustain in being a universe with these characteristics either for a finite or for an infinite period. And, I have argued, all of these characteristics are good, and so, by virtue of His perfect goodness, there is some probability that He will bring them about.

This is basically because, among the good worlds that a God has reason to make are ones in which there are creatures with a limited free choice between good and evil and limited powers to make deeply significant differences to themselves, each other, and their world by means of those choices (including the power to increase their powers and freedom of choice.) The goodness of significant free choice is, I hope, evident. We think it a good gift to give to our own children that they are free to choose their own path in life for good or ill, and to influence the kinds of persons (with what kinds of character and powers) they and others are to be. But good though this is, there is the risk that those who have such free will will make bad choices, form bad characters for themselves, hurt others and influence their characters for evil. For this reason, I suggest that it would not be a good action to create beings with freedom of choice between good and evil and unlimited power to put such choices into effect. If God creates beings with the freedom to choose between good and evil, they must be finite, limited creatures. Even so, the risks are – as we know very well – considerable; and so, I suggest that God would not inevitably bring about such a world. For any world that God could make containing such creatures would be no worse for not containing such creatures. But I suggest that the converse also holds: any world that God could make to which you add such creatures would be none the worse for such an addition. For this reason, there is a probability of $\frac{1}{2}$ that he will make such a world. But my arguments do not depend on giving such a precise probability or such a high probability to God's (if there is a God) making such a world. All that I am claiming is that there is a significant probability that a God would create such a world.

Let us call creatures with limited powers of the kinds just listed free rational creatures. If humans have (libertarian) free will (as is not implausible),[5] evidently our world is a world containing such creatures. We humans make deeply significant choices, choices affecting ourselves, each other, and our world; and our choices include choices to take steps to increase our powers and freedom and to form our characters for good or ill. But our powers in these respects are limited ones. Our world is a world of a kind that God can (with significant probability) be expected to make. Free rational creatures will have to begin life with a limited range of control and the power to choose to extend that range or – alternatively – not to bother to do so. That limited range is their bodies. In order for them to be able to extend their range of control, there must be some procedure that they can utilize – this bodily movement will have this predictable extrabodily effect. That is, the world must be subject to regularities – simple natural laws – that such creatures can choose to try to discover and then choose to utilize in order to influence things distant in space and time. You can learn that if you plant seeds and water them, they will grow into edible plants that will enable you to keep yourself and others alive, or that if you pull the trigger of a gun loaded in a certain way and pointing in a certain direction, it will kill some distant person. And so on. We can choose whether to seek out such knowledge (of how to cure or kill) or not to bother; and we can choose whether to utilize this knowledge for good or for ill. In a chaotic world, that would not be possible – for there would be no recipe for producing effects.

So, given that – as I have argued – there is a significant probability that a God would create free rational creatures (as defined earlier), there is a signficant probability that He will create this necessary condition for the existence of such creatures – a world regular in its conformity to simple natural laws. It is not sufficient that there be natural laws; they must be sufficiently simple to be discoverable by rational creatures. This means that they must be instantiated frequently, and that the simplest extrapolation from their past instantiations will often yield correct predictions. There could be a world with a trillion unconnected laws of nature, each determining that an event of a certain kind would be followed by an event of a certain other kind, but where there were only one or two events of the former kind in the history of the universe. No rational creature could discover such laws. Or there could be laws governing events of a type frequently instantiated but of such enormous mathematical complexity that the simplest extrapolation from the past occurences would never yield correct predictions. The laws must be sufficiently simple and frequently instantiated to be discoverable from a study of past history, at least by a logically omniscient rational being (one who could entertain all possible scientific theories, recognize the simplest, and draw the logical consequences thereof). (The laws, I must add, must not be of a totally deterministic kind and cover all events. They must allow room for free will. However, I shall not discuss that aspect in this chapter.) Also, the conformity of a material world to such laws is beautiful and a good in itself. The simple elegant motions of the stars and of all matter conforming to discoverable laws form a beautiful dance. And that is another reason why, if there is a God, we might expect a law-governed universe, that is, a reason that adds to the probability of there being such a universe, if there is a God.

In order to keep this chapter to a reasonable length, I shall assume that if gods are at work, monotheism of the traditional kind is far more probable than polytheism (that is, the view that many independent gods of finite powers provide the ultimate explanation of things).[6] I shall consider the alternative to which we are contrasting theism to be naturalism, the view that any ultimate explanation of the universe and its properties is of a scientific kind, that is, an explanation in terms of matter–energy and its properties.[7] In this chapter, I seek to investigate further my claim that, given naturalism, even if there is a universe, it is most unlikely that it would be governed by simple laws of nature. My argument in the past has been that if we are confined to scientific explanation, while we can explain lower-level laws by higher level ones, there can be no explanation of the conformity of nature to the most fundamental laws. Yet this conformity consists simply in everything in the universe behaving in exactly the same way. Such a vast coincidence of behaviour, as a vast brute fact, would be a priori extremely improbable. Hence, while simple laws of nature are quite probable if there is a God, they are very improbable otherwise. So their operation is good evidence for the existence of God.

I stand by my argument that, given naturalism, it is vastly improbable that the universe (that is, the one in which we live) would be governed by (simple) laws of nature. But what I had not appreciated before, and what I wish to bring out in this chapter, is that the argument should be phrased as an argument from simple laws of nature (that is, ones discoverable in the sense defined earlier), and that its strength depends on what laws of nature are, and on whether the universe had a temporal beginning, and on what that beginning was like.

The argument is an argument from "the universe" being governed by discoverable laws of nature. By "the universe" I mean that system of physical bodies spatially related to (i.e., at some distance in some direction from) ourselves. I do not rule out the possibility of there being other universes, systems of physical bodies not so related, and we will need to consider that possibility in due course. It is a well-justified extrapolation from study of the spatio-temporal region accessible to our telescopes, a region vastly wider than the region in which we live, that the whole universe is governed by the same laws. They may be the laws of General Relativity, quantum theory, and a few other theories; or the laws of a Grand Unified Theory; or the laws of a Theory of Everything. But what is meant by the claim that it is so governed; what is the truth maker for there being laws of nature? One view, originating from Hume's view of causation, is, of course, the regularity view. "Laws of nature" are simply the ways things behave – have behaved, are behaving, and will behave. "All copper expands when heated" is a law of nature if and only if all bits of copper always have expanded, now expand, and always will expand when heated. We need, however, a distinction between laws of nature and accidental generalizations such as "all spheres of gold are less than one mile in diameter"; and we need to take account of probablistic laws such as "all atoms of C_{14} have a probability of decaying within 5,600 years of $^1/_2$." Regularity theory has reached a developed form that takes account of these matters in the work of David Lewis.

For Lewis, "regularities earn their lawhood not by themselves, but by the joint efforts of a system in which they figure either as axioms or theorems."[8] The best system is the one that has (relative to rivals) the best combination of strength and simplicity. Strength is a matter of how much it successfully predicts (that is, the extent to which it makes many actual events, past, present, or future – whether observed or not – probable, and very few actual events improbable); simplicity is a matter of the laws' fitting together and also having internal simplicity in a way that Lewis does not, but no doubt could, spell out. The true laws are the laws of the best system. So "all spheres of gold are less than one mile in diameter" is probably not a law, because it does not follow from the best system – as is evidenced by the fact that it certainly does not follow from our current best approximation to the ultimate best system – a conjunction of relativity theory and quantum theory. Laws may be probabilistic as well as universal; if "there is a 90 percent probability of an A being B" is a consequence of some theory, it will confer strength on that theory insofar as 90 percent of actual As (past, present, and future) are B. Lewis's account of laws of nature is part of his campaign on behalf of "Humean supervenience," the idea that everything there is supervenes (logically) on "a vast mosaic of local matters of particular fact," which he interprets as a spatio-temporal arrangement of intrinsic properties or "qualities."[9] Laws of nature and causation are, for Lewis, among the things thus supervenient.

Now, there do seem to be overwhelming well-known objections to any Humean account, including Lewis's, if laws of nature are supposed to explain anything – and, in particular, if they are supposed to explain why one thing causes another, as Humeans suppose that they do. Laws explain causation, according to Humeans, because causality reduces to components that include laws of nature. Hume's famous regularity definition of a "cause" describes it as "an object precedent and contiguous to another, and where all the objects resembling the former are placed in a like relation of priority and contiguity to those objects that resemble the latter."[10] "Objects" for Humeans are events or states of affairs, and they are constituted by instantiations of bundles of purely categorical properties (such as, perhaps, being "square" or "red"), in contrast to dispositional properties, whose nature it is to cause or to permit other obects to cause certain effects (such as, perhaps, being "soluble"). For a present day Humean such as Lewis, as I noted earlier, only certain kinds of regularities are laws and so function in an account of causation. On this account, the heating of a particular piece of copper causing its expansion is a matter of the former being followed by the latter, where there is a law that events like the former are followed by events like the latter. But since whether or not some lawlike statement constitutes a law depends, on this account, not merely on what has happened but on what will happen in the whole future history of the universe, it follows that whether A causes B now depends on that future history. Yet how can what is yet to happen (in maybe two billion years' time) make it the case that A now causes B, and thus explain why B happens? Whether A causes B is surely a matter of what happens now, and whether the world ends in two billion years cannot make any difference to whether A now causes B. Events far distant in time cannot make any difference to what is the true explanation of why B occurs (viz., that A occured and caused it) – though, of course, they might make a difference to what we justifiably believe to be the true explanation.

It is because of their role in causation that laws of nature are said to generate counterfactuals. Suppose that I don't heat the copper; it is then fairly evidently the case that "if the copper had been heated, it would have expanded." But if a law simply states what does (or did or will) happen, what grounds does it provide for asserting the counterfactual? It would do that only if there were some kind of necessity built into it.

These seem to me conclusive objections to the regularity account. If, however, despite them, we were to adopt this account, the conformity of all objects to laws of nature being just the fact that they do so conform would have no further cause except from outside the system. If there were no God, it would be a highly improbable coincidence if events in the world fell into kinds in such ways that the simplest extrapolation from the past frequently yielded correct predictions. There are innumerable logically possible ways in which objects could behave today, only one of them being in conformity with the simplest extrapolation from the past. If, on the other hand, God causes the behaviour of physical things, then the coincidence is to be expected, for reasons given earlier. We would, however, need to give some non-Humean account of God's intentional causation – otherwise its universal efficacy would itself constitute a brute coincidence!

So, dismissing Humean accounts of laws, for good reason, let us consider alternative accounts of laws of nature – that is, accounts that represent talk of "laws" as talk about a feature of the world additional to the mere succession of events, a feature of physical necessity that is part of the world. This feature of physical necessity may be thought of either as separate from the objects that are governed by it or as a constituative aspect of those objects. The former approach leads to a picture of the world as consisting of events (constituted perhaps by substances with their properties), on the one hand, and laws of nature, on the other hand; and this approach can be developed so as to allow for the possibility of there being universes in which there are no events but only laws of nature.[11] Laws of nature are thus ontologically concrete entities. The version of this account that has been much discussed recently is the version that claims that laws of nature are logically contingent relations between universals – either Aristotelian instantiated universals (Armstrong) or Platonist not-necesarily-instantiated universals (Tooley). For Armstrong, there being a fundamental law of nature that all Fs are G consists in there being a connection of physical necessity between the universal F and the universal G. There being a fundamental law of nature that "all photons travel at 300,000 km/sec relative to every inertial reference frame" consists in there being such a connection between the universal "being a photon" and the universal "travelling at 300,000 km/sec relative to every inertial reference frame," which we can represent by N (F, G). This relation between universals is itself a (logically) contingently existing universal. The instantiation of F thus inevitably brings with it the instantiation of G. One can perhaps begin to make sense of this suggestion if one thinks of the causing of states of affairs as making properties, which are universals, to be instantiated; and this involving the bringing of them down to Earth from an eternal Heaven, together with whatever is involved with those universals – viz., other universals of (physical) necessity connected thereto. But for Armstrong, there is no such eternal Heaven: "there is nothing to the law except what is instantiated. . . . the law . . . has no existence except in the particular sequences."[12] But in that case, does the relation between universals exist before the law is instantiated for the first time, or not? If so, there is an eternal Heaven in which it exists. If not, what causes it rather than some alternative to exist?

Tooley thinks of the relations between universals as existing in an eternal Heaven prior to their instantiation in this world. This will meet the problem of why they are instantiated on the first occasion, and will also allow for the plausible possibility of there being laws that are never instantiated:

Imagine a world containing ten different types of fundamental particles. Suppose further that the behaviour of particles in interactions depends upon the types of the interacting particles. Considering only interactions involving two particles, there are 55 possibilities with respect to the types of the two particles. Suppose that 54 of these possible interactions have been carefully studied, with the result that 54 laws have been discovered, one for each case, which are not interrelated in any way. Suppose finally that the world is sufficiently deterministic that, given the way particles of the types X and Y are currently distributed, it is impossible for them ever to interact at any time, past, present, or future. In such a situation it would seem very reasonable to believe that there is some *underived* law dealing with the interaction of particles of types X and Y.[13]

If there is such a law, and it consists in a relation between universals, they can only be ones in a Platonist heaven.

But Platonist heavens are very mysterious. God, as an intentional agent, could exercise power over the universe in the way in which we exercise it over our bodies.[14] If there is a God, His causal agency is of a familiar type. But how do universals act on the world? This is a very mysterious causal relation between the non-spatio-temporal world and our world, for which we have no analogue. Thus Lewis: "How can the alleged lawmaker impose a regularity? Why can't we have N(F, G) and still have F's that are not G's?"[15]

If, despite these difficulties, we adopt a relation-between-universals theory, the question then is, if there is no God, why should there be any connections between universals at all, and why should there be universals instantiated frequently enough and mathematical connections sufficiently simple so as to yield discoverable regularities? There might be universals that were instantiated without bringing any other universals with them, so that there was no predictable effect of the instantiation. But on this account, virtually all universals are connected to other universals. And there might be universals, but only ones of kinds instantiated once or twice in the history of the universe, rather than ones such as "photon" and "copper" that are instantiated often and so can be used for useful prediction. And again, the mathematical connections between the universals – for example, between the masses of bodies, their distance apart, and the gravitational attraction between them – might be of such complexity as never to be inferable from past behaviour. Although a priori it is for reasons of simplicity more probable that there will be a universe in which a few particular universals are instantiated, connected in a few particular simple ways, than that there will be a universe in which a certain vast number of particular universals are connected in a particular complicated way, there are so many possible universals and kinds of connection between them, most of which will not yield discoverable laws, that my intuition is that there is a rather low probability that if there is no God, the universe will evince discoverable regularities. Whereas, if there is a God, there is a considerable probability that He would cause the instantiation of a few universals connected in simple ways, if laws of nature consist in connections between universals, for the reasons given earlier. My intuition on the extent to which the simplicity of a theory makes for the prior probability of its truth derives from my asessment of the extent to which we allow it to play a role in determining the relative probability of scientific theories that are equally good at predicting the data.

However, the probability of the existence of God is in no way dependent on this "intuition." If you think that a priori simplicity makes for probability to a far greater extent than I have supposed, then you will indeed suppose that it is a priori probable that in a Godless universe there will be only a few universals connected in simple ways. Still, that leaves a number of concrete entities (universals) and a number of connections between them. And if the simplicity of a supposition makes it as probable as we are supposing, then the simplicity of the supposition that there is a God will make it even more probable a priori that there is a God than that there are a few universals simply connected in a Godless universe. For God is *one* personal being possessing to an infinite degree the properties that are essential to persons; and the notion of an infinite degree of some property is a simpler notion than that of a large finite degree of the property. (It is the notion of zero limits.) In order to be a person, you need to have some power to perform intentional actions and some knowledge of how to perform them. God is supposed to have power and knowledge with zero limits. Persons have some degree of freedom as to which actions to perform – God is supposed to have freedom with zero limits. The supposition that there is a God is thus simpler than the supposition that there are a few particular universals connected in particular ways, or not connected at all. The more you suppose that the relative simplicity of the universe is to be expected if there is no God, the more you must suppose that the existence of God is to be expected a priori. And that will diminish the a priori probability that there is a Godless universe at all. So the more you expect a Godless universe to be orderly, the less you expect there to be a Godless universe at all. Intuitions stronger than mine about the extent to which simplicity is evidence of truth make it a priori probable that there is a God, and so make a posteriori arguments from the character of the universe in favour of the existence of God otiose. So I revert to my "intuition" that while a particular simple theory is more probable than a particular complicated theory, we cannot claim that it is a priori probable (i.e., more probable than not) that the simple theory is the true one.

The alternative to thinking of the physical necessity involved in laws of nature as separate from the objects governed by it is to think of it as a constitutive aspect of those objects. The way in which this is normally developed is what we may call the substances-powers-and-liabilities account of laws of nature. The "objects" that cause effects are individual substances – this planet, those molecules of water. They cause effects in virtue of their powers to do so and their liabilities (deterministic or probabilistic) to exercise those powers under certain conditions. Powers and liabilities are thus among the properties of substances. Laws of nature are then just (logically) contingent regularities – not of mere spatio-temporal succession (as with Hume), but of causal succession, that is, regularities in the causal powers (manifested and unmanifested) of substances of various kinds. The fact that it is a law that heated copper expands is just a matter of every piece of copper having the causal power to expand and the liability to exercise that power when heated. As a matter of contingent fact, substances fall into kinds, such that all objects of the same kind have the same powers and liabilities. The powers and liabilities of large-scale things (e.g., lumps of copper) derive from the powers and liabilities of the small-scale things that compose them (atoms, and ultimately quarks, electrons, etc.). And, given a satisfactory theory integrating all science, all ultimate particulars will have exactly the same powers and liabilities (e.g., the power to cause an effect proportional in a certain way to their mass, charge, spin etc., and the liability to exercise that power under conditions varying with the mass, charge, spin, etc. of other objects). This account of the ultimate determinants of what happens as being merely substances and their causal powers and liabilities does provide explanation of what happens, and in familiar terms. (We ourselves have causal powers that we, unlike inanimate objects, can choose to exercise.) It was the way of explaining things familiar to the ancient and medieval world, before "laws of nature" began to play their role in the sixteenth century. It was revived by Rom Harré and E. H. Madden in *Causal Powers*.[16] "Laws of nature" were originally supposed to be God's laws for nature, and thus to have their natural place in a theistic worldview. The naturalist would seem to me to have difficulty, as was illustrated earlier, in making sense of their operating without a lawgiver. He would do better to adopt the substances-powers-and-liabilities account; and the theist too, unless he is an occasionalist, had better, in my view, endow substances with powers and liabilities so that they act on their own and think of God's "laws" as determining which powers and liabilities substances have, and conserving those powers and liabilities in substances. On this account, causation is an essential component of laws rather than laws being an essential component of causation.[17]

The question then becomes, why do all substances have some powers and liabilities that are identical to each other (e.g., the power to attract each other in accord with a force proportional to mm/r^2 and the liability always to exercise that power), and why with respect to other powers and liabilities do they fall into a small number of kinds (photons, protons, etc.)? The answer provided by this model is given in terms of ancestry. A substance has the powers and liabilities it does because it was produced by another substance exercising its power to produce a substance with just those powers and liabilities. If a proton is produced (together with an electron and an antineutrino) by the decay of a neutron, then the proton's powers and liabilities are caused by the neutron, in virtue of its powers and liabilities. How improbable it will be a priori that all substances fall into kinds in the way described will depend on whether this process had a beginning and on what kind of beginning it was.

Suppose, first, that the universe did have a beginning, a "Big Bang" of some sort. There are two different kinds of theories of a beginning. The first state might have been a spatially extended state or a spatially pointlike state. In the first case, we still have a lot of substances, no doubt crammed into a very small space, but all of them falling into a few kinds in virtue of their different powers to produce the few different kinds of substances we have now. In terms of the Big Bang model, there was not literally a singularity; it was just that as you approach the first instant in the temporally backward direction, you find denser and denser states. But it really all started in a very, but not infinitely, dense state. That state would then still consist of innumerable substances of very few kinds. The alternative first state would be a literally pointlike one. In the first instant, on this theory, there was an unextended point, endowed with the power to decay into innumerable substances of very few kinds, and with the liability to exercise that power at some time or other.

How is a scientific choice to be made between the two theories? We set up a physics that accounts in the simplest way for the present data. Is evolution from a point compatible with that theory? If our theory does not allow infinitely dense matter, or the force necessary for expansion from an infinitely dense state, it is not. We then have the choice of complicating the physics to allow for this possibility, or not. A theory of the latter kind is going to be probable only if not too much complication is required.

What is the a priori probability that the universe began uncaused by God in each of these alternative states? Clearly, if the first state was simply a very condensed version of our present state, there is still the vast coincidence of all the substances' having exactly the same powers and liabilities. Given theism, the coincidence is explained. Suppose now that science supports the theory that the universe began at a point. What is the prior probability that if it did, it would begin with the power to produce the total regularity in the behaviour of observed substances that we find? There are many alternative powers and liabilities with which an initial singularity might be endowed. It might have no powers, or powers with no liability to exercise them unless interfered with, or merely the power and liability to keep itself in being, or the power and liability to produce other substances that themselves have no power to sustain themselves in being for long, or the power and liability to produce other substances that themselves have all kinds of different and unconnected powers and liabilities. And so on. Which is the most probable a priori? Simplicity alone can determine this. My intuition here, based again on what I see as the weight that we give to simplicity in science, is that the most probable state is the zero state (no powers at all), but that a power and liability to produce objects all having the same particular powers (including powers to produce similar objects that would produce similar objects) is more probable than any particular hypothesis of powers and liability to produce objects with chaotic and erratic powers. Yet since there are so many ways of producing objects with chaotic and erratic powers, for each way of producing objects with all the same powers, there is a very low probability a priori that the singularity would have built into it the power to produce innumerable objects all with the same powers and liabilities to go on producing similar objects for billions of years so as to produce discoverable regularities. As before, if you give simplicity a much greater weight than I give it in determining prior probability, the need for a posteriori arguments for the existence of God begins to disappear. However, I suggest that the very low probability of the singularity's having the character just described is not as low as the a priori probability of all of very many substances' beginning their existence uncaused in a Godless universe with the same powers and liabilities. For the former involves a beginning from only one pointlike substance. And so the naturalist should prefer a real to a nominal singularity, if he can have it. (My suspicion is that he cannot – all matter-energy's occupying an unextended point is, I suggest, not a possibility allowed by the current theory of matter–energy, and so not a state that we could justifiably postulate as the cause of the evolution of the universe. But I could well be mistaken about this. And not an enormous amount turns on it.)

Suppose now that the Universe has an infinite age. The properties (of powers and liabilities) of every substance are then caused by those of a preceding substance. So there are substances with exactly the same such properties (including the power to produce substances of the existing kinds), because there always have been. But then this is just as unlikely a priori as that all the substances at an initial moment should have all the same powers and liabilities. A theistic explanation of all substances' having all the same such properties will claim that God conserves substances with their properties from moment to moment – which he can do by virtue of his omnipotence, and which he does for the reasons described earlier.

I have been assuming so far that there is only one universe. But there may be many universes. Our only grounds for believing that there are such will be either that this supposition is a consequence of the simplest explanation of how things are in our universe, or that it is a priori probable. But here we come to an interesting disagreement about which kind of explanation would be the simplest explanation of how things are in our universe. In my view (and in the view of almost all scientists, detectives, etc.), the simplest explanation is one that postulates as few entities as possible behaving in mathematically simple kinds of ways.[18] The most probable explanation of the data is the simplest one that yields the data with high probability. On that view of simplicity, our only grounds for believing that there are other universes would be if extrapolating back from the present state of our universe in accord with the mathematically simplest supposition about its laws were to lead us to a state at which there was a universe split, a state in which those laws dictated that another universe would "bud off" from our universe. But in that case, the other universe would be governed by the laws that govern our universe, in which case we could consider the two universes (or however many universes we learn about) as one multiverse, and the whole preceding structure of argument gives the same results as before.

But it has been suggested that the simplest theory is the one expressible in the shortest number of computational syllables,[19] from which it follows that the simplest theory of universes (compatible with there being at least one unvierse) is the one that claims that all logically possible universes exist. This claim has been put forward by Max Tegmark,[20] but he does not draw out the full consequences of his bold claim. He discusses only universes governed by laws; but clearly, the vast majority of logically possible universes are not law-governed, or, if they are law-governed, their laws are not simple. And he *assumes* that "self-aware substructures," – that is, in my terminology, "rational beings," – are embodied. But since – in my view, fairly clearly – rational beings may be disembodied, if we make Tegmark's assumptions then it remains very improbable that rational beings such as ourselves should find themselves embodied in a physical universe governed by simple laws. And even if rational beings had to be embodied, we live in a universe far more orderly than is needed for our continued existence for a few years. Beyond the area of, say, a small country, the world might be totally chaotic; and what happened more than a few years ago might be undiscoverable. But as far as the telescope can reach and a long way back into history, our universe is totally and discoverably orderly. This sort of universe is not typical of the universes in which we could have existed (whatever our acount of laws), and so a priori it is improbable that we would exist in this universe. But theism can explain why there is *so much* order – viz., we can progresively discover more and more order and extend our range of knowledge and control if we so choose – travel to distant planets, or learn about our remote ancestors. I argued earlier that God had a reason to put us in a universe like this.

So, even on Tegmark's (to my mind) totally mistaken understanding of simplicity, his "simplest explanation" of the existence of our Universe is in fact very improbable, because on his account it is very improbable that we would find ourselves (as we do) in this universe. But theism can explain why we are in a universe with vast discoverable order. Hence I suggest that there is no good reason for believing in the existence of other universes, except perhaps ones governed by the same laws as our own. So it does not affect the issue of why things are law-governed if we suppose that there is more than one universe. And I have argued that whether talk of "laws" is talk of regular successions of events, or of concrete entities determining the behaviour of substances, or of the powers and liabilities of substances, it is a priori improbable that a Godless universe would be governed by simple laws, while there is quite a significant probability that a God-created universe would be governed by simple laws. Hence the operation of laws of nature is evidence – one strand of a cumulative argument – for the existence of God.

Notes

This article is to be published in *Reason, Faith, and History: Essays in Honour of Paul Helm*, ed. Martin Stone (Aldershot, UK: Ashgate, 2005) and is included with permission of Ashgate Publishing.

1. See, especially, my *The Existence of God*, revised edition (Oxford: Clarendon Press, 1991). (For the detailed argument from laws of nature, see Chapter 8 of that work.) See also the simpler version of this, *Is There a God?* (Oxford: Oxford University Press, 1996). (See Chapter 4 for the argument from laws of nature.)

2. See, for example, my *The Coherence of Theism* (Oxford: Clarendon Press, rev. ed. 1993), pp 149–52 and 207–9.

3. For a fuller account of what God's goodness must amount to when there is no best or equal best possible world, see my *The Christian God* (Oxford: Clarendon Press, 1994), pp 65–71 and 134–6.

4. In Plantinga's terminology, God can only weakly actualize such a world, not strongly actualize it. See Alvin Plantinga, *The Nature of Necesity* (Oxford: Clarendon Press, 1974), p. 173. God can, however, make free agents strongly inclined, (though not determined) to choose the good; if he makes such agents, he makes it very probable that they will do good. Or God can make them not so strongly inclined, which will make it less probable that they will do good. But the second sort of freedom may be a freedom more worth having. For the different worths of the different kinds of free will God could create, see my *Providence and the Problem of Evil* (Oxford: Clarendon Press, 1998), pp 82–9.

5. For my arguments in favour of the view that they do have such free will, see my *The Evolution of the Soul* (Oxford: Clarendon Press, rev. ed. 1997), Chapter 13.

6. Mark Wynn has pointed out that there are very many different possible hypotheses, each postulating different numbers of gods with different powers, whereas there is only one hypothesis postulating one God of infinite power. Hence, he claims, although each of the former hypotheses might be less probable a priori than the hypothesis of theism, the disjunction of the former is plausibly more probable than the hypothesis of theism as an explanation of the world's order. (See his "Some Reflections on Richard Swinburne's Argument from Design," *Religious Studies* 29 (1993): 325–35.) But if the order of the world is to be explained by many gods, then some explanation is required for how and why they cooperate in producing the same patterns of order throughout the universe. This becomes a new datum requiring explanation, for the same reason as the fact of order itself. The need for further explanation ends when we postulate one being who is the cause of the existence of all others, and the simplest conceivable such – I urge – is God.

7. I also ignore the claims of John Leslie and Hugh Rice, which are considered seriously by Derek Parfit, that there is at work an inanimate principle producing states of affairs because they are good. For my reasons for ignoring this, see my "Response to Derek Parfit," in *Metaphysics: The Big Questions*, ed. P. Van Inwagen and D. W. Zimmerman (Oxford: Blackwell, 1998).

8. David Lewis, *Philosophical Papers*, vol. 2 (Oxford: Oxford University Press, 1986), "A Subjectivist's Guide to Objective Chance – Postscript," p. 122.

9. Lewis, *Philosophical Papers*, vol. 2, pp. ix–x.

10. David Hume, *A Treatise on Human Nature*, 1.3.14.

11. Thus "I hold . . . that many empty [possible] universes exist. As I see it, there is a world devoid of all material objects and events in which the general principles of Newtonian mechanics are laws; there is another empty world in which the general principles of Aristotelian physics are laws." John W. Carroll, *Law of Nature* (Cambridge: Cambridge University Press, 1994), p. 64 note 4.

12. D. M. Armstrong, *A World of States of Affairs* (Cambridge: Cambridge University Press, 1997), p. 227.

13. Michael Tooley, "The Nature of Laws," *Canadian Journal of Philosophy* 7 (1977): 667–98; see p. 669.

14. Or rather, since we do this by exercising power over our brains, in the way in which we exercise power over our brains. In so doing, we normally think of the power over the brain only in terms of the effect that it causes. But clearly we could – and some people do – train themselves to produce brain states of a kind defined by their internal nature (e.g., to produce α-rhythms) and not in terms of the effects that they normally cause.

15. Lewis, *Philosophical Papers*, vol. 2, p. xii. A similar objection is raised in John Foster, "Regularities, Laws of Nature, and the Existence of God," *Proceedings of the Aristotelian Society* 101 (2000–1): 145–61; see pp. 154–6.

16. R. Harré and E. H. Madden, *Causal Powers* (Oxford: Blackwell, 1975).

17. This allows the logical possibility of singular causation – that is, causation that does not exemplify a pattern captured in a law. I have argued elsewhere that human agency is such causation. When, to take a simple example, I try to lift a weight and succeed, this cannot be represented as an instance of a lawlike succession by virtue of exemplifying some regularity of my trying causing my success in some lawlike fashion. This is because to try to do x just is to exert causal influence in favour of x occurring. "Trying" isn't something separate from "causing"; if it is successful, it just is causing. There is no law at work connecting independent states. Or so I have argued. See my "The Irreducibility of Causation," *Dialectica* 51 (1997): 79–92.

18. For a fuller account of what makes a scientific theory simple, see my *Epistemic Justification* (Oxford: Oxford University Press, 2001), Chaper 4.

19. R. J. Solomonoff, "A Formal Theory of Inductive Inference," *Information and Control* 7 (1964): 1–22.

20. Max Tegmark, "Is 'The Theory of Everything' Merely the Ultimate Ensemble Theory?," *Annals of Physics* 270 (1998): 1–51.

PART IV

INTELLIGENT DESIGN

17

The Logical Underpinnings of Intelligent Design

William A. Dembski

1. RANDOMNESS

For many natural scientists, design, conceived as the action of an intelligent agent, is not a fundamental creative force in nature. Rather, material mechanisms, characterized by chance and necessity and ruled by unbroken laws, are thought to be sufficient to do all nature's creating. Darwin's theory epitomizes this rejection of design.

But how do we know that nature requires no help from a designing intelligence? Certainly, in special sciences ranging from forensics to archaeology to SETI (the Search for Extraterrestrial Intelligence), appeal to a designing intelligence is indispensable. What's more, within these sciences there are well-developed techniques for identifying intelligence. What if these techniques could be formalized and applied to biological systems, and what if they registered the presence of design? Herein lies the promise of Intelligent Design (or ID, as it is now abbreviated).

My own work on ID began in 1988 at an interdisciplinary conference on randomness at Ohio State University. Persi Diaconis, a well-known statistician, and Harvey Friedman, a well-known logician, convened the conference. The conference came at a time when "chaos theory," or "nonlinear dynamics," was all the rage and supposed to revolutionize science. James Gleick, who had written a wildly popular book titled *Chaos*, covered the conference for the *New York Times*.

For all its promise, the conference ended on a thud. No conference proceedings were ever published. Despite a week of intense discussion, Persi Diaconis summarized the conference with one brief concluding statement: "We know what randomness isn't, we don't know what it is." For the conference participants, this was an unfortunate conclusion. The point of the conference was to provide a positive account of randomness. Instead, in discipline after discipline, randomness kept eluding our best efforts to grasp it.

That's not to say that there was a complete absence of proposals for characterizing randomness. The problem was that all such proposals approached randomness through the back door, first giving an account of what was nonrandom and then defining what was random by negating nonrandomness. (Complexity-theoretic approaches to randomness like that of Chaitin [1966] and Kolmogorov [1965] all shared this feature.) For instance, in the case of random number generators, they were good so long as they passed a set of statistical tests. Once a statistical test was found that a random number generator could not pass, the random number generator was discarded as no longer providing suitably random digits.

As I reflected on this asymmetry between randomness and nonrandomness, it became clear that randomness was not an intrinsic property of objects. Instead, randomness was a provisional designation for describing an absence of perceived pattern until such time as a pattern was perceived, at which time the object in question would no longer be considered random. In the case of random number generators, for instance, the statistical tests relative to which their adequacy was assessed constituted a set of patterns. So long as the random number generator passed all these tests, it was considered good, and its output was considered random. But as soon as a statistical test was discovered that the random number generator could not pass, it was no longer good, and its output was considered nonrandom. George Marsaglia, a leading light in random number generation, who spoke at the 1988 randomness conference, made this point beautifully, detailing one failed random number generator after another.

I wrote up these thoughts in a paper titled "Randomness by Design" (1991; see also Dembski 1998a). In that paper, I argued that randomness should properly be thought of as a provisional designation that applies only so long as an object violates all of a set of patterns. Once a pattern is added that the object no longer violates but rather conforms to, the object suddenly becomes nonrandom. Randomness thus becomes a relative notion, relativized to a given set of patterns. As a consequence, randomness is not something fundamental or intrinsic but rather something dependent on and subordinate to an underlying set of patterns or design.

Relativizing randomness to patterns provides a convenient framework for characterizing randomness formally. Even so, it doesn't take us very far in understanding how we distinguish randomness from nonrandomness in practice. If randomness just means violating each pattern from a set of patterns, then anything can be random relative to a suitable set of patterns (each one of which is violated). In practice, however, we tend to regard some patterns as more suitable for identifying randomness than others. This is because we think of randomness not only as patternlessness but also as the output of chance and therefore representative of what we might expect from a chance process.

In order to see this, consider the following two sequences of coin tosses (1 = heads, 0 = tails):

(A) 1100001101011000110111111101000110001101100110011101111
     00011001000010111101110110011111010010100101011110

and

(B) 1111111111111111111111111111111111111111111111111111111
     000000000000000000000000000000000000000000000000000000.

Both sequences are equally improbable (having a probability of 1 in 2^{100}, or approximately 1 in 10^{30}). The first sequence was produced by flipping a fair coin, whereas the second was produced artificially. Yet even if we knew nothing about the causal history of the two sequences, we clearly would regard the first sequence as more random than the second. When tossing a coin, we expect to see heads and tails all jumbled up. We don't expect to see a neat string of heads followed by a neat string of tails. Such a sequence evinces a pattern not representative of chance.

In practice, then, we think of randomness not only in terms of patterns that are alternately violated or conformed to, but also in terms of patterns that are alternately easy or hard to obtain by chance. What, then, are the patterns that are hard to obtain by chance and that in practice we use to eliminate chance? Ronald Fisher's theory of statistical significance testing provides a partial answer. My work on the design inference attempts to round out Fisher's answer.

2. THE DESIGN INFERENCE

In Fisher's (1935, 13–17) approach to significance testing, a chance hypothesis is eliminated provided that an event falls within a prespecified rejection region and provided that the rejection region has sufficiently small probability with respect to the chance hypothesis under consideration. Fisher's rejection regions therefore constitute a type of pattern for eliminating chance. The picture here is of an arrow hitting a target. Provided that the target is small enough, chance cannot plausibly explain the arrow's hitting the target. Of course, the target must be given independently of the arrow's trajectory. Movable targets that can be adjusted after the arrow has landed will not do. (One can't, for instance, paint a target around the arrow after it has landed.)

In extending Fisher's approach to hypothesis testing, the design inference generalizes the types of rejection regions capable of eliminating chance. In Fisher's approach, if we are to eliminate chance because an event falls within a rejection region, that rejection region must be identified prior to the occurrence of the event. This is done in order to avoid the familiar problem known among statisticians as "data snooping" or "cherry picking," in which a pattern is imposed on an event after the fact. Requiring the rejection region to be set prior to the occurrence of an event safeguards against attributing patterns to the event that are factitious and that do not properly preclude its occurrence by chance.

This safeguard, however, is unduly restrictive. In cryptography, for instance, a pattern that breaks a cryptosystem (known as a cryptographic key) is identified after the fact (i.e., after one has listened in and recorded an enemy communication). Nonetheless, once the key is discovered, there is no doubt that the intercepted communication was not random but rather a message with semantic content and therefore designed. In contrast to statistics, which always identifies its patterns before an experiment is performed, cryptanalysis must discover its patterns after the fact. In both instances, however, the patterns are suitable for eliminating chance. Patterns suitable for eliminating chance I call specifications.

Although my work on specifications can, in hindsight, be understood as a generalization of Fisher's rejection regions, I came to this generalization without consciously attending to Fisher's theory (even though, as a probabilist, I was fully aware of it). Instead, having reflected on the problem of randomness and the sorts of patterns we use in practice to eliminate chance, I noticed a certain type of inference that came up repeatedly. These were small probability arguments that, in the presence of a suitable pattern (i.e., specification), did not merely eliminate a single chance hypothesis but rather swept the field clear of chance hypotheses. What's more, having swept the field of chance hypotheses, these arguments inferred to a designing intelligence.

Here is a typical example. Suppose that two parties – call them A and B – have the power to produce exactly the same artifact – call it X. Suppose further that producing X requires so much effort that it is easier to copy X once X has already been produced than to produce X from scratch. For instance, before the advent of computers, logarithmic tables had to be calculated by hand. Although there is nothing esoteric about calculating logarithms, the process is tedious if done by hand. Once the calculation has been accurately performed, however, there is no need to repeat it.

The problem confronting the manufacturers of logarithmic tables, then, was that after expending so much effort to compute logarithms, if they were to publish their results without safeguards, nothing would prevent a plagiarist from copying the logarithms directly and then simply claiming that he or she had calculated the logarithms independently. In order to solve this problem, manufacturers of logarithmic tables introduced occasional – but deliberate – errors into their tables, errors that they carefully noted to themselves. Thus, in a table of logarithms that was accurate to eight decimal places, errors in the seventh and eight decimal places would occasionally be introduced.

These errors then served to trap plagiarists, for even though plagiarists could always claim that they had computed the logarithms correctly by mechanically following a certain algorithm, they could not reasonably claim to have committed the same errors. As Aristotle remarked in his *Nichomachean Ethics* (McKeon 1941, 1106), "It is possible to fail in many ways, . . . while to succeed is possible only in one way." Thus, when two manufacturers of logarithmic tables record identical logarithms that are correct, both receive the benefit of the doubt that they have actually done the work of calculating the logarithms. But when both record the same errors, it is perfectly legitimate to conclude that whoever published second committed plagarism.

To charge whoever published second with plagiarism, of course, goes well beyond merely eliminating chance (chance in this instance being the independent origination of the same errors). To charge someone with plagiarism, copyright infringement, or cheating is to draw a design inference. With the logarithmic table example, the crucial elements in drawing a design inference were the occurrence of a highly improbable event (in this case, getting the same incorrect digits in the seventh and eighth decimal places) and the match with an independently given pattern or specification (the same pattern of errors was repeated in different logarithmic tables).

My project, then, was to formalize and extend our commonsense understanding of design inferences so that they could be rigorously applied in scientific investigation. That my codification of design inferences happened to extend Fisher's theory of statistical significance testing was a happy, though not wholly unexpected, convergence. At the heart of my codification of design inferences was the combination of two things: improbability and specification. Improbability, as we shall see in the next section, can be conceived as a form of complexity. As a consequence, the name for this combination of improbability and specification that has now stuck is *specified complexity* or *complex specified information*.

3. SPECIFIED COMPLEXITY

The term "specified complexity" is about thirty years old. To my knowledge, the origin-of-life researcher Leslie Orgel was the first to use it. In his 1973 book *The Origins of Life*, he wrote: "Living organisms are distinguished by their specified complexity. Crystals such as granite fail to qualify as living because they lack complexity; mixtures of random polymers fail to qualify because they lack specificity" (189). More recently, Paul Davies (1999, 112) identified specified complexity as the key to resolving the problem of life's origin: "Living organisms are mysterious not for their complexity *per se*, but for their tightly specified complexity." Neither Orgel nor Davies, however, provided a precise analytic account of specified complexity. I provide such an account in *The Design Inference* (1998b) and its sequel, *No Free Lunch* (2002). In this section I want briefly to outline my work on specified complexity.

Orgel and Davies used specified complexity loosely. I've formalized it as a statistical criterion for identifying the effects of intelligence. Specified complexity, as I develop it, is a subtle notion that incorporates five main ingredients: (1) a probabilistic version of complexity applicable to events; (2) conditionally independent patterns; (3) probabilistic resources, which come in two forms, replicational and specificational; (4) a specificational version of complexity applicable to patterns; and (5) a universal probability bound. Let's consider these briefly.

Probabilistic Complexity. Probability can be viewed as a form of complexity. In order to see this, consider a combination lock. The more possible combinations of the lock there are, the more complex the mechanism and correspondingly the more improbable it is that the mechanism can be opened by chance. For instance, a combination lock whose dial is numbered from 0 to 39 and that must be turned in three alternating directions will have 64,000 ($= 40 \times 40 \times 40$) possible combinations. This number gives a measure of the complexity of the combination lock, but it also corresponds to a 1/64,000 probability of the lock's being opened by chance. A more complicated combination lock whose dial is numbered from 0 to 99 and that must be turned in five alternating directions will have 10,000,000,000 ($= 100 \times 100 \times 100 \times 100 \times 100$) possible combinations and thus a 1/10,000,000,000 probability of being opened by chance. Complexity and probability therefore vary inversely: the greater the complexity, the smaller the probability. The "complexity" in "specified complexity" refers to this probabilistic construal of complexity.

Conditionally Independent Patterns. The patterns that in the presence of complexity or improbability implicate a designing intelligence must be independent of the event whose design is in question. A crucial consideration here is that patterns not be artificially imposed on events after the fact. For instance, if an archer shoots arrows at a wall and we then paint targets around the arrows so that they stick squarely in the bull's-eyes, we impose a pattern after the fact. Any such pattern is not independent of the arrow's trajectory. On the other hand, if the targets are set up in advance ("specified") and then the archer hits them accurately, we know that it was not by chance but rather by design. The way to characterize this independence of patterns is via the probabilistic notion of conditional independence. A pattern is conditionally independent of an event if adding our knowledge of the pattern to a chance hypothesis does not alter the event's probability. The "specified" in "specified complexity" refers to such conditionally independent patterns. These are the specifications.

Probabilistic Resources. "Probabilistic resources" refers to the number of opportunities for an event to occur or be specified. A seemingly improbable event can become quite probable once enough probabilistic resources are factored in. Alternatively, it may remain improbable even after all the available probabilistic resources have been factored in. Probabilistic resources come in two forms: replicational and specificational. "Replicational resources" refers to the number of opportunities for an event to occur. "Specificational resources" refers to the number of opportunities to specify an event.

In order to see what's at stake with these two types of probabilistic resources, imagine a large wall with N identically sized nonoverlapping targets painted on it, and imagine that you have M arrows in your quiver. Let us say that your probability of hitting any one of these targets, taken individually, with a single arrow by chance is p. Then the probability of hitting any one of these N targets, taken collectively, with a single arrow by chance is bounded by Np, and the probability of hitting any of these N targets with at least one of your M arrows by chance is bounded by MNp. In this case, the number of replicational resources corresponds to M (the number of arrows in your quiver), the number of specificational resources corresponds to N (the number of targets on the wall), and the total number of probabilistic resources corresponds to the product MN. For a specified event of probability p to be reasonably attributed to chance, the number MNp must not be too small.

Specificational Complexity. The conditionally independent patterns that are specifications exhibit varying degrees of complexity. Such degrees of complexity are relativized to personal and computational agents – what I generically refer to as "subjects." Subjects grade the complexity of patterns in light of their cognitive/computational powers and background knowledge. The degree of complexity of a specification determines the number of specificational resources that must be factored in for setting the level of improbability needed to preclude chance. The more complex the pattern, the more specificational resources must be factored in.

In order to see what's at stake, imagine a dictionary of 100,000 ($= 10^5$) basic concepts. There are then 10^5 level-1 concepts, 10^{10} level-2 concepts, 10^{15} level-3 concepts, and so on. If "bidirectional," "rotary," "motor-driven," and "propeller" are basic concepts, then the bacterial flagellum can be characterized as a level-4 concept of the form "bidirectional rotary motor-driven propeller." Now, there are about $N = 10^{20}$ concepts of level 4 or less, which constitute the relevant specificational resources. Given p as the probability for the chance formation of the bacterial flagellum, we think of N as providing N targets for the chance formation of the bacterial flagellum, where the probability of hitting each target is not more than p. Factoring in these N specificational resources, then, amounts to checking whether the probability of hitting any of these targets by chance is small, which in turn amounts to showing that the product Np is small (see the last section on probabilistic resources).

Universal Probability Bound. In the observable universe, probabilistic resources come in limited supply. Within the known physical universe, there are estimated to be around 10^{80} or so elementary particles. Moreover, the properties of matter are such that transitions from one physical state to another cannot occur at a rate faster than 10^{45} times per second. This frequency corresponds to the Planck time, which constitutes the smallest physically meaningful unit of time. Finally, the universe itself is about a billion times younger than 10^{25} seconds old (assuming the universe is between ten and twenty billion years old). If we now assume that any specification of an event within the known physical universe requires at least one elementary particle to specify it and cannot be generated any faster than the Planck time, then these cosmological constraints imply that the total number of specified events throughout cosmic history cannot exceed

$$10^{80} \times 10^{45} \times 10^{25} = 10^{150}.$$

As a consequence, any specified event of probability less than 1 in 10^{150} will remain improbable even after all conceivable probabilistic resources from the observable universe have been factored in. A probability of 1 in 10^{150} is therefore a *universal probability bound* (for the details justifying this universal probability bound, see Dembski 1998b, sec. 6.5). A universal probability bound is impervious to all available probabilistic resources that may be brought against it. Indeed, all the probabilistic resources in the known physical world cannot conspire to render remotely probable an event whose probability is less than this universal probability bound.

The universal probability bound of 1 in 10^{150} is the most conservative in the literature. The French mathematician Emile Borel (1962, 28; see also Knobloch 1987, 228) proposed 1 in 10^{50} as a universal probability bound below which chance could definitively be precluded (i.e., any specified event as improbable as this could never be attributed to chance). Cryptographers assess the security of cryptosystems in terms of brute force attacks that employ as many probabilistic resources as are available in the universe to break a cryptosystem by chance. In its report on the role of cryptography in securing the information society, the National Research Council set 1 in 10^{94} as its universal probability bound for ensuring the security of cryptosystems against chance-based attacks (see Dam and Lin 1996, 380, note 17). The theoretical computer scientist Seth Lloyd (2002) sets 10^{120} as the maximum number of bit operations that the universe could have performed throughout its entire history. That number corresponds to a universal probability bound of 1 in 10^{120}. In his most recent book, *Investigations*, Stuart Kauffman (2000) comes up with similar numbers.

In order for something to exhibit specified complexity, therefore, it must match a conditionally independent pattern (i.e., specification) that corresponds to an event having a probability less than the universal probability bound. Specified complexity is a widely used criterion for detecting design. For instance, when researchers in the SETI project look for signs of intelligence from outer space, they are looking for specified complexity. (Recall the movie *Contact*, in which contact is established when a long sequence of prime numbers comes in from outer space – such a sequence exhibits specified complexity.) Let us therefore examine next the reliability of specified complexity as a criterion for detecting design.

4. RELIABILITY OF THE CRITERION

Specified complexity functions as a criterion for detecting design – I call it the complexity-specification criterion. In general, criteria attempt to classify individuals with respect to a target group. The target group for the complexity-specification criterion comprises all things intelligently caused. How accurate is this criterion in correctly assigning things to this target group and correctly omitting things from it?

The things we are trying to explain have causal histories. In some of those histories intelligent causation is indispensable, whereas in others it is dispensable. An ink blot can be explained without appealing to intelligent causation; ink arranged to form meaningful text cannot. When the complexity-specification criterion assigns something to the target group, can we be confident that it actually is intelligently caused? If not, we have a problem with false positives. On the other hand, when this criterion fails to assign something to the target group, can we be confident that no intelligent cause underlies it? If not, we have a problem with false negatives.

Consider first the problem of false negatives. When the complexity-specification criterion fails to detect design in a thing, can we be sure that no intelligent cause underlies it? No, we cannot. For determining that something is not designed, this criterion is not reliable. False negatives are a problem for it. This problem of false negatives, however, is endemic to design detection in general. One difficulty is that intelligent causes can mimic undirected natural causes, thereby rendering their actions indistinguishable from such unintelligent causes. A bottle of ink happens to fall off a cupboard and spill onto a sheet of paper. Alternatively, a human agent deliberately takes a bottle of ink and pours it over a sheet of paper. The resulting ink blot may look the same in both instances, but in the one case it is the result of natural causes, in the other of design.

Another difficulty is that detecting intelligent causes requires background knowledge on our part. It takes an intelligent cause to recognize an intelligent cause. But if we do not know enough, we will miss it. Consider a spy listening in on a communication channel whose messages are encrypted.

Unless the spy knows how to break the cryptosystem used by the parties on whom she is eavesdropping (i.e., knows the cryptographic key), any messages traversing the communication channel will be unintelligible and might in fact be meaningless.

The problem of false negatives therefore arises either when an intelligent agent has acted (whether consciously or unconsciously) to conceal his actions, or when an intelligent agent, in trying to detect design, has insufficient background knowledge to determine whether design actually is present. This is why false negatives do not invalidate the complexity-specification criterion. This criterion is fully capable of detecting intelligent causes intent on making their presence evident. Masters of stealth intent on concealing their actions may successfully evade the criterion. But masters of self-promotion bank on the complexity-specification criterion to make sure that their intellectual property gets properly attributed, for example. Indeed, intellectual property law would be impossible without this criterion.

And that brings us to the problem of false positives. Even though specified complexity is not a reliable criterion for *eliminating* design, it is a reliable criterion for *detecting* design. The complexity-specification criterion is a net. Things that are designed will occasionally slip past the net. We would prefer that the net catch more than it does, omitting nothing that is designed. But given the ability of design to mimic unintelligent causes and the possibility that ignorance will cause us to pass over things that are designed, this problem cannot be remedied. Nevertheless, we want to be very sure that whatever the net does catch includes only what we intend it to catch – namely, things that are designed. Only things that are designed had better end up in the net. If that is the case, we can have confidence that whatever the complexity-specification criterion attributes to design is indeed designed. On the other hand, if things end up in the net that are not designed, the criterion is in trouble.

How can we see that specified complexity is a reliable criterion for detecting design? Alternatively, how can we see that the complexity-specification criterion successfully avoids false positives – that whenever it attributes design, it does so correctly? The justification for this claim is a straightforward inductive generalization: in every instance where specified complexity obtains and where the underlying causal story is known (i.e., where we are not just dealing with circumstantial evidence but where, as it were, the video camera is running and any putative designer would be caught red-handed), it turns out that design actually is present. Therefore, design actually is present whenever the complexity-specification criterion attributes design.

Although this justification for the complexity-specification criterion's reliability in detecting design may seem a bit too easy, it really isn't. If something genuinely instantiates specified complexity, then it is inexplicable in terms of all material mechanisms (not only those that are known, but all of them). Indeed, to attribute specified complexity to something is to say that the specification to which it conforms corresponds to an event that is highly improbable with respect to all material mechanisms that might give rise to the event. So take your pick – treat the item in question as inexplicable in terms of all material mechanisms or treat it as designed. But since design is uniformly associated with specified complexity when the underlying causal story is known, induction counsels attributing design in cases where the underlying causal story is not known.

To sum up, in order for specified complexity to eliminate chance and detect design, it is not enough that the probability be small with respect to some arbitrarily chosen probability distribution. Rather, it must be small with respect to every probability distribution that might characterize the chance occurrence of the thing in question. If that is the case, then a design inference follows. The use of chance here is very broad and includes anything that can be captured mathematically by a stochastic process. It thus includes deterministic processes whose probabilities all collapse to zero and one (cf. necessities, regularities, and natural laws). It also includes nondeterministic processes, such as evolutionary processes that combine random variation and natural selection. Indeed, chance so construed characterizes all material mechanisms.

5. ASSERTIBILITY

The reliability of specified complexity as a criterion for detecting design is not a problem. Neither is there a problem with specified complexity's coherence as a meaningful concept – specified complexity is well-defined. If there's a problem, it centers on specified complexity's *assertibility*. Assertibility is a term of philosophical use that refers to the epistemic justification or warrant for a claim. Assertibility (with an "i") is distinguished from assertability (with an "a"), where the latter refers to the local factors that in the pragmatics of discourse determine whether asserting a claim is justified (see Jackson 1987, 11). For instance, as a tourist in Iraq prior to the 2003 war with the United States, I would have been epistemically justified in asserting that Saddam Hussein is a monster (in which case the claim would be assertible). Local pragmatic considerations, however, would have told against asserting this remark within Iraqi borders (the claim there would be unassertable). Unlike assertibility, assertability can depend on anything from etiquette and good manners to who happens to hold political power. Assertibility with an "i" is what interests us here.

In order to see what's at stake with specified complexity's assertibility, consider first a mathematical example. It's an open question in mathematics whether the number pi (the ratio of the circumference of a circle to its diameter) is regular – where by "regular" I mean that every number between 0 and 9 appears in the decimal expansion of pi with a limiting relative frequency of 1/10. Regularity is a well-defined mathematical concept. Thus, in asserting that pi is regular, we might be making a true statement. But without a mathematical proof of pi's regularity, we have no justification for asserting that pi is regular. The regularity of pi is, at least for now, unassertible (despite over 200 billion decimal digits of pi having been computed).

But what about the specified complexity of various biological systems? Are there any biological systems whose specified complexity is assertible? Critics of Intelligent Design argue that no attribution of specified complexity to any natural system can ever be assertible. The argument runs as follows. It starts by noting that if some natural system instantiates specified complexity, then that system must be vastly improbable with respect to all purely natural mechanisms that could be operating to produce it. But that means calculating a probability for each such mechanism. This, so the argument runs, is an impossible task. At best, science could show that a given natural system is vastly improbable with respect to known mechanisms operating in known ways and for which the probability can be estimated. But that omits (1) known mechanisms operating in known ways for which the probability cannot be estimated, (2) known mechanisms operating in unknown ways, and (3) unknown mechanisms.

Thus, even if it is true that some natural system instantiates specified complexity, we could never legitimately assert its specified complexity, much less know it. Accordingly, to assert the specified complexity of any natural system is to argue from ignorance. This line of reasoning against specified complexity is much like the standard agnostic line against theism – we can't prove atheism (cf. the total absence of specified complexity from nature), but we can show that theism (cf. the specified complexity of certain natural systems) cannot be justified and is therefore unassertible. This is how skeptics argue that there is no (and indeed can be no) evidence for God or design.

A little reflection, however, makes clear that this attempt by skeptics to undo specified complexity cannot be justified on the basis of scientific practice. Indeed, the skeptic imposes requirements so stringent that they are absent from every other aspect of science. If standards of scientific justification are set too high, no interesting scientific work will ever get done. Science therefore balances its standards of justification with the requirement for self-correction in light of further evidence. The possibility of self-correction in light of further evidence is absent in mathematics, and that accounts for mathematics' need for the highest level of justification, namely, strict logico-deductive proof. But science does not work that way.

Science must work with available evidence, and on that basis (and that basis alone) formulate the best explanation of the phenomenon in question. This means that science cannot explain a phenomenon by appealing to the promise, prospect, or possibility of future evidence. In particular, unknown mechanisms or undiscovered ways by which those mechanisms might operate cannot be invoked in order to explain a phenomenon. If known material mechanisms can be shown to be incapable of explaining a phenomenon, then it is an open question whether any mechanism whatsoever is capable of explaining it. If, further, there are good reasons for asserting the specified complexity of certain biological systems, then design itself becomes assertible in biology. Let's now see how this could be.

6. APPLICATION TO EVOLUTIONARY BIOLOGY

Evolutionary biology teaches that all biological complexity is the result of material mechanisms. These include, principally, the Darwinian mechanism of natural selection and random variation, but they also include other mechanisms (symbiogenesis, gene transfer, genetic drift, the action of regulatory genes in development, self-organizational processes, etc.). These mechanisms are just that: mindless material mechanisms that do what they do irrespective of intelligence. To be sure, mechanisms can be programmed by an intelligence. But any such intelligent programming of evolutionary mechanisms is not properly part of evolutionary biology.

Intelligent Design, by contrast, teaches that biological complexity is not exclusively the result of material mechanisms but also requires intelligence, where the intelligence in question is not reducible to such mechanisms. The central issue, therefore, is not the relatedness of all organisms, or what typically is called common descent. Indeed, Intelligent Design is perfectly compatible with common descent. Rather, the central issue is how biological complexity emerged and whether intelligence played an indispensable (which is not to say exclusive) role in its emergence.

Suppose, therefore, for the sake of argument, that an intelligence – one irreducible to material mechanisms – actually did play a decisive role in the emergence of life's complexity and diversity. How could we know it? Certainly, specified complexity will be required. Indeed, if specified complexity is absent or dubious, then the door is wide open for material mechanisms to explain the object of investigation. Only as specified complexity becomes assertible does the door to material mechanisms start to close.

Nevertheless, evolutionary biology teaches that within biology, the door can never be closed all the way and indeed should not be closed at all. In fact, evolutionary biologists claim to have demonstrated that design is superfluous for understanding biological complexity. The only way actually to demonstrate this, however, is to exhibit material mechanisms that account for the various forms of biological complexity out there. Now, if for every instance of biological complexity some mechanism could readily be produced that accounts for it, Intelligent Design would drop out of scientific discussion. Occam's razor, by proscribing superfluous causes, would in this instance finish off Intelligent Design quite nicely.

But that hasn't happened. Why not? The reason is that there are plenty of complex biological systems for which no biologist can claim to have a clue how they emerged. I'm not talking about hand-waving just-so stories. Biologists have plenty of those. I'm talking about detailed, testable accounts of how such systems could have emerged. In order to see what's at stake, consider how biologists propose to explain the emergence of the bacterial flagellum, a molecular machine that has become the mascot of the Intelligent Design movement.

In public lectures, the Harvard biologist Howard Berg has called the bacterial flagellum "the most efficient machine in the universe." The flagellum is a nano-engineered motor-driven propeller on the backs of certain bacteria. It spins at tens of thousands of rpm, can change direction in a quarter turn, and propels a bacterium through its watery environment. According to evolutionary biology, it had to emerge via some material mechanism(s). Fine, but how?

The usual story is that the flagellum is composed of parts that previously were targeted for different uses and that natural selection then co-opted to form a flagellum. This seems reasonable until we try to fill in the details. The only well-documented examples that we have of successful co-optation come from human engineering. For instance, an electrical engineer might co-opt components from a microwave oven, a radio, and a computer screen in order to create a working television. But in that case, we have an intelligent agent who knows all about electrical gadgets and about televisions in particular.

But natural selection doesn't know a thing about bacterial flagella. So how is natural selection going to take extant protein parts and co-opt them in order to form a flagellum? The problem is that natural selection can select only for preexisting functions. It can, for instance, select for larger finch beaks when the available nuts are harder to open. Here the finch beak is already in place, and natural selection merely enhances its present functionality. Natural selection might even adapt a preexisting structure to a new function; for example, it might start with finch beaks adapted to opening nuts and end with beaks adapted to eating insects.

But in order for co-optation to result in a structure like the bacterial flagellum, it is not a question of enhancing the function of an existing structure or of reassigning an existing structure to a different function, but of reassigning multiple structures previously targeted for different functions to a novel structure exhibiting a novel function. Even the simplest bacterial flagellum requires around forty proteins for its assembly and structure. All these proteins are necessary in the sense that lacking any of them, a working flagellum does not result.

The only way for natural selection to form such a structure by co-optation, then, is for natural selection gradually to enfold existing protein parts into evolving structures whose functions co-evolve with the structures. We might, for instance, imagine a five-part mousetrap consisting of a platform, spring, hammer, holding bar, and catch evolving as follows: it starts as a doorstop (thus consisting merely of the platform), then evolves into a tie clip (by attaching the spring and hammer to the platform), and finally becomes a full mousetrap (by also including the holding bar and catch).

Design critic Kenneth Miller finds such scenarios not only completely plausible but also deeply relevant to biology (in fact, he regularly sports a modified mousetrap cum tie clip). Intelligent Design proponents, by contrast, regard such scenarios as rubbish. Here's why. First, in such scenarios the hand of human design and intention meddles everywhere. Evolutionary biologists assure us that eventually they will discover just how the evolutionary process can take the right and needed steps without the meddling hand of design. All such assurances, however, presuppose that intelligence is dispensable in explaining biological complexity. Yet the only evidence we have of successful co-optation comes from engineering and confirms that intelligence is indispensable in explaining complex structures such as the mousetrap and by implication the flagellum. Intelligence is known to have the causal power to produce such structures. We're still waiting for the promised material mechanisms.

The other reason design theorists are less than impressed with co-optation concerns an inherent limitation of the Darwinian mechanism. The whole point of the Darwinian selection mechanism is that one can get from anywhere in biological configuration space to anywhere else provided one can take small steps. How small? Small enough that they are reasonably probable. But what guarantee is there that a sequence of baby steps connects any two points in configuration space?

The problem is not simply one of connectivity. In order for the Darwinian selection mechanism to connect point A to point B in configuration space, it is not enough that there merely exist a sequence of baby steps connecting the two. In addition, each baby step needs in some sense to be "successful." In biological terms, each step requires an increase in fitness as measured in terms of survival and reproduction. Natural selection, after all, is the motive force behind each baby step, and selection selects only what is advantageous to the organism. Thus, in order for the Darwinian mechanism to connect two organisms, there must be a sequence of successful baby steps connecting the two.

Richard Dawkins (1996) compares the emergence of biological complexity to climbing a mountain – Mount Improbable, as he calls it. According to him, Mount Improbable always has a gradual serpentine path leading to the top that can be traversed in baby steps. But that's hardly an empirical claim. Indeed, the claim is entirely gratuitous. It might be a fact about nature that Mount Improbable is sheer on all sides and that getting to the top from the bottom via baby steps is effectively impossible. A gap like that would reside in nature herself and not in our knowledge of nature (it would not, in other words, constitute a God-of-the-gaps).

Consequently, it is not enough merely to presuppose that a fitness-increasing sequence of baby steps connects two biological systems – it must be demonstrated. For instance, it is not enough to point out that some genes for the bacterial flagellum are the same as those for a type III secretory system (a type of pump) and then to hand-wave that one was co-opted from the other. Anybody can arrange complex systems in series based on some criterion of similarity. But such series do nothing to establish whether the end evolved in Darwinian fashion from the beginning unless (a) the probability of each step in the series can be quantified, (b) the probability at each step turns out to be reasonably large, and (c) each step constitutes an advantage to the evolving system.

Convinced that the Darwinian mechanism must be capable of doing such evolutionary design work, evolutionary biologists rarely ask whether such a sequence of successful baby steps even exists; much less do they attempt to quantify the probabilities involved. I make that attempt in my book *No Free Lunch* (2002, Chapter 5). There, I lay out techniques for assessing the probabilistic hurdles that the Darwinian mechanism faces in trying to account for complex biological structures such as the bacterial flagellum. The probabilities that I calculate – and I try to be conservative – are horrendous and render natural selection utterly implausible as a mechanism for generating the flagellum and structures like it.

Is the claim that the bacterial flagellum exhibits specified complexity assertible? You bet! Science works on the basis of available evidence, not on the promise or possibility of future evidence. Our best evidence points to the specified complexity (and therefore design) of the bacterial flagellum. It is therefore incumbent on the scientific community to admit, at least provisionally, that the bacterial flagellum could be the product of design. Might there be biological examples for which the claim that they exhibit specified complexity is even more assertible? Yes, there might. Unlike truth, assertibility comes in degrees, corresponding to the strength of evidence that justifies a claim. Yet even now, the claim that the bacterial flagellum exhibits specified complexity is eminently assertible.

Evolutionary biology's only recourse for avoiding a design conclusion in instances like this is to look to unknown mechanisms (or known mechanisms operating in unknown ways) to overturn what our best evidence to date indicates is both complex and specified. As far as the evolutionary biologists are concerned, design theorists have failed to take into account indirect Darwinian pathways by which the bacterial flagellum might have evolved through a series of intermediate systems that changed function and structure over time in ways that we do not yet understand. But is it that we do not yet understand the indirect Darwinian evolution of the bacterial flagellum or that it never happened that way in the first place? At this point, there is simply no evidence for such indirect Darwinian evolutionary pathways to account for biological systems such as the bacterial flagellum.

There is further reason to be skeptical of evolutionary biology's general strategy for defeating Intelligent Design by looking to unknown material mechanisms. In the case of the bacterial flagellum, what keeps evolutionary biology afloat is the possibility of indirect Darwinian pathways that might account for it. Practically speaking, this means that even though no slight modification of a bacterial flagellum can continue to serve as a motility structure, a slight modification might serve some other function. But there is now mounting evidence of biological systems for which any slight modification not only destroys the system's existing function but also destroys the possibility of any functioning of the system whatsoever (see Axe 2000). Neither direct nor indirect Darwinian pathways could account for such systems. In that case, we would be dealing with an in-principle argument showing not only that no known material mechanism is capable of accounting for the system but also that no unknown material mechanism is capable of accounting for it. Specified complexity's assertibility in such cases would thus be even greater than in the case of the bacterial flagellum.

It is possible to rule out unknown material mechanisms once and for all provided one has independent reasons for thinking that explanations based on known material mechanisms cannot be overturned by yet-to-be-identified unknown mechanisms. Such independent reasons typically take the form of arguments from contingency that invoke numerous degrees of freedom. Thus, to establish that no material mechanism explains a phenomenon, we must establish that it is compatible with the known material mechanisms involved in its production but that these mechanisms also permit any number of alternatives to the phenomenon. By being compatible with but not required by the known material mechanisms involved in its production, a phenomenon becomes irreducible not only to the known mechanisms but also to any unknown mechanisms. How so? Because known material mechanisms can tell us conclusively that a phenomenon is contingent and allows full degrees of freedom. Any unknown mechanism would therefore have to respect that contingency and allow for the degrees of freedom already discovered.

Consider, for instance, a configuration space comprising all possible character sequences from a fixed alphabet (such spaces model not only written texts but also polymers such as DNA, RNA, and proteins). Configuration spaces such as this are perfectly homogeneous, with one character string geometrically interchangeable with the next. The geometry therefore precludes any underlying mechanisms from distinguishing or preferring some character strings over others. Not material mechanisms but external semantic information (in the case of written texts) or functional information (in the case of biopolymers) is needed to generate specified complexity in these instances. Arguing that this semantic or functional information reduces to material mechanisms is like arguing that Scrabble pieces have inherent in them preferential ways in which they like to be sequenced. They don't. Michael Polanyi (1967, 1968) made such arguments for biological design in the 1960s. Stephen Meyer (2003) has updated them more recently.

7. ELIMINATIVE INDUCTION

To attribute specified complexity to a biological system is to engage in an eliminative induction. Eliminative inductions depend on successfully falsifying competing hypotheses (contrast this with Popperian falsification, where hypotheses are corroborated to the degree that they successfully withstand attempts to falsify them). Now, for many design skeptics, eliminative inductions are mere arguments from ignorance, that is, arguments for the truth of a proposition because it has not been shown to be false. In arguments from ignorance, the lack of evidence for a proposition is used to argue for its truth. A stereotypical argument from ignorance goes something like "Ghosts and goblins exist because you haven't shown me that they don't exist."

But that's clearly not what eliminative inductions are doing. Eliminative inductions argue that competitors to the proposition in question are false. Provided that the proposition and its competitors together form a mutually exclusive and exhaustive class, eliminating all the competitors entails that the proposition is true. This is the ideal case, in which eliminative inductions in fact become deductions. The problem is that, in practice, we don't have a neat ordering of competitors that can then all be knocked down with a few straightforward and judicious blows (like pins in a bowling alley). The philosopher of science John Earman (1992, 165) puts it this way:

The eliminative inductivist [seems to be] in a position analogous to that of Zeno's archer whose arrow can never reach the target, for faced with an infinite number of hypotheses, he can eliminate one, then two, then three, etc., but no matter how long he labors, he will never get down to just one. Indeed, it is as if the arrow never gets half way, or a quarter way, etc. to the target, since however long the eliminativist labors, he will always be faced with an infinite list [of remaining hypotheses to eliminate].

Earman offers these remarks in a chapter titled "A Plea for Eliminative Induction." He himself thinks there is a legitimate and necessary place for eliminative induction in scientific practice. What, then, does he make of this criticism? Here is how he handles it (Earman 1992, 165):

My response on behalf of the eliminativist has two parts. (1) Elimination need not proceed in such a plodding fashion, for the alternatives may be so ordered that an infinite number can be eliminated in one blow. (2) Even if we never get down to a single hypothesis, progress occurs if we succeed in eliminating finite or infinite chunks of the possibility space. This presupposes, of course, that we have some kind of measure, or at least topology, on the space of possibilities.

To this, Earman (1992, 177) adds that eliminative inductions are typically *local inductions*, in which there is no pretense of considering all logically possible hypotheses. Rather, there is tacit agreement on the explanatory domain of the hypotheses as well as on which auxiliary hypotheses may be used in constructing explanations.

In ending this chapter, I want to reflect on Earman's claim that eliminative inductions can be *progressive*. Too often, critics of Intelligent Design charge specified complexity with underwriting a purely negative form of argumentation. But that charge is not accurate. The argument for the specified complexity of the bacterial flagellum, for instance, makes a positive contribution to our understanding of the limitations that natural mechanisms face in trying to account for it. Eliminative inductions, like all inductions and indeed all scientific claims, are fallible. But they must have a place in science. To refuse them, as evolutionary biology tacitly does by rejecting specified complexity as a criterion for detecting design, does not keep science safe from disreputable influences but instead undermines scientific inquiry itself.

The way things stand now, evolutionary biology allows Intelligent Design only to fail but not to succeed. If evolutionary biologists can discover or construct detailed, testable, indirect Darwinian pathways that account for complex biological systems such as the bacterial flagellum, then Intelligent Design will rightly fail. On the other hand, evolutionary biology makes it effectively impossible for Intelligent Design to succeed. According to evolutionary biology, Intelligent Design has only one way to succeed, namely, by showing that complex specified biological structures could not have evolved via any material mechanism. In other words, so long as some unknown material mechanism might have evolved the structure in question, Intelligent Design is proscribed.

Evolutionary theory is thereby rendered immune to disconfirmation in principle, because the universe of unknown material mechanisms can never be exhausted. Furthermore, the evolutionist has no burden of evidence. Instead, the burden of evidence is shifted entirely to the evolutionary skeptic. And what is required of the skeptic? The skeptic must establish a universal negative not by an eliminative induction (such inductions are invariably local and constrained) but by an exhaustive search and elimination of all conceivable possibilities – however remote, however unfounded, however unsupported by evidence. That is not how science is supposed to work.

Science is supposed to give the full range of possible explanations a fair chance to succeed. That's not to say that anything goes; but it is to say that anything might go. In particular, science may not, by a priori fiat, rule out logical possibilities. Evolutionary biology, by limiting itself exclusively to material mechanisms, has settled in advance the question of which biological explanations are true, apart from any consideration of empirical evidence. This is armchair philosophy. Intelligent Design may not be correct. But the only way we could discover that is by admitting design as a real possibility, not by ruling it out a priori. Darwin (1859, 2) himself would have agreed. In the *Origin of Species*, he wrote: "A fair result can be obtained only by fully stating and balancing the facts and arguments on both sides of each question."

References

Axe, D. 2000. Extreme functional sensitivity to conservative amino acid changes on enzyme exteriors. *Journal of Molecular Biology* 301: 585–95.

Borel, E. 1962. *Probabilities and Life*, trans. M. Baudin. New York: Dover.

Chaitin, G. J. 1966. On the length of programs for computing finite binary sequences. *Journal of the Association for Computing Machinery* 13: 547–69.

Dam, K. W., and H. S. Lin (eds.) 1996. *Cryptography's Role in Securing the Information Society*. Washington, DC: National Academy Press.

Darwin, C. [1859] 1964. *On the Origin of Species*, facsimile 1st ed. Cambridge, MA: Harvard University Press.

Davies, P. 1999. *The Fifth Miracle*. New York: Simon and Schuster.

Dawkins, R. 1996. *Climbing Mount Improbable*. New York: Norton.

Dembski, W. A. 1991. Randomness by design. *Nous* 25(1): 75–106.

⎯⎯ 1998a. Randomness. In *The Routledge Encyclopedia of Philosophy*, ed. E. Craig. London: Routledge.

⎯⎯ 1998b. *The Design Inference: Eliminating Chance through Small Probabilities*. Cambridge: Cambridge University Press.

⎯⎯ 2002. *No Free Lunch: Why Specified Complexity Cannot Be Purchased without Intelligence*. Lanham, MD: Rowman and Littlefield.

Earman, J. 1992. *Bayes or Bust? A Critical Examination of Bayesian Confirmation Theory*. Cambridge, MA: MIT Press.

Fisher, R. A. 1935. *The Design of Experiments*. New York: Hafner.

Jackson, F. 1987. *Conditionals*. Oxford: Blackwell.

Kauffman, S. 2000. *Investigations*. New York: Oxford University Press.

Knobloch, E. 1987. Emile Borel as a probabilist. In *The Probabilistic Revolution*, Vol. 1, ed. L. Krüger, L. J. Daston, and M. Heidelberger. 215–33. Cambridge, MA: MIT Press, pp. 215–33.

Kolmogorov, A. 1965. Three approaches to the quantitative definition of information. *Problemy Peredachi Informatsii* (in translation) 1(1): 3–11.

Lloyd, S. 2002. Computational capacity of the universe. *Physical Review Letters* 88(23): 7901–4.

McKeon, R. (ed.) 1941. *The Basic Works of Aristotle*. New York: Random House.

Meyer, S. C. 2003. DNA and the origin of life: Information, specification, and explanation. In *Darwinism, Design and Public Education*, ed. J. A. Campbell and S. C. Meyer, Lansing: Michigan State University Press.

Orgel, L. 1973. *The Origins of Life*. New York: Wiley.

Polanyi, M. 1967. Life transcending physics and chemistry. *Chemical and Engineering News* 45: 54–66.

⎯⎯ 1968. Life's irreducible structure. *Science* 113: 1308–12.

18

Information, Entropy, and the Origin of Life

Walter L. Bradley

1. INTRODUCTION

Darwin's theory of evolution and the development of the Second Law of Thermodynamics by Clausius, Maxwell, Boltzmann, and Gibbs are two of the three major scientific discoveries of the nineteenth century. Maxwell's field equations for electricity and magnetism are the third. The laws of thermodynamics have had a unifying effect in the physical sciences similar to that of the theory of evolution in the life sciences. What is intriguing is that the predictions of one seem to contradict the predictions of the other. The Second Law of Thermodynamics suggests a progression from order to disorder, from complexity to simplicity, in the physical universe. Yet biological evolution involves a hierarchical progression to increasingly complex forms of living systems, seemingly in contradiction to the Second Law of Thermodynamics.

In his great book *The Nature of the Physical World*, Arthur Eddington (1928, 74) says, "If your theory is found to be against the second law of thermodynamics, I can give you no hope; there is nothing for it but to collapse in deepest humiliation." But while nonliving systems dutifully obey the Second Law of Thermodynamics, living systems seem to live in defiance of it. In fact, this is one of the simplest ways of distinguishing living from nonliving systems. Molton (1978, 147) defines life as "regions of order that use energy to maintain their organization against the disruptive force of entropy."

But how is this possible? Lila Gatlin (1972, 1) says, "Life may be defined operationally as an information processing system – a structural hierarchy of functioning units – that has acquired through evolution the ability to store and process the information necessary for its own accurate reproduction." In his classic book *What Is Life?* (1944), Erwin Schroedinger insightfully noted that living systems are characterized by highly ordered, aperiodic structures that survive by continually drawing "negentropy" from their environment and "feeding" on it. Schroedinger used the term "negentropy" to refer to energy that was suitable for utilization by living systems, such as radiant energy and energy-rich compounds. Schroedinger's "highly ordered, aperiodic structures" we recognize today as the informational biopolymers of life – DNA, RNA, and protein. A half-century later, Schroedinger's seminal insights have been confirmed.

If these scientists are right, the characteristic feature of life appears to be its capacity, through the use of information, to survive and exist in a nonequilibrium state, resisting the pull toward equilibrium that is described by the Second Law of Thermodynamics. For them, the origin of life is nothing more or less than the emergence of sufficient biological information to enable a system of biopolymers to (1) store information, (2) replicate with very occasional mistakes, and (3) "feed on negentropy." Unlike biological evolution, where it is fashionable to believe that there is sufficient creative power in mutation combined with natural selection to account for the diversity of life in the biosphere, it is generally recognized that the origin of life is one of the great unsolved mysteries in science (Radetsky1992; Wade 2000).

At the heart of this mystery is the generation of the critical information that is necessary to provide the three life functions just mentioned, in a world in which the Second Law of Thermodynamics seems to naturally move systems in the opposite direction, toward greater randomness. This chapter will begin with a brief introduction to information theory, beginning with the early work of Shannon (1948). This will allow us to quantify the information in biopolymers – especially DNA, RNA, and protein, the molecules that are essential for information storage, replication, and metabolism. Then we will explore the concept of entropy and its ubiquitous increase in nature, usually called the Second Law of Thermodynamics. This will allow us to understand how living systems are able to sustain themselves against the downward pull of the Second Law of Thermodynamics and how thermodynamics affects the origin of information-rich, living systems. Finally, we will explore various scenarios that have been proposed to account for the significant quantity of information that is essential for the emergence of life in a world that so naturally consumes rather than creates information.

2. QUANTIFYING THE INFORMATION IN BIOPOLYMERS

Information theory was developed in 1948 by Claude Shannon of the Bell Laboratories to address issues in communications. However, his approach has found much broader application in many other areas, including the life sciences. Shannon's initial interest was in quantifying the transmission of information, which he considered to be contained in a series of symbols, like letters in an alphabet. For reasons clearly explained in his book, Shannon chose to quantify the information "i" per register (or position) in his message as

$$i = K \log W \tag{1a}$$

where W is the total number of symbols or letters being used to create the message. If each symbol or letter used in his message is equally probable, then the probability of any given symbol is given by $p_i = 1/W$ or $W = 1/p_i$, and

$$i = K \log (1/p_i) = -K \log p_i \tag{1b}$$

In order to express this information in bits, let $K = 1$ and use log to the base 2, or $\log_2$. Equation 1b becomes

$$i = -\log_2 p_i \tag{2}$$

If the probabilities of each symbol are not equal, then Equation 2 becomes

$$i = -\sum p_i \log_2 p_i \tag{3}$$

Shannon Information in DNA. Information in living systems is stored in the DNA molecule, which has four bases called nucleotides that effectively serve as an alphabet of four letters: A-adenine, T-thymine, C-cytosine, and G-guanine. In *E. coli* bacteria, these bases appear equally often, such that $p_i = \frac{1}{4}$ for each one. Thus, using Equation 2, we may calculate the information per nucleotide to be

$$i = -\log_2 (\tfrac{1}{4}) = 2 \text{ bits} \tag{4}$$

Since there are 4×10^6 nucleotides in the DNA of *E. coli* bacteria (Gatlin 1972, 34), the total amount of Shannon information would be

$$I_s = N \bullet i = 4 \times 10^6 \times 2 = 8 \times 10^6 \text{ bits of information} \tag{5}$$

The total Shannon information "I_s" represents the number of binary decisions that must be made in order to get any sequence of base nucleotides in DNA. It is simple (at least in principle) to calculate the number of different messages (or sequences) that one might create in a polynucleotide with 4×10^6 bases such as the polynucleotide in *E. coli*. The total number of unique messages "M" that can be formed from 4×10^6 binary decisions is given by

$$M = 2^{I_s} = 2^{8,000,000} = 10^{2,400,000} \tag{6}$$

For comparison, the typing on this page requires only 10^4 bits of information, so the *E. coli* DNA has information equivalent to $8 \times 10^6/10^4 = 800$ pages like this one. It must be emphasized that each of the $10^{2,400,000}$ alternative sequences or messages in Equation 6 contains the same amount of structural, or syntactic, information – namely, 8,000,000 bits. Yet only a few of

these sequences or messages carry biologically "meaningful" information –
that is, information that can guarantee the functional order of the bacterial
cell (Küppers 1990, 48).

If we consider *Micrococcus lysodeikticus*, the probabilities for the various
nucleotide bases are no longer equal: $p(C) = p(G) = 0.355$ and $p(T) =
p(A) = 0.145$, with the sum of the four probabilities adding to 1.0, as they
must. Using Equation 3, we may calculate the information "i" per nucleotide
as follows:

$$i = -(0.355 \log_2 0.355 + 0.355 \log_2 0.355 + 0.145 \log_2 0.145 +$$
$$0.145 \log_2 0.145) = 1.87 \text{ bits} \qquad (7)$$

Comparing the results from Equation 4 for equally probable symbols and
from Equation 7 for unequally probable symbols illustrates a general point;
namely, that the greatest information is carried when the symbols are equally
probable. If the symbols are not equally probably, then the information per
symbol is reduced accordingly.

Factors Influencing Shannon Information in Any Symbolic Language. The English
language can be used to illustrate this point further. We may consider English
to have twenty-seven symbols – twenty-six letters plus a "space" as a symbol.
If all of the letters were to occur equally frequently in sentences, then the
information per symbol (letter or space) may be calculated, using Equation
2, to be

$$i = -\log_2(1/27) = 4.76 \text{ bits/symbol} \qquad (8)$$

If we use the actual probabilities for these symbols' occurring in sentences
(e.g., space $= 0.2$; E $= 0.105$; A $= 0.63$; Z $= 0.001$), using data from Brillouin
(1962, 5), in Equation 3, then

$$i = 4.03 \text{ bits/symbol} \qquad (9)$$

Since the sequence of letters in English is not random, one can further
refine these calculations by including the nearest-neighbor influences (or
constraints) on sequencing. One finds that

$$i = 3.32 \text{ bits/symbol} \qquad (10)$$

These three calculations illustrate a second interesting point – namely, that
any factors that constrain a series of symbols (i.e., symbols not equally probable, nearest-neighbor influence, second-nearest-neighbor influence, etc.)
will reduce the Shannon information per bit and the number of unique
messages that can be formed in a series of these symbols.

Understanding the Subtleties of Shannon Information. Information can be
thought of in at least two ways. First, we can think of syntactic information,

which has to do only with the structural relationship between characters. Shannon information is only syntactic. Two sequences of English letters can have identical Shannon information "N • i," with one being a beautiful poem by Donne and the other being gibberish. Shannon information is a measure of one's freedom of choice when one selects a message, measured as the $\log_2$ (number of choices). Shannon and Weaver (1964, 27) note,

> The concept of information developed in this theory at first seems disappointing and bizarre – disappointing because it has nothing to do with meaning (or function in biological systems) and bizarre because it deals not with a single message but with a statistical ensemble of messages, bizarre also because in these statistical terms, the two words information and uncertainty find themselves as partners.

Gatlin (1972, 25) adds that Shannon information may be thought of as a measure of information capacity in a given sequence of symbols. Brillouin (1956, 1) describes Shannon information as a measure of the effort to specify a particular message or sequence, with greater uncertainty requiring greater effort. MacKay (1983, 475) says that Shannon information quantifies the uncertainty in a sequence of symbols. If one is interested in messages with meaning – in our case, biological function – then the Shannon information does not capture the story of interest very well.

Complex Specified Information. Orgel (1973, 189) introduced the idea of *complex specified information* in the following way. In order to describe a crystal, one would need only to specify the substance to be used and the way in which the molecules were packed together (i.e., specify the unit cell). A couple of sentences would suffice, followed by the instructions "and keep on doing the same thing," since the packing sequence in a crystal is regular. The instructions required to make a polynucleotide with any random sequence would be similarly brief. Here one would need only to specify the proportions of the four nucleotides to be incorporated into the polymer and provide instructions to assemble them randomly. The crystal is specified but not very complex. The random polymer is complex but not specified. The set of instruction required for each is only a few sentences. It is this set of instructions that we identify as the *complex specified information* for a particular polymer.

By contrast, it would be impossible to produce a correspondingly simple set of instructions that would enable a chemist to synthesize the DNA of *E. coli* bacteria. In this case, the sequence matters! Only by specifying the sequence letter by letter (about 4,600,000 instructions) could we tell a chemist what to make. It would take 800 pages of instructions consisting of typing like that on this page (compared to a few sentences for a crystal or a random polynucleotide) to make such a specification, with no way to shorten it. The DNA of *E. coli* has a huge amount of *complex specified information*.

Brillouin (1956, 3) generalizes Shannon's information to cover the case where the total number of possible messages is W_o and the number of functional messages is W_1. Assuming the complex specified information is effectively zero for the random case (i.e., W_o calculated with no specifications or constraints), Brillouin then calculates the *complex specified information*, I_{CSI}, to be:

$$I_{CSI} = \log_2 (W_o/W_1) \tag{11}$$

For information-rich biological polymers such as DNA and protein, one may assume with Brillouin (1956, 3) that the number of ways in which the polynucleotides or polypeptides can be sequenced is extremely large (W_o). The number of sequences that will provide biological function will, by comparison, be quite small (W_1). Thus, the number of specifications needed to get such a functional biopolymer will be extremely high. The greater the number of specifications, the greater the constraints on permissible sequences, ruling out most of the possibilities from the very large set of random sequences that give no function, and leaving W_1 necessarily small.

Calculating the Complex Specified Information in the Cytochrome c Protein Molecule. If one assembles a random sequence of the twenty common amino acids in proteins into a polymer chain of 110 amino acids, each with $p_i = .05$, then the average information "I" per amino acid is given by Equation 2; it is $\log_2(20) = 4.32$. The total Shannon information is given by $I = N \cdot i = 110 \cdot 4.32 = 475$. The total number of unique sequences that are possible for this polypeptide is given by Equation 6 to be

$$M = 2^1 = 2^{475} \cong 10^{143} = W_o \tag{12}$$

It turns out that the amino acids in cytochrome c are not equiprobable ($p_i = 0.05$) as assumed earlier. If one takes the actual probabilities of occurrence of the amino acids in cytochrome c, one may calculate the average information per residue (or link in our 110-link polymer chain) to be 4.139, using Equation 3, with the total information being given by $I = N \cdot i = 4.139 \times 110 = 455$. The total number of unique sequences that are possible for this case is given by Equation 6 to be

$$M = 2^{455} = 1.85 \times 10^{137} = W_o \tag{13}$$

Comparison of Equation 12 to Equation 13 illustrates again the principle that the maximum number of sequences is possible when the probabilities of occurrence of the various amino acids in the protein are equal.

Next, let's calculate the number of sequences that actually give a functional cytochrome c protein molecule. One might be tempted to assume that only one sequence will give the requisite biological function. However, this is not so. Functional cytochrome c has been found to allow more than one amino acid to occur at some residue sites (links in my 110-link polymer chain). Taking this flexibility (or interchangeability) into account, Yockey (1992, 242–58) has provided a rather more exacting calculation of the information required to make the protein cytochrome c. Yockey calculates the total Shannon information for these functional cytochrome c proteins to be 310 bits, from which he calculates the number of sequences of amino acids that give a functional cytochrome c molecule:

$$M = 2^{310} = 2.1 \times 10^{93} = W_1 \tag{14}$$

This result implies that, on average, there are approximately three amino acids out of twenty that can be used interchangeably at each of the 110 sites and still give a functional cytochrome c protein. The chance of finding a functional cytochrome c protein in a prebiotic soup of randomly sequenced polypeptides would be:

$$W_1/W_o = 2.1 \times 10^{93}/1.85 \times 10^{137} = 1.14 \times 10^{-44} \tag{15}$$

This calculation assumes that there is no intersymbol influence – that is, that sequencing is not the result of dipeptide bonding preferences. Experimental support for this assumption will be discussed in the next section (Kok, Taylor, and Bradley 1988; Yeas 1969). The calculation also ignores the problem of chirality, or the use of exclusively left-handed amino acids in functional protein. In order to correct this shortcoming, Yockey repeats his calculation assuming a prebiotic soup with thirty-nine amino acids, nineteen with a left-handed and nineteen with a right-handed structures, assumed to be of equal concentration, and glysine, which is symmetric. W_1 is calculated to be 4.26×10^{62} and $P = W_1/W_o = 4.26 \times 10^{62}/1.85 \times 10^{137} = 2.3 \times 10^{-75}$. It is clear that finding a functional cytochrome c molecule in the prebiotic soup is an exercise in futility.

Two recent experimental studies on other proteins have found the same incredibly low probabilities for accidental formation of a functional protein that Yockey found; namely, 1 in 10^{75} (Strait and Dewey 1996) and 1 in 10^{63} (Bowie et al. 1990). All three results argue against any significant nearest-neighbor influence in the sequencing of amino acids in proteins, since this would make the sequencing much less random and the probability of formation of a functional protein much higher. In the absence of such intrinsic sequencing, the probability of accidental formation of a functional protein is incredibly low. The situation for accidental formation of functional polynucleotides (RNA or DNA) is much worse than for proteins, since the total information content is much higher (e.g., $\sim 8 \times 10^6$ bits for *E. coli* DNA versus 455 bits for the protein cytochrome c).

Finally, we may calculate the complex specified information, I_{CSI}, necessary to produce a functional cytochrome c by utilizing the results of Equation

15 in Equation 11, as follows:

$$I_{CSI} = \log_2 (1.85 \times 10^{137}/2.1 \times 10^{93}) = 146 \text{ bits of information, or}$$

$$I_{CSI} = \log_2 (1.85 \times 10^{137}/4.26 \times 10^{62}) = 248 \text{ bits of information} \qquad (16)$$

The second of these equations includes chirality in the calculation. It is this huge amount of complex specified information, I_{CSI}, that must be accounted for in many biopolymers in order to develop a credible origin-of-life scenario.

Summary. Shannon information, I_s, is a measure of the complexity of a biopolymer and quantifies the maximum capacity for complex specified information, I_{CSI}. Complex specified information measures the essential information that a biopolymer must have in order to store information, replicate, and metabolize. The complex specified information in a modest-sized protein such as cytochrome c is staggering, and one protein does not a first living system make. A much greater amount of information is encoded in DNA, which must instruct the production of all the proteins in the menagerie of molecules that constitute a simple living system. At the heart of the origin-of-life question is the source of this very, very significant amount of complex specified information in biopolymers. The role of the Second Law of Thermodynamics in either assisting or resisting the formation of such biopolymers that are rich in information will be considered next.

3. THE SECOND LAW OF THERMODYNAMICS AND THE ORIGIN OF LIFE

Introduction. "The law that entropy always increases – the 2nd Law of Thermodynamics – holds I think the supreme position among the laws of nature." So said Sir Arthur Eddington (1928, 74). If entropy is a measure of the disorder or disorganization of a system, this would seem to imply that the Second Law hinders if not precludes the origin of life, much like gravity prevents most animals from flying. At a minimum, the origin of life must be shown somehow to be compatible with the Second Law. However, it has recently become fashionable to argue that the Second Law is actually the driving force for abiotic as well as biotic evolution. For example, Wicken (1987, 5) says, "The emergence and evolution of life are phenomena causally connected with the Second Law." Brooks and Wiley (1988, xiv) indicate, "The axiomatic behavior of living systems should be increasing complexity and self-organization as a result of, not at the expense of increasing entropy." But how can this be?

What Is Entropy Macroscopically? The First Law of Thermodynamics is easy to understand: energy is always conserved. It is a simple accounting exercise.

When I burn wood, I convert chemical energy into thermal energy, but the total energy remains unchanged. The Second Law is much more subtle in that it tells us something about the nature of the available energy (and matter). It tells us something about the flow of energy, about the availability of energy to do work. At a macroscopic level, entropy is defined as

$$\Delta S = Q/T \qquad (17)$$

where S is the entropy of the system and Q is the heat or thermal energy that flows into or out of the system. In the wintertime, the Second Law of Thermodynamics dictates that heat flows from inside to outside your house. The resultant entropy change is

$$\Delta S = -Q/T_1 + Q/T_2 \qquad (18)$$

where T_1 and T_2 are the temperatures inside and outside your house. Conservation of energy, the First Law of Thermodynamics, tell us that the heat lost from your house $(-Q)$ must exactly equal the heat gained by the surroundings $(+Q)$. In the wintertime, the temperature inside the house is greater than the temperature outside $(T_1 > T_2)$, so that $\Delta S > 0$, or the entropy of the universe increases. In the summer, the temperature inside your house is lower than the temperature outside, and thus, the requirement that the entropy of the universe must increase means that heat must flow from the outside to the inside of your house. That is why people in Texas need a large amount of air conditioning to neutralize this heat flow and keep their houses cool despite the searing temperature outside. When people combust gasoline in their automobiles, chemical energy in the gasoline is converted into thermal energy as hot, high-pressure gas in the internal combustion engine, which does work and releases heat at a much lower temperature to the surroundings. The total energy is conserved, but the residual capacity of the energy that is released to do work on the surroundings is virtually nil.

Time's Arrow. In reversible processes, the entropy of the universe remains unchanged, while in irreversible processes, the entropy of the universe increases, moving from a less probable to a more probable state. This has been referred to as "time's arrow" and can be illustrated in everyday experience by our perceptions as we watch a movie. If you were to see a movie of a pendulum swinging, you could not tell the difference between the movie running forward and the movie running backward. Here potential energy is converted into kinetic energy in a completely reversible way (no increase in entropy), and no "arrow of time" is evident. But if you were to see a movie of a vase being dropped and shattered, you would readily recognize the difference between the movie running forward and running backward, since the shattering of the vase represents a conversion of kinetic energy into the surface energy of the many pieces into which the vase is broken, a quite irreversible and energy-dissipative process.

What Is Entropy Microscopically? Boltzmann, building on the work of Maxwell, was the first to recognize that entropy can also be expressed microscopically, as follows:

$$S = k \log_e \Omega \qquad (19)$$

where k is Boltzmann's constant and Ω is the number of ways in which the system can be arranged. An orderly system can be arranged in only one or possibly a few ways, and thus would be said to have a small entropy. On the other hand, a disorderly system can be disorderly in many different ways and thus would have a high entropy. If "time's arrow" says that the total entropy of the universe is always increasing, then it is clear that the universe naturally goes from a more orderly to a less orderly state in the aggregate, as any housekeeper or gardener can confirm. The number of ways in which energy and/or matter can be arranged in a system can be calculated using statistics, as follows:

$$\Omega = N!/(a!b!c!......) \qquad (20)$$

where $a+b+c+ ..... = N$. As Brillouin (1956, 6) has demonstrated, starting with Equation 20 and using Stirling's approximation, it may be easily shown that

$$\log \Omega = -\sum p_i \log p_i \qquad (21)$$

where $p_1 = a/N$, $p_2 = b/N$, ... A comparison of Equations 19 and 21 for Boltzmann's thermodynamic entropy to Equations 1 and 3 for Shannon's information indicate that they are essentially identical, with an appropriate assignment of the constant K. It is for this reason that Shannon information is often referred to as Shannon entropy. However, K in Equation 1 should not to be confused with the Boltzmann's constant k in Equation 19. K is arbitrary and determines the unit of information to be used, whereas k has a value that is physically based and scales thermal energy in much the same way that Planck's constant "h" scales electromagnetic energy. Boltzmann's entropy measures the amount of uncertainty or disorder in a physical system – or, more precisely, the lack of information about the actual structure of the physical system. Shannon information measures the uncertainty in a message. Are Boltzmann entropy and Shannon entropy causally connected in any way? It is apparent that they are not.

The probability space for Boltzmann entropy, which is a measure of the number of ways in which mass and energy can be arranged in biopolymers, is quite different from the probability space for Shannon entropy, which focuses on the number of different messages that might be encoded on the biopolymer. According to Yockey (1992, 70), in order for Shannon and Boltzmann entropies to be causally connected, their two probability spaces would need to be either isomorphic or related by a code, which they are not. Wicken (1987, 21–33) makes a similar argument that these two entropies are conceptually distinct and not causally connected. Thus the Second Law cannot be the proximate cause for any observed changes in the Shannon information (or entropy) that determines the complexity of the biopolymer (via the polymerized length of the polymer chain) or the complex specified information having to do with the sequencing of the biopolymer.

Thermal and Configurational Entropy. The total entropy of a system is a measure of the number of ways in which the mass and the energy in the system can be distributed or arranged. The entropy of any living or nonliving system can be calculated by considering the total number of ways in which the energy and the matter can be arranged in the system, or

$$S = k \ln (\Omega_{th}\Omega_{conf}) = k \ln \Omega_{th} + k \ln \Omega_{conf} = \Delta S_{th} + \Delta S_c \quad (22)$$

with ΔS_{th} and ΔS_c equal to the thermal and configurational entropies, respectively. The atoms in a perfect crystal can be arranged in only one way, and thus it has a very low configurational entropy. A crystal with imperfections can be arranged in a variety of ways (i.e., various locations of the imperfections), and thus it has a higher configurational entropy. The Second Law would lead us to expect that crystals in nature will always have some imperfections, and they do. The change in configurational entropy is a force driving chemical reactions forward, though a relatively weak one, as we shall see presently. Imagine a chemical system that is comprised of fifty amino acids of type A and fifty amino acids of type B. What happens to the configurational entropy if two of these molecules chemically react? The total number of molecules in the systems drops from 100 to 99, with 49 A molecules, 49 B molecules, and a single A-B bipeptide. The change in configurational entropy is given by

$$S_{cf} - S_{co} = \Delta S_c = k \ln [99!/(49!49!1!)] - k \ln [100!/50!50!]$$
$$= k \ln (25) \quad (23)$$

The original configurational entropy S_{co} for this reaction can be calculated to be $k \ln 10^{29}$, so the driving force due to changes in configuration entropy is seen to be quite small. Furthermore, it decreases rapidly as the reaction goes forward, with $\Delta S_c = k \ln (12.1)$ and $\Delta S_c = k \ln (7.84)$ for the formation of the second and third dipeptides in the reaction just described. The thermal entropy also decreases as such polymerization reactions take place owing to the significant reduction in the availability of translational and rotational modes of thermal energy storage, giving a net decrease in the total entropy (configuration plus thermal) of the system. Only at the limit, as the yield goes to zero in a large system, does the entropic driving force for configurational entropy overcome the resistance to polymerization provided by the concurrent decrease in thermal entropy.

Wicken (1987) argues that configurational entropy is the driving force responsible for increasing the complexity, and therefore the information capacity, of biological polymers by driving polymerization forward and thus making longer polymer chains. It is in this sense that he argues that the Second Law is a driving force for abiotic as well as biotic evolution. But as noted earlier, this is only true for very, very trivial yields. The Second Law is at best a trivial driving force for complexity!

Thermodynamics of Isolated Systems. An isolated system is one that does not exchange either matter or energy with its surroundings. An idealized thermos jug (i.e., one that loses no heat to its surroundings), filled with a liquid and sealed, would be an example. In such a system, the entropy of the system must either stay constant or increase due to irreversible energy-dissipative processes taking place inside the thermos. Consider a thermos containing ice and water. The Second Law requires that, over time, the ice melts, which gives a more random arrangement of the mass and thermal energy, which is reflected in an increase in the thermal and configurational entropies.

The gradual spreading of the aroma of perfume in a room is an example of the increase in configurational entropy in a system. Your nose processes the gas molecules responsible for the perfume aroma as they spread spontaneously throughout the room, becoming randomly distributed. Note that the reverse does not happen. The Second Law requires that processes that are driven by an increase in entropy are not reversible.

It is clear that life cannot exist as an isolated system that monotonically increases its entropy, losing its complexity and returning to the simple components from which it was initially constructed. An isolated system is a dead system.

Thermodynamics of Open Systems. Open systems allow the free flow of mass and energy through them. Plants use radiant energy to convert carbon dioxide and water into sugars that are rich in chemical energy. The system of chemical reactions that gives photosynthesis is more complex, but effectively gives

$$6CO_2 + 6H_2O + \text{radiant energy} \rightarrow 6C_6H_{12}O_6 + 6O_2 \qquad (24)$$

Animals consume plant biomass and use this energy-rich material to maintain themselves against the downward pull of the Second Law. The total entropy change that takes place in an open system such as a living cell must be consistent with the Second Law of Thermodynamics and can be described as follows:

$$\Delta S_{cell} + \Delta S_{surroundings} > 0 \qquad (25)$$

The change in the entropy of the surroundings of the cell may be calculated as Q/T, where Q is positive if energy is released to the surroundings by exothermic reactions in the cell and Q is negative if heat is required from the surroundings due to endothermic reactions in the cell. Equation 25, which is a statement of the Second Law, may now be rewritten using Equation 22 to be

$$\Delta S_{th} + \Delta S_{conf} + Q/T > 0 \tag{26}$$

Consider the simple chemical reaction of hydrogen and nitrogen to produce ammonia. Equation 26, which is a statement of the Second Law, has the following values, expressed in entropy units, for the three terms:

$$-14.95 - 0.79 + 23.13 > 0 \tag{27}$$

Note that the thermal entropy term and the energy exchange term Q/T are quite large compared to the configurational entropy term, which in this case is even negative because the reaction is assumed to have a high yield. It is the large exothermic chemical reaction that drives this reaction forward, despite the resistance provided by the Second Law. This is why making amino acids in Miller-Urey-type experiments is as easy as getting water to run downhill, if and only if one uses energy-rich chemicals such as ammonia, methane, and hydrogen that combine in chemical reactions that are very exothermic (50–250 kcal/mole). On the other hand, attempts to make amino acids from water, nitrogen, and carbon dioxide give at best minuscule yields because the necessary chemical reactions collectively are endothermic, requiring an increase in energy of more than 50 kcal/mole, akin to getting water to run uphill. Electrical discharge and other sources of energy used in such experiments help to overcome the kinetic barriers to the chemical reaction but do not change the thermodynamic direction dictated by the Second Law.

Energy-Rich Chemical Reactants and Complexity. Imagine a pool table with a small dip or cup at the center of the table. In the absence of such a dip, one might expect the pool balls to be randomly positioned on the table after one has agitated the table for a short time. However, the dip will cause the pool balls to assume a distinctively nonrandom arrangement – all of them will be found in the dip at the center of the table. When we use the term "energy-rich" to describe molecules, we generally mean double covalent bonds that can be broken to give two single covalent bonds, with a more negative energy of interaction or a larger absolute value for the bonding energy. Energy-rich chemicals function like the dip in the pool table, causing a quite nonrandom outcome to the chemistry as reaction products are attracted into this chemical bonding energy "well," so to speak.

The formation of ice from water is a good example of this principle, with $Q/T = 80\,cal/gm$ and $\Delta S_{th} + \Delta S_{conf} = 0.29\ cal/K$ for the transition from water to ice. The randomizing influence of thermal energy drops sufficiently low at 273K to allow the bonding forces in water to draw the water molecules into a crystalline array. Thus water goes from a random to an orderly state due to a change in the bonding energy between water and ice – a bonding potential-energy well, so to speak. The release of the heat of fusion to the surroundings gives a greater increase in the entropy of the surroundings than the entropy decrease associated with the ice formation. So the entropy of the universe does increase as demanded by the Second Law, even as ice freezes.

Energy-rich Biomass. Polymerization of biopolymers such as DNA and protein in living systems is driven by the consumption of energy-rich reactants (often in coupled chemical reactions). The resultant biopolymers themselves are less rich than the reactants, but still much more energy-rich than the equilibrium chemical mixture to which they can decompose – and will decompose, if cells or the whole system dies. Sustaining living systems in this nonequilibrium state is analogous to keeping a house warm on a cold winter's night. Living systems also require a continuous source of energy, either from radiation or from biomass, and metabolic "machinery" that functions in a way analogous to the heater in a house. Morowitz (1968) has estimated that *E. coli* bacteria have an average energy from chemical bonding of .27eV/atom greater (or richer) than the simple compounds from which the bacteria is formed. As with a hot house on a cold winter's night, the Second Law says that living systems are continuously being pulled toward equilibrium. Only the continuous flow of energy through the cell (functioning like the furnace in a house) can maintain cells at these higher energies.

Summary. Informational biopolymers direct photosynthesis in plants and the metabolism of energy-rich biomass in animals that make possible the cell's "levitation" above chemical equilibrium and physical death. Chemical reactions that form biomonomers and biopolymers require exothermic chemical reactions in order to go forward, sometimes assisted in a minor way by an increase in the configurational entropy (also known as the law of mass action) and resisted by much larger decreases in the thermal entropy. At best, the Second Law of Thermodynamics gives an extremely small yield of unsequenced polymers that have no biological function. Decent yields required exothermic chemical reactions, which are not available for some critical biopolymers. Finally, Shannon (informational) entropy and Boltzmann (thermodynamic) entropy are not causally connected, meaning in practice that the sequencing needed to get functional biopolymers is not facilitated by the Second Law, a point that Wicken (1987) and Yockey (1992) have both previously made.

The Second Law is to the emergence of life what gravity is to flight, a challenge to be overcome. Energy flow is necessary to sustain the levitation of life above thermodynamic equilibrium but is not a sufficient cause for the formation of living systems. I find myself in agreement with Yockey's

(1977) characterization of thermodynamics as an "uninvited (and probably unwelcome) guest in emergence of life discussions." In the next section, we will critique the various proposals for the production of complex specified information in biopolymers that are essential to the origin of life.

4. CRITIQUE OF VARIOUS ORIGIN-OF-LIFE SCENARIOS

In this final section, we will critique major scenarios of how life began, using the insights from information theory and thermodynamics that have been developed in the preceding portion of this chapter. Any origin-of-life scenario must somehow explain the origin of molecules encoded with the necessary minimal functions of life. More specifically, the scenario must explain two major observations: (1) how very complex molecules such as polypeptides and polynucleotides that have large capacities for information came to be, and (2) how these molecules are encoded with complex specified information. All schemes in the technical literature use some combination of chance and necessity, or natural law. But they differ widely in the magnitude of chance that is invoked and in which natural law is emphasized as guiding or even driving the process part of this story. Each would seek to minimize the degree of chance that is involved. The use of the term "emergence of life," which is gradually replacing "origin of life," reflects this trend toward making the chance step(s) as small as possible, with natural processes doing most of the "heavy lifting."

Chance Models and Jacques Monod (1972). In his classic book *Chance and Necessity* (1972), Nobel laureate Jacques Monod argues that life began essentially by random fluctuations in the prebiotic soup that were subsequently acted upon by selection to generate information. He readily admits that life is such a remarkable accident that it is almost certainly occurred only once in the universe. For Monod, life is just a quirk of fate, the result of a blind lottery, much more the result of chance than of necessity. But in view of the overwhelming improbability of encoding DNA and protein to give functional biopolymers, Monod's reliance on chance is simply believing in a miracle by another name and cannot in any sense be construed as a rational explanation for the origin of life.

Replicator-first Models and Eigen and Winkler-Oswatitsch (1992). In his book *Steps toward Life*, Manfred Eigen seeks to demonstrate that the laws of nature can be shown to reduce significantly the improbability of the emergence of life, giving life a "believable" chance. Eigen and Winkler-Oswatitsch (1992, 11) argue that

[t]he genes found today cannot have arisen randomly, as it were by the throw of a dice. There must exist a process of optimization that works towards functional

efficiency. Even if there were several routes to optimal efficiency, mere trial and error cannot be one of them. . . . It is reasonable to ask how a gene, the sequence of which is one out of 10^{600} possible alternatives of the same length, copies itself spontaneously and reproducibly.

It is even more interesting to wonder how such a molecule emerged in the first place. Eigen's answer is that the emergence of life began with a self-replicating RNA molecule that, through mutation/natural selection over time, became increasingly optimized in its biochemical function. Thus, the information content of the first RNA is assumed to have been quite low, making this "low-tech start" much less chancy. The reasonableness of Eigen's approach depends entirely on how "low-tech" one can go and still have the necessary biological functions of information storage, replication with occasional (but not too frequent) replicating mistakes, and some meaningful basis for selection to guide the development of more molecular information over time.

Robert Shapiro, a Harvard-trained DNA chemist, has recently critiqued all RNA-first replicator models for the emergence of life (2000). He says,

A profound difficulty exists, however, with the idea of RNA, or any other replicator, at the start of life. Existing replicators can serve as templates for the synthesis of additional copies of themselves, but this device cannot be used for the preparation of the very first such molecule, which must arise spontaneously from an unorganized mixture. The formation of an information-bearing homopolymer through undirected chemical synthesis appears very improbable.

Shapiro then addresses various assembly problems and the problem of even getting all the building blocks, which he addresses elsewhere (1999).

Potentially an even more challenging problem than making a polynucleotide that is the precursor to a functional RNA is encoding it with enough information to direct the required functions. What kind of selection could possibly guide the encoding of the initial information required to "get started"? In the absence of some believable explanation, we are back to Monod's unbelievable chance beginning. Bernstein and Dillion (1997) have recently addressed this problem as follows.

Eigen has argued that natural selection itself represents an inherent form of self-organization and must necessarily yield increasing information content in living things. While this is a very appealing theoretical conclusion, it suffers, as do most reductionist theories, from the basic flaw that Eigen is unable to identify the source of the natural selection during the origin of life. By starting with the answer (an RNA world), he bypasses the nature of the question that had to precede it.

Many models other than Eigen's begin with replication first, but few address the origins of metabolism (see Dyson 1999), and all suffer from the same shortcomings as Eigen's hypercycle, assuming too complicated a starting point, too much chance, and not enough necessity. The fundamental question remains unresolved – namely, is genetic replication a necessary prerequisite for the emergence of life or just a consequence of it?

Metabolism-first Models of Wicken (1987), Fox (1984), and Dyson (1999). Sidney Fox has made a career of making and studying proteinoid microspheres. By heating dry amino acids to temperatures that drive off the water that is released as a byproduct of polymerization, he is able to polymerize amino acids into polypeptides, or polymer chains of amino acids. Proteinoid molecules differ from actual proteins in at least three significant (and probably critical) ways: (1) a significant percentage of the bonds are not the peptide bonds found in modern proteins; (2) proteinoids are comprised of a mixture of L and D amino acids, rather than of all L amino acids (like actual proteins); and (3) their amino acid sequencing gives little or no catalytic activity. It is somewhat difficult to imagine how such a group of "protein wannabes" that have attracted other "garbage" from solution and formed a quasi-membrane can have sufficient encoded information to provide any biological function, much less sufficient biological function to benefit from any imaginable kind of selection. Again, we are back to Monod's extreme dependence on chance.

Fox and Wicken have proposed a way out of this dilemma. Fox (1984, 16) contends that "[a] guiding principle of non-randomness has proved to be essential to understanding origins. . . . As a result of the new protobiological theory the neo-Darwinian formulation of evolution as the natural selection of random variations should be modified to the natural selection of non-random variants resulting from the synthesis of proteins and assemblies thereof." Wicken (1987) appeals repeatedly to inherent nonrandomness in polypeptides as the key to the emergence of life. Wicken recognizes that there is little likelihood of sufficient intrinsic nonrandomness in the sequencing of bases in RNA or DNA to provide any basis for biological function. Thus his hope is based on the possibility that variations in steric interference in amino acids might give rise to differences in the dipeptide bonding tendencies in various amino acid pairs. This could potentially give some nonrandomness in amino acid sequencing. But it is not just nonrandomness but complex specificity that is needed for function.

Wicken bases his hypothesis on early results published by Steinman and Cole (1967) and Steinman (1971), who claimed to show that dipeptide bond frequencies measured experimentally were nonrandom (some amino acids reacted preferentially with other amino acids) and that these nonrandom chemical bonding affinities are reflected in the dipeptide bonding frequencies in actual proteins, based on a study of the amino acid sequencing in ten protein molecules. Steinman subsequently coauthored a book with Kenyon (1969) titled *Biochemical Predestination* that argued that the necessary information for functional proteins was encoded in the relative chemical reactivities of the various amino acid "building blocks" themselves, which directed them to self-assemble into functional proteins. However, a much more comprehensive study by Kok, Taylor, and Bradley (1988), using the same approach but studying 250 proteins and employing a more rigorous statistical analysis, concluded that there was absolutely no correlation between the dipeptide bond frequencies measured by Steinman and Cole and the dipeptide bond frequencies found in actual proteins.

The studies by Yockey (1992), Strait and Dewey (1996), Sauer and Reidhaar-Olson (1990), and Kok, Taylor, and Bradley (1988) argue strongly from empirical evidence that one cannot explain the origin of metabolic behavior by bonding preferences. Their work argues convincingly that the sequencing that gives biological function to proteins cannot be explained by bipeptide bonding preferences. Thus the metabolism-first approach would appear to be back to the "chance" explanation of Monod (1972). It is also interesting to note that Kenyon has now repudiated his belief in biochemical predestination (Thaxton, Bradley, and Olsen 1992).

Self-organization in Systems Far from Equilibrium – Prigogine. Prigogine has received a Nobel Prize for his work on the behavior of chemical systems far from equilibrium. Using mathematical modeling and thoughtful experiments, he has demonstrated that systems of chemical reactions that have autocatalytic behavior resulting in nonlinear kinetics can have surprising self-organization (Nicolis and Prigogine 1977; Prigogine 1980; Prigogine and Stengers 1984). There is also a tendency for such systems to bifurcate when the imposed gradients reach critical values. However, the ordering produced in Prigogine's mathematical models and experiments seems to be of the same order of magnitude as the information implicit in the boundary conditions, proving once again that it is hard to get something for nothing. Second, the ordering observed in these systems has no resemblance to the specified complexity characteristic of living systems. Put another way, the complex specified information in such systems is quite modest compared to that of living systems. Thus Prigogine's approach also seems to fall short of providing any "necessity" for the emergence of life, leaving us again with Monod's chance explanation.

Complexity and the Work of Kauffman and the Sante Fe Institute. Kauffman defines "life" as a closed network of catalyzed chemical reactions that reproduce each molecule in the network – a self-maintaining and self-reproducing metabolism that does not require self-replicating molecules. Kauffman's ideas are based on computer simulations alone, without any experimental support. He claims that when a system of simple chemicals reaches a critical level of diversity and connectedness, it undergoes a dramatic transition, combining to create larger molecules of increasing complexity and catalytic capability – Kauffman's definition of life.

Such computer models ignore important aspects of physical reality that, if included in the models, would make the models not only more complicated but also incapable of the self-organizing behavior that is desired by the modelers. For example, Kauffman's origin-of-life model requires a critical diversity of molecules so that there is a high probability that the production of each molecule is catalyzed by another molecule. For example, he posits $1/1,000,000$ as the probability that a given molecule catalyzes the production of another molecule (which is too optimistic a probability, based on catalyst chemistry). If one has a system of 1,000,000 molecules, then in theory it becomes highly probable that most molecules are catalyzed in their production, at which point this catalytic closure causes the system to "catch fire" – in effect to come to life (Kauffman 1995, 64).

Einstein said that we want our models to be as simple as possible, but not too simple (i.e., ignoring important aspects of physical reality). Kauffman's model for the origin of life ignores critical thermodynamic and kinetic issues that, if included in his model, would kill his "living system." For example, there are huge kinetic transport issues in taking Kauffman's system with 1,000,000 different types of molecules, each of which can be catalyzed in its production by approximately one type of molecule, and organizing it in such a way that the catalyst that produces a given molecule will be in the right proximity to the necessary reactants to allow it to be effective. Kauffman's simple computer model ignores this enormous organizational problem that must precede the "spontaneous self-organization" of the system. Here he is assuming away (not solving) a system-level configurational entropy problem that is completely analogous to the molecular-level configurational entropy problem discussed in Thaxton, Bradley, and Olsen (1984). The models themselves seem to represent reality poorly, and the lack of experimental support makes Kauffman's approach even more speculative than the previous four, none of which seemed to be particularly promising.

5. SUMMARY

Biological life requires a system of biopolymers of sufficient specified complexity to store information, replicate with very occasional mistakes, and utilize energy flow to maintain the levitation of life above thermodynamic equilibrium and physical death. And there can be no possibility of information generation by the Maxwellian demon of natural selection until this significant quantity of complex specified information has been provided a priori. A quotation from Nicholas Wade, writing in the *New York Times* (June 13, 2000), nicely summarizes the dilemma of the origin of life:

The chemistry of the first life is a nightmare to explain. No one has yet developed a plausible explanation to show how the earliest chemicals of life – thought to be RNA – might have constructed themselves from the inorganic chemicals likely to have been around on early earth. The spontaneous assembly of a small RNA molecule on the primitive earth "would have been a near miracle," two experts in the subject helpfully declared last year.

The origin of life seems to be the ultimate example of irreducible complexity. I believe that cosmology and the origin of life provide the most compelling examples of Intelligent Design in nature. I am compelled to agree with the eloquent affirmation of design by Harold Morowitz (1987): "I find it hard not to see design in a universe that works so well. Each new scientific discovery seems to reinforce that vision of design. As I like to say to my friends, the universe works much better than we have any right to expect."

References

Bernstein, Robert S., and Patrick F. Dillon. 1997. Molecular complementarity I: The complementarity theory of the origin and evolution of life. *Journal of Theoretical Biology* 188: 447–79.

Bowie, J. U., J. F. Reidhaar-Olson, W. A. Lim, and R. T. Sauer. 1990. Deciphering the message in protein sequences: Tolerance to amino acid substitution. *Science* 247: 1306–10.

Brillouin, Leon. 1956. *Science and Information Theory*. New York: Academic Press.

Brooks, Daniel R., and E. O. Wiley. 1988. *Evolution as Entropy*, 2nd ed. Chicago: University of Chicago Press.

Dyson, Freeman. 1999. *Origins of Life*, 2nd ed. Cambridge: Cambridge University Press.

Eddington, A. S. 1928. *The Nature of the Physical World*. New York: Macmillan.

Eigen, Manfred, and Ruthild Winkler-Oswatitsch. 1992. *Steps toward Life*. Oxford: Oxford University Press.

Fox, Sidney. 1984. Proteinoid experiments and evolutionary theory. In *Beyond Neo-Darwinism: An Introduction to the New Evolutionary Paradigm*, ed. Mae-Wan Ho and Peter T. Saunders. New York: Academic Press.

Gatlin, Lila L. 1972. *Information Theory and the Living System*. New York: Columbia University Press.

Kenyon, Dean H., and Gary Steinman. 1969. *Biochemical Predestination*. New York: McGraw-Hill.

Kok, Randall, John A. Taylor, and Walter L. Bradley. 1988. A statistical examination of self ordering of amino acids in proteins. *Origins of Life and Evolution of the Biosphere* 18: 135–42.

Küppers, Bernd-Olaf. 1990. *Information and the Origin of Life*. Cambridge, MA: MIT Press.

Kauffman, Stuart. 1995. *At Home in the Universe: The Search for the Laws of Self-Organization and Complexity*. New York: Oxford University Press.

MacKay, Donald M. 1983. The wider scope of information theory. In *The Study of Information: Interdisciplinary Messages*, ed. Fritz Machlup and Una Mansfield. New York: Wiley.

Molton, Peter M. 1978. Polymers to living cells: Molecules against entropy. *Journal of the British Interplanetary Society* 31: 147.

Monod, J. 1972. *Chance and Necessity.* New York: Vintage.

Morowitz, Harold J. 1968. *Energy Flow in Biology.* New York: Academic Press.

1987. *Cosmic Joy and Local Pain.* New York: Scribners.

Nicolis, G., and I. Prigogine. 1977. *Self-Organization in Nonequilibrium Systems.* New York: Wiley.

Orgel, Leslie E. 1973. *The Origins of Life.* New York: Wiley.

Prigogine, I. 1980. *From Being to Becoming.* San Francisco: W. H. Freeman.

Prigogine, I., and I. Strenger. 1984. *Order out of Chaos.* New York: Bantam.

Radetsky, Peter. 1992. How did life start? *Discover* 10 (November): 74.

Sauer, R. T., and J. F. Reidhaar-Olson. 1990. Functionally acceptable substitutions in two-helical regions of repressor. *Proteins: Structure, Function and Genetics* 7: 306–16.

Schroedinger, E. 1945. *What Is Life?* Cambridge: Cambridge University Press.

Shannon, Claude, and Warren Weaver. [1948] 1964. *The Mathematical Theory of Communication.* Urbana: University of Illinois Press.

Shapiro, Robert. 1999. Prebiotic cytosine synthesis: A critical analysis and implications for the origin of life. *Proceedings of the National Academy of Sciences (USA)* 96: 4396–401.

2000. A replicator was not involved in the origin of life. *IUBMB Life* 49: 173–6.

Steinman, G. 1971. Non-enzymic synthesis of biologically pertinent peptides. In *Prebiotic and Biochemical Evolution,* ed. A. Kimball and J. Oro. New York: Elsevier.

Steinman, G., and M. Cole. 1967. *Proceedings of the National Academy of Sciences (USA)* 58: 735.

Strait, Bonnie J., and Gregory T. Dewey. 1996. The Shannon information entropy of protein sequences. *Biophysical Journal* 71: 148–55.

Thaxton, Charles B., Walter L. Bradley, and Roger L. Olsen. [1984] 1992. *The Mystery of Life's Origin: Reassessing Current Theories.* Dallas: Lewis and Stanley.

Wicken, Jeffrey S. 1987. *Evolution, Thermodynamics, and Information.* New York: Oxford University Press.

Yeas, M. 1969. *The Biological Code.* Amsterdam: North-Holland.

Yockey, H. 1977. A calculation of the probability of spontaneous biogenesis by information theory. *Journal of Theoretical Biology* 67: 377–98.

1992. *Information Theory in Molecular Biology.* Cambridge: Cambridge University Press.

19

Irreducible Complexity

Obstacle to Darwinian Evolution

Michael J. Behe

A SKETCH OF THE INTELLIGENT DESIGN HYPOTHESIS

In his seminal work *On the Origin of Species*, Darwin hoped to explain what
no one had been able to explain before – how the variety and complexity of
the living world might have been produced by simple natural laws. His idea
for doing so was, of course, the theory of evolution by natural selection. In a
nutshell, Darwin saw that there was variety in all species. For example, some
members of a species are bigger than others, some faster, some brighter
in color. He knew that not all organisms that are born will survive to reproduce, simply because there is not enough food to sustain them all. So
Darwin reasoned that the ones whose chance variation gives them an edge
in the struggle for life would tend to survive and leave offspring. If the variation could be inherited, then over time the characteristics of the species
would change, and over great periods of time, perhaps great changes could
occur.

It was an elegant idea, and many scientists of the time quickly saw that
it could explain many things about biology. However, there remained an
important reason for reserving judgment about whether it could actually
account for all of biology: the basis of life was as yet unknown. In Darwin's
day, atoms and molecules were still theoretical constructs – no one was sure
if such things actually existed. Many scientists of Darwin's era took the cell to
be a simple glob of protoplasm, something like a microscopic piece of Jell-O.
Thus the intricate molecular basis of life was utterly unknown to Darwin and
his contemporaries.

In the past hundred years, science has learned much more about the cell
and, especially in the past fifty years, much about the molecular basis of life.
The discoveries of the double helical structure of DNA, the genetic code,
the complicated, irregular structure of proteins, and much else have given
us a greater appreciation for the elaborate structures that are necessary to
sustain life. Indeed, we have seen that the cell is run by machines – literally,

machines made of molecules. There are molecular machines that enable the cell to move, machines that empower it to transport nutrients, machines that allow it to defend itself.

In light of the enormous progress made by science since Darwin first proposed his theory, it is reasonable to ask if the theory still seems to be a good explanation for life. In *Darwin's Black Box: The Biochemical Challenge to Evolution* (Behe 1996), I argued that it is not. The main difficulty for Darwinian mechanisms is that many systems in the cell are what I termed "irreducibly complex." I defined an irreducibly complex system as: a single system that is necessarily composed of several well-matched, interacting parts that contribute to the basic function, and where the removal of any one of the parts causes the system to effectively cease functioning (Behe 2001). As an example from everyday life of an irreducibly complex system, I pointed to a mechanical mousetrap such as one finds in a hardware store. Typically, such traps have a number of parts: a spring, a wooden platform, a hammer, and other pieces. If one removes a piece from the trap, it can't catch mice. Without the spring, or hammer, or any of the other pieces, one doesn't have a trap that works half as well as it used to, or a quarter as well; one has a broken mousetrap, which doesn't work at all.

Irreducibly complex systems seem very difficult to fit into a Darwinian framework, for a reason insisted upon by Darwin himself. In the *Origin*, Darwin wrote that "[i]f it could be demonstrated that any complex organ existed which could not possibly have been formed by numerous, successive, slight modifications, my theory would absolutely break down. But I can find out no such case" (Darwin 1859, 158). Here Darwin was emphasizing that his was a gradual theory. Natural selection had to improve systems by tiny steps, over a long period of time, because if things improved too rapidly, or in large steps, then it would begin to look as if something other than natural selection were driving the process. However, it is hard to see how something like a mousetrap could arise gradually by something akin to a Darwinian process. For example, a spring by itself, or a platform by itself, would not catch mice, and adding a piece to the first nonfunctioning piece wouldn't make a trap either. So it appears that irreducibly complex biological systems would present a considerable obstacle to Darwinian evolution.

The question then becomes, are there any irreducibly complex systems in the cell? Are there any irreducibly complex molecular machines? Yes, there are many. In *Darwin's Black Box*, I discussed several biochemical systems as examples of irreducible complexity: the eukaryotic cilium, the intracellular transport system, and more. Here I will just briefly describe the bacterial flagellum (DeRosier 1998; Shapiro 1995), since its structure makes the difficulty for Darwinian evolution easy to see (Figure 19.1). The flagellum can be thought of as an outboard motor that bacteria use to swim. It was the first truly rotary structure discovered in nature. It consists of a long filamentous

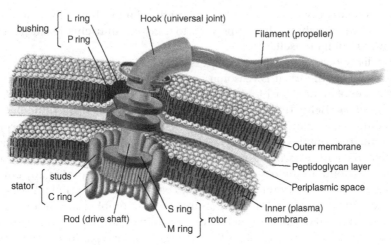

FIGURE 19.1. The bacterial flagellum. Reproduced from D. Voet and J. G. Voet, *Biochemistry*, 2nd ed. (New York: Wiley, 1995), Figure 34–84, with permission of John Wiley Publishers and Donald Voet, who wished to emphasize that "this is an artist-drawn representation of the flagellum rather than a photo or drawing of an actual flagellum."

tail that acts as a propeller; when it is spun, it pushes against the liquid medium and can propel the bacterium forward. The propeller is attached to the drive shaft indirectly through something called the hook region, which acts as a universal joint. The drive shaft is attached to the motor, which uses a flow of acid or sodium ions from the outside to the inside of the cell to power rotation. Just as an outboard motor has to be kept stationary on a motorboat while the propeller turns, there are proteins that act as a stator structure to keep the flagellum in place. Other proteins act as bushings to permit the drive shaft to pass through the bacterial membrane. Studies have shown that thirty to forty proteins are required to produce a functioning flagellum in the cell. About half of the proteins are components of the finished structure, while the others are necessary for the construction of the flagellum. In the absence of almost any of the proteins – in the absence of the parts that act as the propeller, drive shaft, hook, and so forth – no functioning flagellum is built.

As with the mousetrap, it is quite difficult to see how Darwin's gradualistic process of natural selection sifting random mutations could produce the bacterial flagellum, since many pieces are required before its function appears. A hook by itself, or a driveshaft by itself, will not act as a propulsive device. But the situation is actually much worse than it appears from this cursory description, for several reasons. First, there is associated with the functioning of the flagellum an intricate control system, which tells the flagellum when to rotate, when to stop, and sometimes when to reverse itself and rotate in the opposite direction. This allows the bacterium to swim toward or away from an appropriate signal, rather than in a random direction that could much more easily take it the wrong way. Thus the problem of accounting for the origin of the flagellum is not limited to the flagellum itself but extends to associated control systems as well.

Second, a more subtle problem is how the parts assemble themselves into a whole. The analogy to an outboard motor fails in one respect: an outboard motor is generally assembled under the direction of a human – an intelligent agent who can specify which parts are attached to which other parts. The information for assembling a bacterial flagellum, however (or, indeed, for assembling any biomolecular machine), resides in the component proteins of the structure itself. Recent work shows that the assembly process for a flagellum is exceedingly elegant and intricate (Yonekura et al. 2000). If that assembly information is absent from the proteins, then no flagellum is produced. Thus, even if we had a hypothetical cell in which proteins homologous to all of the parts of the flagellum were present (perhaps performing jobs other than propulsion) but were missing the information on how to assemble themselves into a flagellum, we would still not get the structure. The problem of irreducibility would remain.

Because of such considerations, I have concluded that Darwinian processes are not promising explanations for many biochemical systems in the cell. Instead, I have noted that, if one looks at the interactions of the components of the flagellum, or cilium, or other irreducibly complex cellular system, they look like they were designed – purposely designed by an intelligent agent. The features of the systems that indicate design are the same ones that stymie Darwinian explanations: the specific interaction of multiple components to accomplish a function that is beyond the individual components. The logical structure of the argument to design is a simple inductive one: whenever we see such highly specific interactions in our everyday world, whether in a mousetrap or elsewhere, we unfailingly find that the systems were intentionally arranged – that they were designed. Now we find systems of similar complexity in the cell. Since no other explanation has successfully addressed them, I argue that we should extend the induction to subsume molecular machines, and hypothesize that they were purposely designed.

MISCONCEPTIONS ABOUT WHAT A HYPOTHESIS OF DESIGN ENTAILS

The hypothesis of Intelligent Design (ID) is quite controversial, mostly because of its philosophical and theological overtones, and in the years since *Darwin's Black Box* was published a number of scientists and philosophers have tried to refute its main argument. I have found these rebuttals to be unpersuasive, at best. Quite the opposite, I think that some putative counterexamples to design are unintentionally instructive. Not only do they fail to make their case for the sufficiency of natural selection, they show clearly the obstacle that irreducible complexity poses to Darwinism. They also show that Darwinists have great trouble recognizing problems with their own theory. I will examine two of those counterexamples in detail a little later in this chapter. Before I do, however, I will first address a few common misconceptions that surround the biochemical design argument.

First of all, it is important to understand that a hypothesis of Intelligent Design has no quarrel with evolution per se – that is, evolution understood simply as descent with modification, but leaving the mechanism open. After all, a designer may have chosen to work that way. Rather than common descent, the focus of ID is on the *mechanism* of evolution – how did all this happen, by natural selection or by purposeful Intelligent Design?

A second point that is often overlooked but should be emphasized is that Intelligent Design can happily coexist with even a large degree of natural selection. Antibiotic and pesticide resistance, antifreeze proteins in fish and plants, and more may indeed be explained by a Darwinian mechanism. The critical claim of ID is not that natural selection doesn't explain *anything*, but that it doesn't explain *everything*.

My book, *Darwin's Black Box*, in which I flesh out the design argument, has been widely discussed in many publications. Although many issues have been raised, I think the general reaction of scientists to the design argument is well and succinctly summarized in the recent book *The Way of the Cell*, published by Oxford University Press and authored by the Colorado State University biochemist Franklin Harold. Citing my book, Harold writes, "We should reject, as a matter of principle, the substitution of intelligent design for the dialogue of chance and necessity (Behe 1996); but we must concede that there are presently no detailed Darwinian accounts of the evolution of any biochemical system, only a variety of wishful speculations" (Harold 2001, 205).

Let me emphasize, in reverse order, Harold's two points. First, as other reviewers of my book have done,[1] Harold acknowledges that Darwinists have no real explanation for the enormous complexity of the cell, only hand-waving speculations, more colloquially known as "just-so stories." I had claimed essentially the same thing six years earlier in *Darwin's Black Box* and encountered fierce resistance – mostly from internet fans of Darwinism who claimed that, why, there were hundreds or thousands of research papers describing the Darwinian evolution of irreducibly complex biochemical systems, and who set up web sites to document them.[2]

As a sufficient response to such claims, I will simply rely on Harold's statement quoted here, as well as the other reviewers who agree that there is a dearth of Darwinian explanations. After all, if prominent scientists who are no fans of Intelligent Design agree that the systems remain unexplained, then that should settle the matter. Let me pause, however, to note that I find this an astonishing admission for a theory that has dominated biology for so long. That Darwinian theory has borne so little fruit in explaining the molecular basis of life – despite its long reign as the fundamental theory of biology – strongly suggests that it is not the right framework for understanding the origin of the complexity of life.

Harold's second point is that there is some principle that forbids us from investigating Intelligent Design, even though design is an obvious idea that quickly pops into your mind when you see a drawing of the flagellum (Figure 19.1) or other complex biochemical system. What principle is that? He never spells it out, but I think the principle probably boils down to this: design appears to point strongly beyond nature. It has philosophical and theological implications, and that makes many people uncomfortable. Because they think that science should avoid a theory that points so strongly beyond nature, they want to rule out intelligent design from the start.

I completely disagree with that view and find it fainthearted. I think science should follow the evidence wherever it seems to lead. That is the only way to make progress. Furthermore, not only Intelligent Design, but *any* theory that purports to explain how life occurred will have philosophical and theological implications. For example, the Oxford biologist Richard Dawkins has famously said that "Darwin made it possible to be an intellectually-fulfilled atheist" (Dawkins 1986, 6). A little less famously, Kenneth Miller has written that "[God] used evolution as the tool to set us free" (Miller 1999, 253). Stuart Kauffman, a leading complexity theorist, thinks Darwinism cannot explain all of biology: "Darwinism is not enough. . . . [N]atural selection cannot be the sole source of order we see in the world" (Kauffman 1995, viii). But Kauffman thinks that his theory will somehow show that we are "at home in the universe." The point, then, is that all theories of origins carry philosophical and theological implications. There is no way to avoid them in an explanation of life.

Another source of difficulty for some people concerns the question, how could biochemical systems have been designed? A common misconception is that designed systems would have to be created from scratch in a puff of smoke. But that isn't necessarily so. The design process may have been much more subtle. In fact, it may have contravened no natural laws at all. Let's consider just one possibility. Suppose the designer is indeed God, as most people would suspect. Well, then, as Kenneth Miller points out in his book, *Finding Darwin's God*:

The indeterminate nature of quantum events would allow a clever and subtle God to influence events in ways that are profound, but scientifically undetectable to us. Those events could include the appearance of mutations . . . and even the survival of individual cells and organisms affected by the chance processes of radioactive decay. (Miller 1999, 241)

Miller doesn't think that guidance is necessary in evolution, but if it were (as I believe), then a route would be open for a subtle God to design life without overriding natural law. If quantum events such as radioactive decay are not governed by causal laws, then it breaks no law of nature to influence such events. As a theist like Miller, that seems perfectly possible to me. I would add, however, that such a process would amount to Intelligent Design, not Darwinian evolution. Further, while we might not be able to detect quantum manipulations, we may nevertheless be able to conclude confidently that the final structure was designed.

MISCONCEPTIONS CONCERNING SUPPOSED WAYS AROUND THE IRREDUCIBILITY OF BIOCHEMICAL SYSTEMS

Consider a hypothetical example where proteins homologous to all of the parts of an irreducibly complex molecular machine first had other individual functions in the cell. Might the irreducible system then have been put together from individual components that originally worked on their own, as some Darwinists have proposed? Unfortunately, this picture greatly oversimplifies the difficulty, as I discussed in *Darwin's Black Box* (Behe 1996, 53). Here analogies to mousetraps break down somewhat, because the parts of a molecular system have to find each other automatically in the cell. They can't be arranged by an intelligent agent, as a mousetrap is. In order to find each other in the cell, interacting parts have to have their surfaces shaped so that they are very closely matched to each other, as pictured in Figure 19.2. Originally, however, the individually acting components would not have had complementary surfaces. So all of the interacting surfaces of

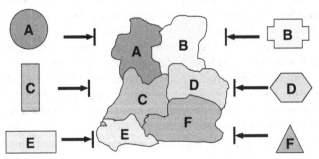

FIGURE 19.2. The parts of an irreducibly complex molecular machine must have surfaces that are closely matched to each other to allow specific binding. This drawing emphasizes that even if individually acting proteins homologous to parts of a complex originally had separate functions, their surfaces would not be complementary to each other. Thus the problem of irreducibility remains even if the separate parts originally had individual functions. (The blocked arrows indicate that the original protein shapes are not suitable to bind other proteins in the molecular machine.)

all of the components would first have to be adjusted before they could function together. And only then would the new function of the composite system appear. Thus, I emphasize strongly, *the problem of irreducibility remains, even if individual proteins homologous to system components separately and originally had their own functions.*

Another area where one has to be careful is in noticing that some systems that have extra or redundant components may have an irreducibly complex *core*. For example, a car with four spark plugs might get by with three or two, but it certainly can't get by with none. Rat traps often have two springs, to give them extra strength. The trap can still work if one spring is removed, but it can't work if both springs are removed. Thus in trying to imagine the origin of a rat trap by Darwinian means, we still have all the problems we had with a mousetrap. A cellular example of redundancy is the hugely complex eukaryotic cilium, which contains about 250 distinct protein parts (Dutcher 1995). The cilium has multiple copies of a number of components, including multiple microtubules and dynein arms. Yet a working cilium needs at least one copy of each in order to work, as I pictured in my book (Behe 1996, 60). Thus, like the rat trap's, its gradual Darwinian production remains quite difficult to envision. Kenneth Miller has pointed to the redundancy of the cilium as a counterexample to my claim of its irreducibility (Miller 1999, 140–3). But redundancy only delays irreducibility; it does not eliminate it.

Finally, rather than showing how their theory could handle the obstacle, some Darwinists are hoping to get around irreducible complexity by verbal tap dancing. At a debate between proponents and opponents of Intelligent Design sponsored by the American Museum of Natural History in April 2002, Kenneth Miller actually claimed (the transcript is available at the web site of the National Center for Science Education) that a mousetrap isn't irreducibly complex because subsets of a mousetrap, and even each individual part, could still "function" on their own. The holding bar of a mousetrap, Miller observed, could be used as *a toothpick*, so it still has a "function" outside the mousetrap. Any of the parts of the trap could be used as a paperweight, he continued, so they all have "functions." And since any object that has mass can be a paperweight, then any part of anything has a function of its own. *Presto*, there is no such thing as irreducible complexity! Thus the acute problem for gradualism that any child can see in systems like the mousetrap is smoothly explained away.

Of course, the facile explanation rests on a transparent fallacy, a brazen equivocation. Miller uses the word "function" in two different senses. Recall that the definition of irreducible complexity notes that removal of a part "causes the *system* to effectively cease functioning." Without saying so, in his exposition Miller shifts the focus from the separate function of the intact *system* itself to the question of whether we can find a different use (or "function") for some of the *parts*. However, if one removes a part from the mousetrap that I have pictured, it can no longer catch mice. The *system* has indeed effectively ceased functioning, so the *system* is irreducibly complex, just as I have written. What's more, the functions that Miller glibly assigns to the parts – paperweight, toothpick, key chain, and so forth – have little or nothing to do with the function of the system – catching mice (unlike the mousetrap series proposed by John McDonald, to be discussed later) – so they give us no clue as to how the system's function could arise gradually. Miller has explained precisely nothing.

With the problem of the mousetrap behind him, Miller then moved on to the bacterial flagellum – and again resorted to the same fallacy. If nothing else, one has to admire the breathtaking audacity of verbally trying to turn another severe problem for Darwinism into an advantage. In recent years, it has been shown that the bacterial flagellum is an even more sophisticated system than had been thought. Not only does it act as a rotary propulsion device, it also contains within itself an elegant mechanism used to transport the proteins that make up the outer portion of the machine from the inside of the cell to the outside (Aizawa 1996). Without blinking, Miller asserted that the flagellum is not irreducibly complex because some proteins of the flagellum could be missing and the remainder could still transport proteins, perhaps independently. (Proteins similar – but not identical – to some found in the flagellum occur in the type III secretory system of some bacteria. See Hueck 1998). Again, he was equivocating, switching the focus from the function of the system, acting as a rotary propulsion machine, to the ability of a subset of the system to transport proteins across a membrane. However, taking away the parts of the flagellum certainly destroys the ability of the system to act as a rotary propulsion machine, as I have argued. Thus, contra Miller, the flagellum is indeed irreducibly complex. What's more, the function of transporting proteins has as little directly to do with the function of rotary propulsion as a toothpick has to do with a mousetrap. So discovering the supportive function of transporting proteins tells us precisely nothing about how Darwinian processes might have put together a rotary propulsion machine.

THE BLOOD CLOTTING CASCADE

Having dealt with some common misconceptions about intelligent design, in the next two sections I will examine two systems that were proposed as serious counterexamples to my claim of irreducible complexity. I will show not only that they fail, but also how they highlight the seriousness of the obstacle of irreducible complexity.

In *Darwin's Black Box*, I argued that the blood clotting cascade is an example of an irreducibly complex system (Behe 1996, 74–97). At first glance, clotting seems to be a simple process. A small cut or scrape will bleed for a while and then slow down and stop as the visible blood congeals. However, studies over the past fifty years have shown that the visible simplicity is undergirded by a system of remarkable complexity (Halkier 1992). In all, there are over a score of separate protein parts involved in the vertebrate clotting system. The concerted action of the components results in the formation of a weblike structure at the site of the cut, which traps red blood cells and stops the bleeding. Most of the components of the clotting cascade are involved not in the structure of the clot itself, but in the control of the timing and placement of the clot. After all, it would not do to have clots forming at inappropriate times and places. A clot that formed in the wrong place, such as in the heart or brain, could lead to a heart attack or stroke. Yet a clot that formed even in the right place, but too slowly, would do little good.

The insoluble weblike fibers of the clot material itself are formed of a protein called fibrin. However, an insoluble web would gum up blood flow before a cut or scrape happened, so fibrin exists in the bloodstream initially in a soluble, inactive form called fibrinogen. When the closed circulatory system is breached, fibrinogen is activated by having a piece cut off from one end of two of the three proteins that comprise it. This exposes sticky sites on the protein, which allows them to aggregate. Because of the shape of the fibrin, the molecules aggregate into long fibers that form the meshwork of the clot. Eventually, when healing is completed, the clot is removed by an enzyme called plasmin.

The enzyme that converts fibrinogen to fibrin is called thrombin. Yet the action of thrombin itself has to be carefully regulated. If it were not, then thrombin would quickly convert fibrinogen to fribrin, causing massive blood clots and rapid death. It turns out that thrombin exists in an inactive form called prothrombin, which has to be activated by another component called Stuart factor. But by the same reasoning, the activity of Stuart factor has to be controlled, too, and it is activated by yet another component. Ultimately, the component that usually begins the cascade is tissue factor, which occurs on cells that normally do not come in contact with the circulatory system. However, when a cut occurs, blood is exposed to tissue factor, which initiates the clotting cascade.

Thus in the clotting cascade, one component acts on another, which acts on the next, and so forth. I argued that the cascade is irreducibly complex because, if a component is removed, the pathway is either immediately turned on or permanently turned off. It would not do, I wrote, to postulate that the pathway started from one end, fibrinogen, and then added components, since fibrinogen itself does no good. Nor is it plausible even to start with something like fibrinogen and a nonspecific enzyme that might cleave it, since the clotting would not be regulated and would be much more likely to do harm than good.

So said I. But Russell Doolittle – an eminent protein biochemist, a professor of biochemistry at the University of California–San Diego, a member of the National Academy of Sciences, and a lifelong student of the blood clotting system – disagreed. As part of a symposium discussing my book and Richard Dawkins' *Climbing Mount Improbable* in the *Boston Review*, which is published by the Massachusetts Institute of Technology, Doolittle wrote an essay discussing the phenomenon of gene duplication – the process by which a cell may be provided with an extra copy of a functioning gene. He then conjectured that the components of the blood clotting pathway, many of which have structures that are similar to each other, arose by gene duplication and gradual divergence. This is the common view among Darwinists. Professor Doolittle went on to describe a then-recent experiment that, he thought, showed that the cascade is not irreducible after all. Professor Doolittle cited a paper by Bugge and colleagues (1996a) entitled "Loss of Fibrinogen Rescues Mice from the Pleiotropic Effects of Plasminogen Deficiency." Of that paper, he wrote:

Recently the gene for plaminogen [sic] was knocked out of mice, and, predictably, those mice had thrombotic complications because fibrin clots could not be cleared away. Not long after that, the same workers knocked out the gene for fibrinogen in another line of mice. Again, predictably, these mice were ailing, although in this case hemorrhage was the problem. And what do you think happened when these two lines of mice were crossed? For all practical purposes, the mice lacking both genes were normal! Contrary to claims about irreducible complexity, the entire ensemble of proteins is not needed. Music and harmony can arise from a smaller orchestra. (Doolittle 1997)

(Again, fibrinogen is the precursor of the clot material itself. Plasminogen is the precursor of plasmin, which removes clots once their purpose is accomplished.) So if one knocks out either one of those genes of the clotting pathway, trouble results; but, Doolittle asserted, if one knocks out both, then the system is apparently functional again. That would be a very interesting result, but it turns out to be incorrect. Doolittle misread the paper.

The abstract of the paper states that "[m]ice deficient in plasminogen and fibrinogen are phenotypically indistinguishable from fibrinogen-deficient mice." In other words, the double mutants have all the problems that the mice lacking just fibrinogen have. Those problems include inability to clot, hemorrhaging, and death of females during pregnancy. Plasminogen deficiency leads to a different suite of symptoms – thrombosis, ulcers, and high mortality. Mice missing both genes were "rescued" from the ill effects of plasminogen deficiency only to suffer the problems associated with fibrinogen deficiency.[3] The reason for this is easy to see. Plasminogen is needed to remove clots that, left in place, interfere with normal functions. However, if the gene for fibrinogen is also knocked out, then clots can't form in the first place, and their removal is not an issue. Yet if clots can't form, then there is no functioning clotting system, and the mice suffer the predictable consequences.

TABLE 19.1. *Effects of knocking out genes for blood clotting components*

Missing Protein	Symptoms	Reference
Plasminogen	Thrombosis, high mortality	Bugge et al. 1995
Fibrinogen	Hemorrhage, death in pregnancy	Suh et al. 1995
Plasminogen/fibrinogen	Hemorrhage, death in pregnancy	Bugge et al. 1996a
Prothrombin	Hemorrhage, death in pregnancy	Sun et al. 1998
Tissue factor	Hemorrhage, death in pregnancy	Bugge et al. 1996b

Clearly, the double-knockout mice are not "normal." They are not promising evolutionary intermediates.

The same group that produced the mice missing plasminogen and fibrinogen has also produced mice individually missing other components of the clotting cascade – prothrombin and tissue factor. In each case, the mice are severely compromised, which is *exactly* what one would expect if the cascade is irreducibly complex (Table 19.1).

What lessons can we draw from this incident? The point is certainly not that Russell Doolittle misread a paper, which anyone might do. (Scientists, as a rule, are not known for their ability to write clearly, and Bugge and colleagues were no exception.) Rather, the main lesson is that irreducible complexity seems to be a much more severe problem than Darwinists recognize, since the experiment Doolittle himself chose to demonstrate that "music and harmony can arise from a smaller orchestra" showed exactly the opposite. A second lesson is that gene duplication is not the panacea that it is often made out to be. Professor Doolittle knows as much about the structures of the clotting proteins and their genes as anyone on Earth, and he is convinced that many of them arose by gene duplication and exon shuffling. Yet that knowledge did not prevent him from proposing utterly nonviable mutants as possible examples of evolutionary intermediates. A third lesson is that, as I had claimed in *Darwin's Black Box*, there are no papers in the scientific literature detailing how the clotting pathway could have arisen by Darwinian means. If there were, Doolittle would simply have cited them.

Another significant lesson that we can draw is that, while the majority of academic biologists and philosophers place their confidence in Darwinism, that confidence rests on no firmer grounds than Professor Doolittle's. As an illustration, consider the words of the philosopher Michael Ruse:

For example, Behe is a real scientist, but this case for the impossibility of a small-step natural origin of biological complexity has been trampled upon contemptuously by the scientists working in the field. They think his grasp of the pertinent science is weak and his knowledge of the literature curiously (although conveniently) outdated.

For example, far from the evolution of clotting being a mystery, the past three decades of work by Russell Doolittle and others has thrown significant light on the ways in which clotting came into being. More than this, it can be shown that the clotting mechanism does not have to be a one-step phenomenon with everything already in place and functioning. One step in the cascade involves fibrinogen, required for clotting, and another, plaminogen [*sic*], required for clearing clots away. (Ruse 1998)

And Ruse goes on to quote Doolittle's passage from the *Boston Review* that I quoted earlier. Now, Ruse is a prominent Darwinist and has written many books on various aspects of Darwiniana. Yet, as his approving quotation of Doolittle's mistaken reasoning shows (complete with his copying of Doolittle's typo-misspelling of "plaminogen"), Ruse has no independent knowledge of how natural selection could have put together complex biochemical systems. As far as the scientific dispute is concerned, Ruse has nothing to add.

Another such example is seen in a recent essay in *The Scientist*, "Not-So-Intelligent Design," by Neil S. Greenspan, a professor of pathology at Case Western Reserve University, who writes (Greenspan 2002), "The Design advocates also ignore the accumulating examples of the reducibility of biological systems. As Russell Doolittle has noted in commenting on the writings of one ID advocate . . ." Greenspan goes on to cite approvingly Doolittle's argument in the *Boston Review*. He concludes, with unwitting irony, that "[t]hese results cast doubt on the claim by proponents of ID that they know which systems exhibit irreducible complexity and which do not." But since the results are precisely the opposite of what Greenspan supposed, the shoe is now on the other foot. This incident casts grave doubt on the claim by Darwinists – both biologists and philosophers – that they know that complex cellular systems are explainable in Darwinian terms. It demonstrates that Darwinists either cannot or will not recognize difficulties for their theory.

THE MOUSETRAP

The second counterargument to irreducibility I will discuss here concerns not a biological example but a conceptual one. In *Darwin's Black Box*, I pointed to a common mechanical mousetrap as an example of irreducible complexity. Almost immediately after the book's publication, some Darwinists began proposing ways in which the mousetrap could be built step by step. One proposal that has gotten wide attention, and that has been endorsed by some prominent scientists, was put forward by John McDonald, a professor of biology at the University of Delaware, and can be seen on his web site.[4] His series of traps is shown in Figure 19.3. McDonald's main point was that the trap that I pictured in my book consisted of five parts, yet he could build a trap with fewer parts.

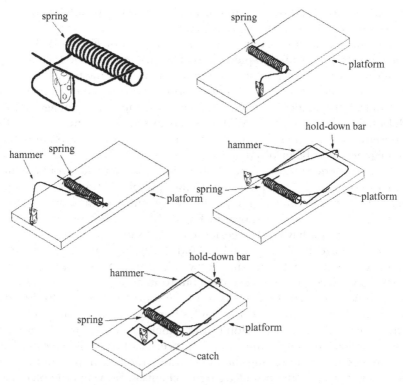

FIGURE 19.3. A series of mousetraps with an increasing number of parts, as proposed by John McDonald <http://udel.edu/~mcdonald/oldmousetrap.html> and reproduced here with his permission. Yet intelligence is still required to construct one trap from another, as described in the text.

I agree. In fact, I said exactly the same thing in my book. I wrote:

We need to distinguish between a *physical* precursor and a *conceptual* precursor. The trap described above is not the only system that can immobilize a mouse. On other occasions my family has used a glue trap. In theory at least, one can use a box propped open with a stick that could be tripped. Or one can simply shoot the mouse with a BB gun. However, these are not physical precursors to the standard mousetrap since they cannot be transformed, step-by-Darwinian-step, into a trap with a base, hammer, spring, catch, and holding bar. (Behe 1996, 43)

Thus the point is not that mousetraps can be built in different ways, with different numbers of pieces. (My children have a game at home called "Mousetrap," which has many, many pieces and looks altogether different from the common mechanical one.) Of course they can. The only question is whether a particular trap can be built by "numerous, successive, slight modifications" to a simple starting point – without the intervention of intelligence – as Darwin insisted that his theory required.

The McDonald traps cannot. Shown at the top of Figure 19.3 are his one-piece trap and his two-piece trap. The structure of the second trap, however, is not a single, small, random step away from the first. First notice that the one-piece trap is not a simple spring – it is shaped in a very special way. In fact, the shape was deliberately chosen by an intelligent agent, John McDonald, to act as a trap. Well, one has to start somewhere. But if the mousetrap series is to have any relevance at all to Darwinian evolution, then intelligence can't be involved at any further point.

Yet intelligence saturates the whole series. Consider what would be necessary to convert the one-piece trap to the "two-piece" trap. One can't just place the first trap on a simple piece of wood and have it work as the second trap does. Rather, as shown in Figure 19.3, the two protruding ends of the spring first have to be reoriented. What's more, two staples (barely visible in Figure 19.3) are added to hold the spring onto the platform so that it can be under tension in the two-piece trap. So we have gone not from a one-piece to a two-piece trap, but from a one-piece to a four-piece trap. Notice also that the placement of the staples in relation to the edge of the platform is critical. If the staples were moved a quarter-inch from where they are, the trap wouldn't work. Finally, consider that, in order to have a serious analogy to the robotic processes of the cell, we can't have an intelligent human setting the mousetrap – the first trap would have to be set by some unconscious charging mechanism. So, when the pieces are rearranged, the charging mechanism too would have to change for the second trap.

It's easy for us intelligent agents to overlook our role in directing the construction of a system, but nature cannot overlook any step at all, so the McDonald mousetrap series completely fails as an analogy to Darwinian evolution. In fact, the second trap is best viewed not as some Darwinian descendant of the first but as a completely different trap, designed by an intelligent agent, perhaps using a refashioned part or two from the first trap.

Each of the subsequent steps in the series suffers from analogous problems, which I have discussed elsewhere.[5]

In his endorsement of the McDonald mousetrap series, Kenneth Miller wrote: "If simpler versions of this mechanical device [the mousetrap] can be shown to work, then simpler versions of biochemical machines could work as well . . . and this means that complex biochemical machines could indeed have had functional precursors."[6] But that is exactly what it doesn't show – if by "precursor" Miller means "Darwinian precursor." On the contrary, McDonald's mousetrap series shows that even if one does find a simpler system to perform some function, that gives one no reason to think that a more complex system performing the same function could be produced by a Darwinian process starting with the simpler system. Rather, the difficulty in doing so for a simple mousetrap gives us compelling reason to think it cannot be done for complex molecular machines.

FUTURE PROSPECTS OF THE INTELLIGENT DESIGN HYPOTHESIS

The misconceived arguments by Darwinists that I have recounted here offer strong encouragement to me that the hypothesis of Intelligent Design is on the right track. After all, if well-informed opponents of an idea attack it by citing data that, when considered objectively, actually demonstrate its force, then one is entitled to be confident that the idea is worth investigating.

Yet it is not primarily the inadequacy of Darwinist responses that bodes well for the design hypothesis. Rather, the strength of design derives mainly from the work-a-day progress of science. In order to appreciate this fact, it is important to realize that the idea of Intelligent Design arose not from the work of any individual but from the collective work of biology, particularly in the last fifty years. Fifty years ago, the cell seemed much simpler, and in our innocence it was easier then to think that Darwinian processes might have accounted for it. But as biology progressed and the imagined simplicity vanished, the idea of design became more and more compelling. That trend is continuing inexorably. The cell is not getting any simpler; it is getting much more complex. I will conclude this chapter by citing just one example, from the relatively new area of proteomics.

With the successful sequencing of the entire genomes of dozens of microorganisms and one vertebrate (us), the impetus has turned toward analyzing the cellular interactions of the proteins that the genomes code for, taken as a whole. Remarkable progress has already been made. Early in 2002, an exhaustive study of the proteins comprising the yeast proteome was reported. Among other questions, the investigators asked what proportion of yeast proteins work as groups. They discovered that nearly fifty percent of proteins work as complexes of a half-dozen or more, and many as complexes of ten or more (Gavin et al. 2002).

This is not at all what Darwinists had expected. As Bruce Alberts wrote earlier in the article "The Cell as a Collection of Protein Machines":

We have always underestimated cells. Undoubtedly we still do today. But at least we are no longer as naive as we were when I was a graduate student in the 1960s. Then most of us viewed cells as containing a giant set of second-order reactions. . . .

But, as it turns out, we can walk and we can talk because the chemistry that makes life possible is much more elaborate and sophisticated than anything we students had ever considered. Proteins make up most of the dry mass of a cell. But instead of a cell dominated by randomly colliding individual protein molecules, we now know that nearly every major process in a cell is carried out by assemblies of 10 or more protein molecules. And, as it carries out its biological functions, each of these protein assemblies interacts with several other large complexes of proteins.

Indeed, the entire cell can be viewed as a factory that contains an elaborate network of interlocking assembly lines, each of which is composed of a set of large protein machines. (Alberts 1998)

The important point here for a theory of Intelligent Design is that molecular machines are not confined to the few examples that I discussed in *Darwin's Black Box*. Rather, most proteins are found as components of complicated molecular machines. Thus design might extend to a large fraction of the features of the cell, and perhaps beyond that into higher levels of biology.

Progress in twentieth-century science has led us to the design hypothesis. I expect progress in the twenty-first century to confirm and extend it.

Notes

1. For example, the microbiologist James Shapiro of the University of Chicago declared in *National Review* that "[t]here are no detailed Darwinian accounts for the evolution of any fundamental biochemical or cellular system, only a variety of wishful speculations" (Shapiro 1996, 65). In *Nature*, the University of Chicago evolutionary biologist Jerry Coyne stated, "There is no doubt that the pathways described by Behe are dauntingly complex, and their evolution will be hard to unravel.... [W]e may forever be unable to envisage the first proto-pathways" (Coyne 1996, 227). In a particularly scathing review in *Trends in Ecology and Evolution*, Tom Cavalier-Smith, an evolutionary biologist at the University of British Columbia, nonetheless wrote, "For none of the cases mentioned by Behe is there yet a comprehensive and detailed explanation of the probable steps in the evolution of the observed complexity. The problems have indeed been sorely neglected – though Behe repeatedly exaggerates this neglect with such hyperboles as 'an eerie and complete silence' " (Cavalier-Smith 1997, 162). The Evolutionary biologist Andrew Pomiankowski, writing in *New Scientist*, agreed: "Pick up any biochemistry textbook, and you will find perhaps two or three references to evolution. Turn to one of these and you will be lucky to find anything better than 'evolution selects the fittest molecules for their biological function' " (Pomiankowski 1996, 44). In *American Scientist*, the Yale molecular biologist Robert Dorit averred, "In a narrow sense, Behe is correct when he argues that we do not yet fully understand the evolution of the flagellar motor or the blood clotting cascade" (Dorit 1997, 474).
2. A good example is found on the "World of Richard Dawkins" web site, maintained by a Dawkins fan named John Catalano at <www.world-of-dawkins.com/Catalano/box/published.htm>. It is to this site that the Oxford University physical chemist Peter Atkins was referring when he wrote in a review of *Darwin's Black Box* for the "Infidels" web site: "Dr. Behe claims that science is largely silent on the details of molecular evolution, the emergence of complex biochemical pathways and processes that underlie the more traditional manifestations of evolution at the level of organisms. Tosh! There are hundreds, possibly thousands, of scientific papers that deal with this very subject. For an entry into this important and flourishing field, and an idea of the intense scientific effort that it represents (see the first link above) [*sic*]" (Atkins 1998).

3. Bugge and colleagues (1996a) were interested in the question of whether plasminogen had any role in metabolism other than its role in clotting, as had been postulated. The fact that the direct effects of plasminogen deficiency were ameliorated by fibrinogen deficiency showed that plasminogen probably had no other role.
4. <http://udel.edu/~mcdonald/oldmousetrap.html>. Professor McDonald has recently designed a new series of traps that can be seen at <http://udel.edu/~mcdonald/mousetrap.html>. I have examined them and have concluded that they involve his directing intelligence to the same degree.
5. M. J. Behe, "A Mousetrap Defended: Response to Critics." <www.crsc.org>
6. <http://biocrs.biomed.brown.edu/Darwin/DI/Mousetrap.html>

References

Aizawa, S. I. 1996. Flagellar assembly in Salmonella typhimurium. *Molecular Microbiology* 19: 1–5.

Alberts, B. 1998. The cell as a collection of protein machines: Preparing the next generation of molecular biologists. *Cell* 92: 291–4.

Atkins, P. W. 1998. Review of Michael Behe's *Darwin's Black Box*. <*www.infidels.org/library/modern/peter_atkins/behe.html*>.

Behe, M. J. 1996. *Darwin's Black Box: The Biochemical Challenge to Evolution.* New York: The Free Press.

2001. Reply to my critics: A response to reviews of *Darwin's Black Box: The Biochemical Challenge to Evolution. Biology and Philosophy* 16: 685–709.

Bugge, T. H., M. J. Flick, C. C. Daugherty, and J. L. Degen. 1995. Plasminogen deficiency causes severe thrombosis but is compatible with development and reproduction. *Genes and Development* 9: 794–807.

Bugge, T. H., K. W. Kombrinck, M. J. Flick, C. C. Daugherty, M. J. Danton, and J. L. Degen. 1996a. Loss of fibrinogen rescues mice from the pleiotropic effects of plasminogen deficiency. *Cell* 87: 709–19.

Bugge, T. H., Q. Xiao, K. W. Kombrinck, M. J. Flick, K. Holmback, M. J. Danton, M. C. Colbert, D. P. Witte, K. Fujikawa, E. W. Davie, and J. L. Degen. 1996b. Fatal embryonic bleeding events in mice lacking tissue factor, the cell-associated initiator of blood coagulation. *Proceedings of the National Academy of Sciences (USA)* 93: 6258–63.

Cavalier-Smith, T. 1997. The blind biochemist. *Trends in Ecology and Evolution* 12: 162–3.

Coyne, J. A. 1996. God in the details. *Nature* 383: 227–8.

Darwin, C. 1859. *The Origin of Species.* New York: Bantam Books.

Dawkins, R. 1986. *The Blind Watchmaker.* New York: Norton.

DeRosier, D. J. 1998. The turn of the screw: The bacterial flagellar motor. *Cell* 93: 17–20.

Doolittle, R. F. A delicate balance. *Boston Review*, February/March 1997, pp. 28–9.

Dorit, R. 1997. Molecular evolution and scientific inquiry, misperceived. *American Scientist* 85: 474–5.

Dutcher, S. K. 1995. Flagellar assembly in two hundred and fifty easy-to-follow steps. *Trends in Genetics* 11: 398–404.

Gavin, A. C., et al. 2002. Functional organization of the yeast proteome by systematic analysis of protein complexes. *Nature* 415: 141–7.

Greenspan, N. S. 2002. Not-so-intelligent design. *The Scientist* 16: 12.

Halkier, T. 1992. *Mechanisms in Blood Coagulation Fibrinolysis and the Complement System.* Cambridge: Cambridge University Press.

Harold, F. M. 2001. *The Way of the Cell.* Oxford: Oxford University Press.

Hueck, C. J. 1998. Type III protein secretion systems in bacterial pathogens of animals and plants. *Microbiology and Molecular Biology Reviews* 62: 379–433.

Kauffman, S. A. 1995. *At Home in the Universe: The Search for Laws of Self-Organization and Complexity.* New York: Oxford University Press.

Miller, K. R. 1999. *Finding Darwin's God: A Scientist's Search for Common Ground between God and Evolution.* New York: Cliff Street Books.

Pomiankowski, A. 1996. The God of the tiny gaps. *New Scientist,* September 14, pp. 44–5.

Ruse, M. 1998. Answering the creationists: Where they go wrong and what they're afraid of. *Free Inquiry,* March 22, p. 28.

Shapiro, J. 1996. In the details . . . what? *National Review,* September 16, pp. 62–5.

Shapiro, L. 1995. The bacterial flagellum: From genetic network to complex architecture. *Cell* 80: 525–7.

Suh, T. T., K. Holmback, N. J. Jensen, C. C. Daugherty, K. Small, D. I. Simon, S. Potter, and J. L. Degen. 1995. Resolution of spontaneous bleeding events but failure of pregnancy in fibrinogen-deficient mice. *Genes and Development* 9: 2020–33.

Sun, W. Y., D. P. Witte, J. L. Degen, M. C. Colbert, M. C. Burkart, K. Holmback, Q. Xiao, T. H. Bugge, and S. J. Degen. 1998. Prothrombin deficiency results in embryonic and neonatal lethality in mice. *Proceedings of the National Academy of Sciences USA* 95: 7597–602.

Yonekura, K., S. Maki, D. G. Morgan, D. J. DeRosier, F. Vonderviszt, K. Imada, and K. Namba. 2000. The bacterial flagellar cap as the rotary promoter of flagellin self-assembly. *Science* 290: 2148–52.

20

The Cambrian Information Explosion

Evidence for Intelligent Design

Stephen C. Meyer

INTRODUCTION

In his book *The Philosophy of Biology*, Elliott Sober (2000) notes that many evolutionary biologists regard the design hypothesis as inherently untestable and, therefore, unscientific in principle simply because it no longer commands scientific assent. He notes that while logically unbeatable versions of the design hypothesis have been formulated (involving, for example, a "trickster God" who creates a world that appears to be undesigned), design hypotheses in general need not assume an untestable character. A design hypothesis could, he argues, be formulated as a fully scientific "inference to the best explanation." He notes that scientists often evaluate the explanatory power of a "hypothesis by testing it against one or more competing hypotheses" (44). Thus, he argues that William Paley's design hypothesis was manifestly testable but was rejected precisely because it could not explain the relevant evidence of contemporary biology as well as the fully naturalistic theory of Charles Darwin. Sober then casts his lot with modern neo-Darwinism on evidential grounds. But the possibility remains, he argues, "that there is some other version of the design hypothesis that both disagrees with the hypothesis of evolution and also is a more likely explanation of what we observe. No one, to my knowledge, has developed such a version of the design hypothesis. But this does not mean that no one ever will" (46).

In recent essays (Meyer 1998, 2003), I have advanced a design hypothesis of the kind that Sober acknowledges as a scientific possibility. Specifically, I have argued that the hypothesis of Intelligent Design can be successfully formulated as "an inference to the best explanation" for the origin of the information necessary to produce the first life. Such a design hypothesis stands, not as a competitor to *biological* evolutionary theory (i.e., neo-Darwinism), but instead as a competitor to *chemical* evolutionary theories of how life first arose from nonliving chemicals.

In order to make this argument, I show that considerations of causal adequacy (Hodge 1977, 239; Lipton 1991, 32–88) typically determine which among a group of competing explanations qualify as *best*. I then argue *against* the causal adequacy of each of the main categories of naturalistic explanation – chance, necessity, and their combination – for the origin of biological information. Further, in order to avoid formulating a purely negative "argument from ignorance," I also argue *for* the positive adequacy of intelligent agency as a cause of information. I note, in the words of the information theorist Henry Quastler, that the "creation of new information is habitually associated with conscious activity" (1964, 16). Thus, I conclude that Intelligent Design stands as the best – most causally adequate – explanation for the origin of the information necessary to produce the first life.

In this volume, Professors Dembski and Bradley amplify the two complementary aspects of this argument – Dembski, by suggesting that living systems possess a reliable positive indicator of the activity of an intelligent cause, namely, "complex specified information"; Bradley, by challenging the causal adequacy of naturalistic explanations for the origin of the information necessary to the first life. Jointly, these two chapters provide both a negative case against the adequacy of naturalistic theories and a positive case for the causal adequacy of Intelligent Design, thereby supporting Intelligent Design as the best explanation for the information necessary to the first life.

THESIS

This chapter extends this line of reasoning by formulating another, more radical design hypothesis. Rather than positing Intelligent Design solely as an explanation for the origin of the information necessary to the first life, this chapter will offer Intelligent (or purposive) Design as an explanation for the information necessary to produce the novel animal body plans that arise during the history of life. This design hypothesis thus competes directly with neo-Darwinism in two respects. First, it seeks to explain the origin of the novel biological form (and the information necessary to produce it) that emerges *after* the origin of the first life. Second, it posits the action of a purposive intelligence, not just a purposeless or undirected process, in the history of life.

Many scientists now openly acknowledge the fundamental difficulties facing chemical evolutionary theories of the origin of life, including the problem of explaining the origin of biological information from nonliving chemistry. Nevertheless, many assume that theories of biological evolution do not suffer from a similar information problem. While many scientists recognize that invoking natural selection at the pre-biotic level remains theoretically problematic (since natural selection presumably acts only on self-replicating organisms), neo-Darwinists assume that natural selection acting on random mutations within already living organisms can generate the information needed to produce fundamentally new organisms from preexisting forms. I will dispute this claim. I will argue that explaining the origin of novel biological information is not a problem confined to origin-of-life research, but rather one that afflicts specifically biological theories of evolution as well.

In order to this make this case, I will examine a paradigm example of a discrete increase in biological information during the history of life: the Cambrian explosion. I will then compare the explanatory power of three competing models – neo-Darwinism, self-organization, and Intelligent Design – with respect to the origin of the information that arises during the Cambrian.

THE CAMBRIAN EXPLOSION

The "Cambrian explosion" refers to the geologically sudden appearance of many new animal body plans about 530 million years ago. At this time, at least nineteen and perhaps as many as thirty-five phyla (of forty total phyla) made their first appearance on Earth within a narrow five-million-year window of geologic time (Meyer et al. 2003; Bowring et al. 1993). Phyla constitute the highest categories in the animal kingdom, with each phylum exhibiting a unique architecture, blueprint, or structural body plan. Familiar examples of basic animal body plans are mollusks (squids and shellfish), arthropods (crustaceans, insects, and trilobites), and chordates, the phylum to which all vertebrates belong.

An especially dramatic feature of the Cambrian explosion was the first appearance of invertebrate phyla with mineralized exoskeletons, including members of the phyla *Mollusca, Echinodermata,* and *Arthropoda.* Many well-preserved animals with soft tissues also first appeared, including representatives of *Ctenophora, Annelida, Onycophora, Phoronida,* and *Priapulida.* Fossil discoveries from the Lower Cambrian Yuanshan Formation in China have also shown the presence of animals from the phylum *Chordata,* including two fish fossils, *Myllokunmingia fengjiaoa* and *Haikouichthys ercaicunensis,* suggesting an earlier appearance for vertebrates than previously thought (Shu et al. 1999).

To say that the fauna of the Cambrian period appeared in a geologically sudden manner also implies the absence of clear transitional intermediate forms connecting Cambrian animals with simpler pre-Cambrian forms. And indeed, in almost all cases, the Cambrian animals have no clear morphological antecedents. Debate now exists about the extent to which this pattern of evidence can be reconciled with the theory of universal common descent. This essay will not address that question but will instead analyze whether the neo-Darwinian *mechanism* of natural selection acting on random mutations can generate the information necessary to produce the animals that arise in the Cambrian.

DEFINING BIOLOGICAL INFORMATION

Before proceeding, I must define the term "information" as used in biology. In classical Shannon information theory, the amount of information in a system is inversely related to the probability of the arrangement of constituents in a system or the characters along a communication channel (Shannon 1948). The more improbable (or complex) the arrangement, the more Shannon information, or information-carrying capacity, a string or system possesses.

Since the 1960s, mathematical biologists have realized that Shannon's theory could be applied to the analysis of DNA and proteins to measure their information-carrying capacity. Since DNA contains the assembly instructions for building proteins, the information-processing system in the cell represents a kind of communication channel (Yockey 1992, 110). Further, DNA conveys information via specifically arranged sequences of four different chemicals – called nucleotide bases – that function as alphabetic or digital characters in a linear array. Since each of the four bases has a roughly equiprobable chance of occurring at each site along the spine of the DNA molecule, biologists can calculate the probability, and thus the information-carrying capacity, of any particular sequence **n** bases long.

The ease with which information theory applies to molecular biology has created confusion about the type of information that DNA and proteins possess. Sequences of nucleotide bases in DNA, or amino acids in a protein, are highly improbable and thus have a large information-carrying capacity. But, like meaningful sentences or lines of computer code, genes and proteins are also *specified* with respect to function. Just as the meaning of a sentence depends upon the specific arrangement of the letters in the sentence, so too does the function of a gene sequence depend upon the specific arrangement of the nucleotide bases in the gene. Thus, as Sarkar points out, molecular biologists beginning with Francis Crick have equated *information* not only with complexity but also with "specificity," where "specificity" or "specified" has meant "necessary to function" (1996, 191).

Similarly, this chapter poses a question, not about the origin of Shannon information – mere complexity of arrangement – but about the origin of the "specified complexity" or "complex specified information" (CSI) that characterizes living systems and their biomolecular components.

THE CAMBRIAN INFORMATION EXPLOSION

The Cambrian explosion represents a remarkable jump in the specified complexity or CSI of the biological world. For over three billion years, the biological realm included little more than bacteria and algae. Then, beginning about 570 mya, the first complex multicellular organisms appeared in the rock strata, including sponges, cnidarians, and the peculiar Ediacaran biota. Forty million years later, the Cambrian explosion occurred. The emergence of the Ediacaran biota (570 mya), and then to a much greater extent the Cambrian explosion (530 mya), represented steep climbs up the biological complexity gradient.

One way to measure the increase in CSI that appears with the Cambrian animals is to assess the number of new cell types that emerge (Valentine 1995, 91–3). Studies of modern animals suggest that the sponges that appeared in the late Precambrian, for example, would have required five cell types, whereas the more complex animals that appeared in the Cambrian (such as representatives of *Arthropoda*) would have required fifty or more cell types. Functionally more complex animals require more cell types to perform their more diverse functions. New cell types require many new and specialized proteins. New proteins, in turn, require new genetic information. Thus an increase in the number of cell types implies (at minimum) a considerable increase in the amount of specified genetic information. Molecular biologists have recently estimated that a minimally complex single-celled organism would require between 318 and 562 kilobase pairs of DNA to produce the proteins necessary to maintain life (Koonin 2001). More complex single cells might require upward of a million base pairs. Yet to build the proteins necessary to sustain a complex arthropod such as a trilobite would require orders of magnitude more coding instructions. The genome size of the modern fruitfly *Drosophila melanogaster* (an arthropod) is approximately 120 million base pairs (Gerhart and Kirschner 1997, 121). Transitions from a single cell to colonies of cells to complex animals represent significant (and, in principle, measurable) increases in CSI.

Building a new animal from a single-celled organism requires a vast amount of new genetic information. It also requires a way of arranging gene products – proteins – into higher levels of organization. New proteins are required to service new cell types. But new proteins must be organized into new systems within the cell; new cell types must be organized into new tissues, organs, and body parts (Müller and Newman 2003). These, in turn, must be organized to form body plans. New animals, therefore, embody hierarchically organized systems of lower-level parts within a functional whole. Such hierarchical organization itself represents a type of information, since body plans comprise both highly improbable and functionally specified arrangements of lower-level parts. The specified complexity of new body plans requires explanation in any account of the Cambrian explosion.

Can neo-Darwinism explain the discontinuous increase in CSI that appears in the Cambrian explosion – either in the form of new genetic information or in the form of hierarchically organized systems of parts? We will now examine the two parts of this question.

NOVEL GENES AND PROTEINS

Many scientists and mathematicians have questioned the ability of mutation and selection to generate information in the form of novel genes and proteins. Such skepticism often derives from consideration of the extreme improbability (and specificity) of functional genes and proteins.

A typical gene contains over one thousand precisely arranged bases. For any specific arrangement of four nucleotide bases of length n, there is a corresponding number of possible arrangements of bases, 4^n. For any protein, there are 20^n possible arrangements of protein-forming amino acids. A gene 999 bases in length represents one of 4^{999} possible nucleotide sequences; a protein of 333 amino acids is one of 20^{333} possibilities.

Since the 1960s, biologists have generally thought functional proteins to be rare among the set of possible amino acid sequences (of corresponding length). Some have used an analogy with human language to illustrate why this should be the case. Denton, for example, has shown that meaningful words and sentences are extremely rare among the set of possible combinations of English letters, especially as sequence length grows. (The ratio of meaningful 12-letter words to 12-letter sequences is $1/10^{14}$; the ratio of 100-letter sentences to possible 100-letter strings is roughly $1/10^{100}$.) Further, Denton shows that most meaningful sentences are *highly isolated* from one another in the space of possible combinations, so that random substitutions of letters will, after a very few changes, inevitably degrade meaning. Apart from a few closely clustered sentences accessible by random substitution, the overwhelming majority of meaningful sentences lie, probabilistically speaking, beyond the reach of random search.

Denton and others have argued that similar constraints apply to genes and proteins (1986, 301–24). They have questioned whether an undirected search via mutation/selection would have a reasonable chance of locating new islands of function – representing fundamentally new genes or proteins – within the time available (Schuetzenberger 1967; Løvtrup 1979; Berlinski 1996). Some have also argued that alterations in sequencing would likely result in loss of protein function before fundamentally new function could arise. Nevertheless, neither the sensitivity of genes and proteins to functional loss as a result of sequence change, nor the extent to which functional proteins are isolated within sequence space, has been fully known.

Recently, experiments in molecular biology have shed light on these questions. A variety of "mutagenesis" techniques have shown that proteins (and thus the genes that produce them) are indeed highly specified relative to biological function (Bowie and Sauer 1989; Reidhaar-Olson and Sauer 1990; Taylor et al. 2001). Mutagenesis research tests the sensitivity of proteins (and, by implication, DNA) to functional loss as a result of alterations in sequencing. Studies of protein mutations have long shown that amino acid residues at many active site positions cannot vary without functional loss (Perutz and Lehmann 1968). More recent protein studies (including mutagenesis experiments) have shown that functional requirements place significant constraints on sequencing even at nonactive site positions (Bowie and Sauer 1989; Reidhaar-Olson and Sauer 1990; Chothia, Gelfland, and Kister 1998;

Axe 2000; Taylor et al. 2001). In particular, Axe (2000) has shown that multiple as opposed to single amino acid substitutions inevitably result in loss of protein function, even when these changes occur at sites that allow variation when altered in isolation. Cumulatively, these constraints imply that proteins are highly sensitive to functional loss as a result of alterations in sequencing, and that functional proteins represent highly isolated and improbable arrangements of amino acids – arrangements that are far more improbable, in fact, than would be likely to arise by chance, even given our multibillion-year-old universe (Kauffman 1995, 44; Dembski 1998, 175–223).

Of course, neo-Darwinists do not envision a completely random search through the space of possible nucleotide sequences. They see natural selection acting to preserve small advantageous variations in genetic sequences and their corresponding protein products. Richard Dawkins (1996), for example, likens an organism to a high mountain peak. He compares climbing the sheer precipice up the front side of the mountain to building a new organism by chance. He acknowledges that this approach up "Mount Improbable" will not succeed. Nevertheless, he suggests that there is a gradual slope up the backside of the mountain that could be climbed in small incremental steps. In his analogy, the backside climb up "Mount Improbable" corresponds to the process of *natural selection* acting on random changes in the genetic text. What chance alone cannot accomplish blindly or in one leap, selection (acting on mutations) can accomplish through the cumulative effect of many slight successive steps.

Yet the extreme specificity and complexity of proteins presents a difficulty not only for the chance origin of specified biological information (i.e., for random mutations acting alone), but also for selection and mutation acting in concert. Indeed, mutagenesis experiments cast doubt on each of the two scenarios by which neo-Darwinists envision new information arising from the mutation/selection mechanism. For neo-Darwinists, new functional genes either arise from noncoding sections in the genome or from preexisting genes. Both scenarios are problematic.

In the first scenario, neo-Darwinists envision new genetic information arising from those sections of the genetic text that can presumably vary freely without consequence to the organism. According to this scenario, noncoding sections of the genome, or duplicated sections of coding regions, can experience a protracted period of "neutral evolution" during which alterations in nucleotide sequences have no discernible effect on the function of the organism. Eventually, however, a new gene sequence will arise that can code for a novel protein. At that point, natural selection can favor the new gene and its functional protein product, thus securing the preservation and heritability of both.

This scenario has the advantage of allowing the genome to vary through many generations, as mutations "search" the space of possible base sequences. The scenario has an overriding problem, however: the size of the combinatorial space and the extreme rarity and isolation of the functional sequences within that space of possibilities. Since natural selection can do nothing to help *generate* new functional sequences, but rather can only preserve such sequences once they have arisen, chance alone – random variation – must do the work of information generation – that is, of finding the exceedingly rare functional sequences within a combinatorial universe of possibilities. Yet the probability of randomly assembling (or "finding," in the previous sense) a functional sequence is vanishingly small even on a scale of billions of years. Robert Sauer's mutagenesis experiments imply that the probability of attaining (at random) the correct sequencing for a short protein 100 amino acids long is about 1 in 10^{65} (Reidhaar-Olson and Sauer 1990; Behe 1992, 65–9). More recent mutagenesis research suggests that Sauer's methods imply probability measures that are, if anything, too optimistic (Axe 2000).

Other considerations imply additional improbabilities. First, new Cambrian animals would require proteins much longer than 100 residues to perform necessary specialized functions. Susumu Ohno (1996) has noted that Cambrian animals would have required complex proteins such as lysyl oxidase in order to support their stout body structures. Lysyl oxidase molecules in extant organisms comprise over 400 amino acids. These molecules represent highly complex (nonrepetitive) and tightly specified arrangements of matter. Reasonable extrapolation from mutagenesis experiments done on shorter protein molecules suggests that the probability of producing functionally sequenced proteins of this length at random is far smaller than 1 in 10^{150} – the point at which, according to Dembski's calculation of the universal probability bound, appeals to chance become absurd, given the time and other probabilistic resources of the entire universe (1998, 175–223). Second, the Cambrian explosion took far less time (5×10^6 years) than the duration of the universe (2×10^{10} years) assumed by Dembski in his calculation. Third, DNA mutation rates are far too low to generate the novel genes and proteins necessary to building the Cambrian animals, given the duration of the explosion. As Susumo Ohno has explained:

Assuming a spontaneous mutation rate to be a generous 10^{-9} per base pair per year ... it still takes 10 million years to undergo 1% change in DNA base sequences. It follows that 6–10 million years in the evolutionary time scale is but a blink of an eye. The Cambrian explosion ... within the time span of 6–10 million years can't possibly be explained by mutational divergence of individual gene functions. (1996, 8475)

The selection/mutation mechanism faces another probabilistic obstacle. The animals that arise in the Cambrian exhibit structures that would have required many new *types* of cells, each of which would have required many novel proteins to perform their specialized functions. Further, new cell types require *systems* of proteins that must, as a condition of function, act in close coordination with one another. The unit of selection in such systems ascends to the system as a whole. Natural selection selects for functional advantage. But new cell types require whole systems of proteins to perform their distinctive functions. In such cases, natural selection cannot contribute to the process of information generation until *after* the information necessary to build the requisite *system* of proteins has arisen. Thus random variations must, again, do the work of information generation – and now not simply for one protein, but for many proteins arising at nearly the same time. Yet the odds of this occurring by chance are far smaller than the odds of the chance origin of a single gene or protein.

As Richard Dawkins has acknowledged, "we can accept a certain amount of luck in our explanations, but not too much" (1986, 139). The neutral theory of evolution, which, by its own logic, prevents natural selection from playing a role in generating genetic information until after the fact, relies on entirely "too much luck." The sensitivity of proteins to functional loss as the result of random changes in sequencing, the need for long proteins to build new cell types and animals, the need for whole new *systems* of proteins to service new cell types, the brevity of the Cambrian explosion relative to mutation rates – all suggest the immense improbability (and implausibility) of any scenario for the origin of Cambrian genetic information that relies upon chance alone unassisted by natural selection.

Yet the neutral theory requires novel genes and proteins to arise – essentially – by random mutation alone. Adaptive advantage accrues *after* the generation of new functional genes and proteins. Thus, natural selection cannot play a role *until* new information-bearing molecules have independently arisen. Thus the neutral theory envisions the need to scale the steep face of a Dawkins-style precipice on which there is *no* gradually sloping backside – a situation that, by Dawkins' own logic, is probabilistically untenable.

In the second scenario, neo-Darwinists envision novel genes and proteins arising by numerous successive mutations in the preexisting genetic text that codes for proteins. To adapt Dawkins's metaphor, this scenario envisions gradually climbing down one functional peak and then ascending another. Yet mutagenesis experiments again suggest a difficulty. Recent experiments performed by Douglas Axe at Cambridge University show that, even when exploring a region of sequence space populated by proteins of a single fold and function, most multiple-position changes quickly lead to loss of function (Axe 2000). Yet to turn one protein into another with *a completely novel* structure and function requires specified changes at many more sites. Given this reality, the probability of escaping total functional loss during a random search for the changes needed to produce a new function is vanishingly small – and this probability diminishes exponentially with each additional requisite change. Thus, Axe's results imply that, in all probability, random searches for novel proteins (through sequence space) will result in functional loss long before any novel functional protein will emerge.

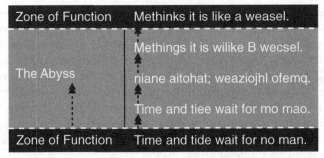

FIGURE 20.1. Multiple undirected changes in the arrangement of letters in a sentence will destroy meaning before a new sentence can arise. Mutagenesis experiments suggest that a similar problem applies to sequence-specific genes and proteins.

Francisco Blanco at the European Molecular Biology Laboratory has come to a similar conclusion. Using directed mutagenesis, his team has found that the sequence space between two natural protein domains is not populated by folded or functional conformations (i.e., biologically-relevant proteins). Instead, mutant sequences "lack a well defined three-dimensional structure." They conclude:

[B]oth the hydrophobic core residues and the surface residues are important in determining the structure of the proteins, and suggest that the appearance of a completely new fold from an existing one *is unlikely to occur by evolution through a route of folded intermediate sequences* [emphasis added]. (Blanco, Angrand, and Serrano 1999, 741)

Thus, although this second neo-Darwinian scenario has the advantage of starting with functional genes and proteins, it also has a lethal disadvantage: any process of random mutation or rearrangement in the genome would in all probability generate nonfunctional intermediate sequences before fundamentally new functional genes or proteins would arise (Figure 20.1). Clearly, nonfunctional intermediate sequences confer no survival advantage on their host organisms. Yet natural selection favors *only* functional advantage. It cannot select or favor nucleotide sequences or polypeptide chains that do not yet perform biological functions, and still less will it favor sequences that efface or destroy preexisting function.

Evolving genes and proteins will almost inevitably range through a series of nonfunctional intermediate sequences that natural selection will not favor or preserve but will, in all probability, eliminate (Blanco et al. 1999; Axe, 2000). When this happens, selection-driven evolution will cease. At this point, neutral evolution of the genome (unhinged from selective pressure) may ensue, but, as we have seen, such a process must overcome immense probabilistic hurdles, even granting cosmic time.

Thus, whether one envisions the evolutionary process beginning with a noncoding region of the genome or a preexisting functional gene, the functional specificity and complexity of proteins impose very stringent limitations on the efficacy of mutation and selection. In the first case, function must arise first, before natural selection can act to favor a novel variation. In the second case, function must be continuously maintained in order to prevent deleterious (or lethal) consequences to the organism and to allow further evolution. Yet the complexity and functional specificity of proteins implies that both these conditions will be extremely difficult to meet. Therefore, the neo-Darwinian mechanism appears to be inadequate to generate the new information present in the novel genes and proteins that arise with the Cambrian animals.

NOVEL BODY PLANS

The problems with the neo-Darwinian mechanism run deeper still. In order to explain the origin of the Cambrian animals, one must account not only for new proteins and cell types, but also for the origin of new body plans. Within the past decade, developmental biology has dramatically advanced our understanding of how body plans are built during ontogeny. In the process, it has also uncovered a profound difficulty for neo-Darwinism.

Significant morphological change in organisms requires attention to timing. Mutations in genes that are expressed late in the development of an organism will not affect the body plan. Mutations expressed early in development, however, could conceivably produce significant morphological change (Arthur 1997, 21). Thus, events expressed early in the development of organisms have the only realistic chance of producing large-scale macroevolutionary change (Thomson 1992). As John and Miklos explain, "macroevolutionary change" requires changes in "very early embryogenesis" (1988, 309).

Yet recent studies in developmental biology make clear that mutations expressed early in development typically have deleterious (or, at best, neutral) effects (Arthur 1997, 21), including mutations in crucially important "master regulator" or hox genes. For example, when early-acting body plan molecules, or morphogens such as bicoid (which helps to set up the anterior–posterior head-to-tail axis in *Drosophila*), are perturbed, development shuts down (Nusslein-Volhard and Wieschaus 1980; Lawrence and Struhl 1996). The resulting embryos die. Moreover, there is a good reason for this. If an engineer modifies the length of the piston rods in an internal combustion engine without modifying the crankshaft accordingly, the engine won't start. Similarly, processes of development are tightly integrated spatially and temporally in such a way that changes early in development will require a host of other coordinated changes in separate but functionally interrelated developmental processes downstream. Thus, as Stuart Kuaffman explains, "A mutation disrupting formation of a spinal column and cord is more likely to be lethal than one affecting the number of fingers . . ." (1995, 200).

This problem has led to what the geneticist John F. McDonald has called "a great Darwinian paradox" (1983, 93). He notes that genes that vary within natural populations affect only minor aspects of form and function, while genes that govern major changes – the very stuff of macroevolution – apparently do not vary, or vary only to the detriment of the organism. As he puts it, "those [genetic] *loci* that are obviously variable within natural populations do not seem to lie at the basis of many major adaptive changes, while those *loci* that seemingly do constitute the foundation of many if not most major adaptive changes are not variable . . ." (93). In other words, mutations of the kind that macroevolution *doesn't* need (namely, viable genetic mutations in DNA expressed late in development) do occur, but those that it *does* need (namely, beneficial *Bauplan* mutations expressed early in development) *don't occur*.

Darwin wrote that "nothing can be effected" by natural selection "unless favorable variations occur" (1859, 108). Yet discoveries about the genetic regulation of development suggest that variations of the kind required by neo-Darwinism – favorable *Bauplan* mutations – do not occur.

Developmental biology has raised another formidable problem for the mutation/selection mechanism. Embryological evidence has long shown that DNA does not wholly determine morphological form (Goodwin 1985; Sapp 1987; Nijhout 1990), suggesting that mutations in DNA alone cannot account for the morphological changes required to build a new body plan (Müller and Newman 2003).

DNA directs protein synthesis. It also helps to regulate the timing and expression of the synthesis of various proteins within cells. Nevertheless, DNA alone does not determine how individual proteins assemble themselves into larger systems of proteins; still less does it solely determine how cell types, tissue types, and organs arrange themselves into body plans. Instead, other factors – such as the structure and organization of the cell membrane and cytoskeleton – play important roles in determining developmental pathways that determine body plan formation during embryogenesis.

For example, the shape and location of microtubules in the cytoskeleton influence the "patterning" of embryos. Arrays of microtubules help to distribute the essential proteins used during development to their correct locations in the cell. Of course, microtubules themselves are made of many protein subunits. Nevertheless, the protein subunits in the cell's microtubules are identical to one another. Neither they nor the genes that produce them account for the different shapes and locations of microtubule arrays that distinguish different kinds of embryos and developmental pathways. As Jonathan Wells explains, "What matters in development is the shape and location of microtubule arrays, and the shape and location of a microtubule array is not determined by its units" (1999, 52).

Two analogies may help to clarify the point. At a building site, builders will make use of many materials: lumber, wires, nails, drywall, piping, and windows. Yet building materials do not determine the floor plan of the house, or the arrangement of houses in a neighborhood. Similarly, electronic circuits are composed of many components, such as resistors, capacitors, and transistors. But such lower-level components do not determine their own arrangement in an integrated circuit. Biological systems also depend on hierarchical arrangements of parts. Genes and proteins are made from simple building blocks – nucleotide bases and amino acids – arranged in specific ways. Cell types are made of, among other things, systems of specialized proteins. Organs are made of specialized arrangements of cell types and tissues. And body plans comprise specific arrangements of specialized organs. Yet, clearly, the properties of individual proteins[1] (or, indeed, the lower-level parts in the hierarchy generally) do not determine the organization of these higher-level structures and organizational patterns. It follows, therefore, that the genetic information that codes for proteins does not determine these higher-level structures either.

These considerations pose another challenge to the sufficiency of the neo-Darwinian mechanism. Neo-Darwinism seeks to explain the origin of new information, form, and structure as a result of selection acting on randomly arising variation at a very low level within the biological hierarchy – namely, within the genetic text. Yet major morphological innovations depend on a specificity of arrangement at a much higher level of the organizational hierarchy, a level that DNA alone does not determine. Yet if DNA is not wholly responsible for body plan morphogenesis, then DNA sequences can mutate indefinitely, without regard to realistic probabilistic limits, and still not produce a new body plan. Thus, the mechanism of natural selection acting on random mutations in DNA cannot *in principle* generate novel body plans, including those that first arose in the Cambrian explosion.

SELF-ORGANIZATIONAL MODELS

Of course, neo-Darwinism is not the only naturalistic model for explaining the origin of novel biological form. Stuart Kauffman, for example, also doubts the efficacy of the mutation/selection mechanism. Nevertheless, he has advanced a self-organizational model to account for the emergence of new form, and presumably the information necessary to generate it. Whereas neo-Darwinism attempts to explain new form as the consequence of selection acting on random mutation, Kauffman suggests that selection acts, not mainly on random variations, but on emergent patterns of order that self-organize via the laws of nature.

Kauffman illustrates how this might work using various model systems in a computer environment (1995, 47–92). In one, he conceives a system of buttons connected by strings. Buttons represent novel genes or gene products; strings represent the lawlike forces of interaction that obtain between gene products – that is, proteins. Kauffman suggests that when the complexity of the system (as represented by the number of buttons and strings) reaches a critical threshold, new modes of organization can arise in the system "for free" – that is, without intelligent guidance – after the manner of a phase transition in chemistry.

Another model that Kauffman develops is a system of interconnected lights. Each light can flash in a variety of states – on, off, twinkling, and so on. Since there is more than one possible state for each light, and many lights, there is a vast number of possible states that the system can adopt. Further, in his system, rules determine how past states will influence future states. Kauffman asserts that, as a result of these rules, the system will, if properly tuned, eventually produce a kind of order in which a few basic patterns of light activity recur with greater-than-random frequency. Since these actual patterns of light activity represent a small portion of the total number of possible states in which the system can reside, Kaufman suggests that self-organizational laws might similarly result in highly improbable biological outcomes – perhaps even sequences (of bases or amino acids) within a much larger sequence space of possibilities.

Do these simulations of self-organizational processes accurately model the origin of novel genetic information? It is hard to think so.

First, in both examples, Kaufmann presupposes but does not explain significant sources of preexisting information. In his buttons-and-strings system, the buttons represent proteins – themselves packets of CSI and the result of preexisting genetic information. Where does this information come from? Kauffman doesn't say, but the origin of such information is an essential part of what needs to be explained in the history of life. Similarly, in his light system, the order that allegedly arises for "for free" – that is, apart from any intelligent input of information – actually arises only if the programmer of the model system "tunes" it in such a way as to keep it from either (a) generating an excessively rigid order or (b) devolving into chaos (86–8). Yet this necessary tuning involves an intelligent programmer selecting certain parameters and excluding others – that is, inputting information.

Second, Kauffman's model systems are not constrained by functional considerations and thus are not analogous to biological systems. A system of interconnected lights governed by pre-programmed rules may well settle into a small number of patterns within a much larger space of possibilities. But because these patterns have no function, and need not meet any functional requirement, they have no specificity analogous to that present in actual organisms. Instead, examination of Kauffman's model systems shows that they produce sequences or systems characterized not by *specified* complexity, but instead by large amounts of symmetrical order or internal redundancy interspersed with aperiodicity or (mere) complexity (53, 89, 102). Getting a law-governed system to generate repetitive patterns of flashing lights, even with a certain amount of variation, is clearly interesting, but it is not biologically relevant. On the other hand, a system of lights flashing "Eat at Joe's" would model a biologically relevant self-organizational process, at least if such messages arose without agents previously programming the system with equivalent amounts of CSI. In any case, Kauffman's systems do not produce *specified* complexity, and thus they do not offer promising models for explaining the new genes and proteins that arose in the Cambrian.

Even so, Kauffman suggests that his self-organizational models can specifically elucidate aspects of the Cambrian explosion. According to Kauffman, new Cambrian animals emerged as the result of "long jump" mutations that established new body plans in a discrete rather than gradual fashion (199–201). He also recognizes that mutations affecting early development are almost inevitably harmful. Thus he concludes that body plans, once established, will not change, and that any subsequent evolution must occur within an established *Bauplan*. And indeed, the fossil record does show a curious (from a Darwinian point of view) top-down pattern of appearance, in which higher taxa (and the body plans they represent) appear first, only later to be followed by the multiplication of lower taxa representing variations within those original body designs. Further, as Kauffman expects, body plans appear suddenly and persist without significant modification over time.

But here, again, Kauffman begs the most important question, which is: *what produces the new Cambrian body plans in the first place?* Granted, he invokes "long jump mutations" to explain this, but he identifies no specific self-organizational process that can produce such mutations. Moreover, he concedes a principle that undermines the plausibility of his own proposal. Kauffman acknowledges that mutations that occur early in development are almost inevitably deleterious. Yet developmental biologists know that mutations of this kind are the only ones that have a realistic chance of producing large-scale evolutionary change – that is, the big jumps that Kauffman invokes. Though Kauffman repudiates the neo-Darwinian reliance upon random mutations in favor of self-organizing order, in the end he must invoke the most implausible kind of random mutation in order to provide a self-organizational account of the new Cambrian body plans. Clearly, his model is not sufficient.

DESIGN WITHOUT A DESIGNER?

Neo-Darwinists such as Francisco Ayala, Richard Dawkins, and Richard Lewontin acknowledge that organisms appear to have been designed. As Dawkins notes, "biology is the study of complicated things that give the appearance of having been designed for a purpose" (1986, 1). Of course, neo-Darwinists assert that what Ayala calls the "obvious design" of living things is *only* apparent. As Ayala explains:

The functional design of organisms and their features would therefore seem to argue for the existence of a designer. It was Darwin's greatest accomplishment to show that the directive organization of living beings can be explained as the result of a natural process, natural selection, without any need to resort to ... [an] external agent. ... (1994, 5)

According to neo-Darwinists, mutation and selection – and perhaps other similar (though less significant) naturalistic mechanisms – are fully sufficient to explain the appearance of design in biological systems. Self-organizational theorists modify this claim but affirm its essential tenet. They argue that natural selection acting on self-organizing order can explain the complexity of living things – again, without any appeal to design.

Most biologists now acknowledge that the Darwinian mechanism can explain micro-evolutionary adaptation, such as cyclical variations in the size of Galapagos finch beaks. But can it explain *all* appearances of design, including the genetic and other forms of CSI necessary to produce morphological innovations in the history of life? As Dawkins has noted, "the machine code of the genes is uncannily computer like. Apart from differences in jargon, the pages of a molecular-biology journal might be interchanged with those of a computer-engineering journal" (1995, 11). Certainly, the presence of CSI in living organisms, and the discontinuous increases of CSI that occurred during the Cambrian explosion, are at least suggestive of design. Can any fully naturalistic model of evolutionary change explain *these* appearances of design without reference to actual design?

This chapter has argued that neither neo-Darwinism nor self-organization provides an adequate explanation of the origin of the information that arises in the Cambrian. If this is the case, could the appearance of design – as specifically manifest in new information-rich genes, proteins, cell types and body plans – have resulted from Intelligent Design rather than from a purposeless process that merely mimics the powers of a designing intelligence? Perhaps what Sober has conceded as a possibility can now be advanced as a reality. Perhaps a design hypothesis that competes with neo-Darwinism can be defended as an inference to the best explanation. In this concluding section, I will argue as much.

Studies in the history and philosophy of science have shown that many scientific theories, particularly in the historical sciences, are formulated and justified as inferences to the best explanation (Lipton 1991, 32–88; Sober 2000, 44). Historical scientists, in particular, assess competing hypotheses by evaluating which hypothesis would, if true, provide the best explanation of some set of relevant data. Those with greater explanatory power are typically judged to be better – more probably true – theories. Darwin himself used this method of reasoning in defending his theory of universal common descent (Darwin 1896, 437). Moreover, contemporary studies on the method of "inference to the best explanation" have shown that determining which among a set of competing possible explanations constitutes the best one depends upon judgments about the causal adequacy, or "causal powers," of competing explanatory entities (Lipton, 32–88).

I have argued that the two most widely held naturalistic mechanisms for generating biological form are not causally adequate to produce the discontinuous increases of CSI that arose in the Cambrian. Do intelligent agents have causal powers sufficient to produce such increases in CSI, either in the form of sequence-specific lines of code or hierarchically arranged systems of parts? Clearly, they do.

In the first place, intelligent human agents have demonstrated the power to produce linear sequence-specific arrangements of characters. Indeed, experience affirms that information of this type routinely arises from the activity of intelligent agents. A computer user who traces the information on a screen back to its source invariably comes to a *mind* – that of a software engineer or programmer. The information in a book or inscription ultimately derives from a writer or scribe – from a mental, rather than a strictly material, cause. Our experience-based knowledge of information flow confirms that systems with large amounts of specified complexity (especially codes and languages) invariably originate from an intelligent source – from a mind or personal agent. To quote Henry Quastler again: the "creation of new information is habitually associated with conscious activity" (1964, 16). Experience teaches this obvious truth.

Further, intelligent agents have just those necessary powers that natural selection lacks as a condition of its causal adequacy. Recall that at several points in our previous analysis, we demonstrated that natural selection lacks the ability to generate novel information precisely because it can act only after the fact – that is, after new functional CSI has already arisen. Natural selection can favor new proteins and genes, but only after they provide some function. The job of generating new functional genes, proteins, and systems of proteins therefore falls to entirely random mutations. Yet without functional criteria to guide a search through the space of possible sequences, random variation is probabilistically doomed. What is needed is not just a source of variation (i.e., the freedom to search a space of possibilities) or a mode of selection that can operate after the fact of a successful search, but instead a means of selection that (a) operates during a search – before success – and that (b) is guided by information about, or knowledge of, a functional target.

Demonstration of this requirement has come from an unlikely quarter: genetic algorithms. Genetic algorithms are programs that allegedly simulate the creative power of mutation and selection. Richard Dawkins and Bernd-Olaf Kuppers, for example, have developed computer programs that putatively simulate the production of genetic information by mutation and natural selection (Dawkins 1986, 47–9; Kuppers 1987, 355–69). Nevertheless, as I have shown elsewhere (Meyer 1998b, 127–8), these programs succeed only by the illicit expedient of providing the computer with a "target sequence" and then treating relatively greater proximity to *future* function (i.e., the target sequence), not actual present function, as a selection criterion. As David Berlinski (2000) has argued, genetic algorithms need something akin to a "forward-looking memory" in order to succeed. Yet such foresighted selection has no analogue in nature. In biology, where differential survival depends upon maintaining function, selection cannot occur before new functional sequencing arises. Natural selection lacks foresight.

What natural selection lacks, intelligent selection – purposive or goal-directed design – provides. Agents can arrange matter with distant goals in mind. In their use of language, they routinely "find" highly isolated and improbable functional sequences amid vast spaces of combinatorial possibilities. Analysis of the problem of the origin of biological information exposes a deficiency in the causal powers of natural selection that corresponds precisely to powers that agents are uniquely known to possess. Intelligent agents have foresight. Agents can select functional goals *before* they exist. They can devise or select material means to accomplish those ends from among an array of possibilities and then actualize those goals in accord with a *pre*conceived design and/or independent set of functional requirements. The causal powers that natural selection lacks – almost by definition – are associated with the attributes of consciousness and rationality – with purposive intelligence. Thus, by invoking Intelligent Design to explain the origin of new information, design theorists are not positing an arbitrary explanatory element unmotivated by a consideration of the evidence. Instead, design theorists are positing an entity with precisely the attributes and causal powers that the phenomenon in question requires as a condition of its production and explanation.

Secondly, the highly specified hierarchical arrangements of parts in animal body plans also bespeak *design*. At every level of the biological hierarchy, organisms require specified and highly improbable arrangements of lower-level constituents in order to maintain their form and function. Genes require specified arrangements of nucleotide bases; proteins require specified arrangements of amino acids; new cell types require specified arrangements of systems of proteins; body plans require specialized arrangements of cell types and organs. Not only do organisms contain information-rich components (such as proteins and genes), they also comprise information-rich arrangements of those components and the systems that comprise them.

Based on experience, we know that human agents possessing rationality, consciousness, and foresight have, as a consequence of these attributes, the ability to produce information-rich hierarchies in which both individual modules and the arrangements of those modules exhibit complexity and specificity – information so defined. Individual transistors, resistors, and capacitors exhibit considerable complexity and specificity of design; at a higher level of organization, their specific arrangement within an integrated circuit represents additional information and reflects further design.

Conscious and rational agents have, as part of their powers of purposive intelligence, the capacity to design information-rich parts and to organize those parts into functional information-rich systems and hierarchies. Further, we know of no other causal entity or process that has this capacity. Clearly, we have good reason to doubt that either mutation and selection or self-organizational processes can produce the information-rich components, systems, and body plans that arose in the Cambrian. Instead, explaining the origin of such information requires causal powers that we uniquely associate with conscious and rational activity – with intelligent causes, not purely natural processes or material mechanisms. Thus, based on our experience and analysis of the causal powers of various explanatory entities, we can infer Intelligent Design as the best – most causally adequate – explanation for the origin of the complex specified information required to build the Cambrian animals. In other words, the remarkable explosion of Cambrian information attests to the power and activity of a purposive intelligence in the history of life.

Note

1. Of course, many proteins bind chemically with each other to form complexes and structures within cells. Nevertheless, these "self-organizational" properties do not fully account for higher levels of organization in cells, organs, or body plans.

References

Arthur, W. 1997. *The Origin of Animal Body Plans.* Cambridge: Cambridge University Press.

Axe, D. D. 2000. Biological function places unexpectedly tight constraints on protein sequences. *Journal of Molecular Biology* 301 (3): 585–96.

Ayala, F. 1994. Darwin's revolution. In *Creative Evolution?!*, ed. J. Campbell and J. Schopf. Boston: Jones and Bartlett, pp. 1–17.

Behe, M. 1992. Experimental support for regarding functional classes of proteins to be highly isolated from each other. In *Darwinism: Science or Philosophy?*, ed. J. Buell and G. Hearn. Richardson, Tx: Foundation for Thought and Ethics, pp. 60–71.

Berlinski, D. 1996. The deniable Darwin. *Commentary* (June): 19–29.
    2000. On assessing genetic algorithms. Lecture delivered to the Science and Evidence of Design in the Universe conference, Yale University, November 4.

Blanco, F., I. Angrand, and L. Serrano. 1999. Exploring the confirmational properties of the sequence space between two proteins with different folds: An experimental study. *Journal of Molecular Biology* 285: 741–53.

Bowie, J., and R. Sauer. 1989. Identifying determinants of folding and activity for a protein of unknown sequences: Tolerance to amino acid substitution. *Proceedings of the National Academy of Sciences (USA)* 86: 2152–6.

Bowring, S. A., J. P. Grotzinger, C. E. Isachsen, A. H. Knoll, S. M. Pelechaty, and P. Kolosov. 1993. Calibrating rates of Early Cambrian evolution. *Science* 261: 1293–8.

Chothia, C., I. Gelfland, and A. Kister. 1998. Structural determinants in the sequences of immunoglobulin variable domain. *Journal of Molecular Biology* 278: 457–79.

Darwin, C. 1859. *On the Origin of Species.* London: John Murray.

Darwin, F. (ed.). 1896. *Life and Letters of Charles Darwin*, vol. 1. London: D. Appleton.

Dawkins, R. 1986. *The Blind Watchmaker.* London: Penguin.

    1995. *River Out of Eden.* New York: Basic Books.

    1996. *Climbing Mount Improbable.* New York: Norton.

Dembski, W. A. 1998. *The Design Inference.* Cambridge: Cambridge University Press.

Denton, M. 1986. *Evolution: A Theory in Crisis.* London: Adler and Adler.

Gerhart J., and M. Kirschner. 1997. *Cells, Embryos, and Evolution.* London: Blackwell Science.

Goodwin, B. C. 1985. What are the causes of morphogenesis? *BioEssays* 3: 32–6.

Hodge, M. J. S. 1977. The structure and strategy of Darwin's long argument. *British Journal for the History of Science* 10: 237–45.

John, B., and G. Miklos. 1988. *The Eukaryote Genome in Development and Evolution.* London: Allen and Unwin.

Kauffman, S. 1995. *At Home in the Universe.* Oxford: Oxford University Press.

Koonin, E. 2000. How many genes can make a cell?: The minimal genome concept. *Annual Review of Genomics and Human Genetics* 1: 99–116.

Kuppers, B.-O. 1987. On the prior probability of the existence of life. In *The Probabilistic Revolution*, ed. L. Kruger et al. Cambridge, MA.: MIT Press, pp. 355–69.

Lawrence, P. A., and G. Struhl. 1996. Morphogens, compartments and pattern: Lessons from Drosophila? *Cell* 85: 951–61.

Lipton, P. 1991. *Inference to the Best Explanation.* New York: Routledge.

Løvtrup, S. 1979. Semantics, logic and vulgate neo-Darwinism. *Evolutionary Theory* 4: 157–72.

McDonald, J. F. 1983. The molecular basis of adaptation: A critical review of relevant ideas and observations. *Annual Review of Ecology and Systematics* 14: 77–102.

Meyer, S. C. 1998. DNA by design: An inference to the best explanation for the origin of biological information. *Rhetoric and Public Affairs* 1 (4): 519–55. Lansing: Michigan State University Press.

    2003. DNA and the origin of life: Information, specification and explanation. In *Darwinism, Design and Public Education*, ed. J. A. Campbell and S. C. Meyer. Lansing: Michigan State University Press, pp. 223–85.

Meyer, S. C., M. Ross, P. Nelson, and P. Chien. 2003. The Cambrian explosion: biology's big bang. In *Darwinism, Design and Public Education*, ed. J. A. Campbell and S. C. Meyer. Lansing: Michigan State University Press, pp. 323–402. See also: Appendix C, Stratigraphic first appearance of phyla body plans, pp. 593–8.

Müller, G. B., and S. A. Newman. 2003. *Origination of Organismal Form: The Forgotten Cause in Evolutionary Theory*, ed. Gerd B. Müller and Stuart A. Newman. Cambridge, MA: MIT Press, pp. 3–10.

Nijhout, H. F. 1990. Metaphors and the role of genes in development. *BioEssays* 12: 441–6.

Nusslein-Volhard, C., and E. Wieschaus. 1980. Mutations affecting segment number and polarity in Drosophila. *Nature* 287: 795–801.

Ohno, S. 1996. The notion of the Cambrian pananimalia genome. *Proceedings of the National Academy of Sciences (USA)* 93: 8475–8.

Perutz, M. F., and H. Lehmann. 1968. Molecular pathology of human hemoglobin. *Nature* 219: 902–9.

Quastler, H. 1964. *The Emergence of Biological Organization*. New Haven, CT: Yale University Press.

Reidhaar-Olson, J. and R. Sauer. 1990. Functionally acceptable solutions in two alpha-helical regions of lambda repressor. *Proteins, Structure, Function, and Genetics* 7: 306–16.

Sapp, J. 1987. *Beyond the Gene.* New York: Oxford University Press.

Sarkar, S. 1996. Biological information: A skeptical look at some central dogmas of molecular biology. In *The Philosophy and History of Molecular Biology: New Perspectives* ed. S. Sarkar. Dordrecht: Kluwer, pp. 187–233.

Schuetzenberger, M. 1967. Algorithms and the neo-Darwinian theory of evolution. In *Mathematical Challenges to the Darwinian Interpretation of Evolution*, ed. P. S. Morehead and M. M. Kaplan. New York: Allen R. Liss Publishing.

Shannon, C. 1948. A mathematical theory of communication. *Bell System Technical Journal* 27: 379–423, 623–56.

Shu, D. G., H. L. Lou, S. Conway Morris, X. L. Zhang, S. X. Hu, L. Chen, J. Han, M. Zhu, Y. Li, and L. Z. Chen. 1999. Lower Cambrian vertebrates from south China. *Nature* 402: 42–6.

Sober, E. 2000. *The Philosophy of Biology*, 2nd ed. San Francisco: Westview Press.

Taylor, S. V., K. U. Walter, P. Kast, and D. Hilvert. 2001. Searching sequence space for protein catalysts. *Proceedings of the National Academy of Science (USA)* 98: 10596–601.

Thomson, K. S. 1992. Macroevolution: The morphological problem. *American Zoologist* 32: 106–12.

Valentine, J. W. 1995. Late Precambrian bilaterians: Grades and clades. In *Tempo and Mode in Evolution: Genetics and Paleontology 50 Years after Simpson*, ed. W. M. Fitch, and F. J. Ayala. Washington, DC: National Academy Press, pp. 87–107.

Wells, J. 1999. Making sense of biology: The evidence for development by design. *Touchstone* (July/August): 51–5.

Yockey, H. P. 1992. *Information Theory and Molecular Biology*. Cambridge: Cambridge University Press.

Index

Printed in the United States
By Bookmasters